P9-DWA-303

Stormy Weather

Guy Dauncey with Patrick Mazza

Foreword by Ross Gelbspan

NEW SOCIETY PUBLISHERS

Cataloguing in Publication Data:
A catalog record for this publication is available from the National Library of Canada.

Cover design by Diane McIntosh. Photo: Lightning image Don Farrall.

Printed in Canada by Friesens Inc.

New Society Publishers acknowledges the support of the Government of Canada through the Book Publishing Industry Development Program (BPIDP) for our publishing activities, and the assistance of the Province of British Columbia through the British Columbia Arts Council.

Paperback ISBN: 0-86571-421-5

To order directly from the publishers, please add $4.50 shipping to the price of the first copy, and $1.00 for each additional copy (plus GST in Canada). Send check or money order to:

New Society Publishers
P.O. Box 189, Gabriola Island, BC V0R 1X0, Canada

New Society Publishers aims to publish books for fundamental social change through nonviolent action. We are committed to building an ecologically sustainable and just society not just through education, but through action. We are acting on our commitment to the world's remaining ancient forests by phasing out our paper supply from ancient forests worldwide. This book is one step towards ending global deforestation and climate change. It is printed on acid-free paper that is 100% old growth forest-free (100% post-consumer recycled), processed chlorine free — supplied by New Leaf Paper — and printed with vegetable based, low VOC inks. For further information, or to browse our full list of books and purchase securely, visit our website at: www.newsociety.com

NEW SOCIETY PUBLISHERS
www.newsociety.com

CONTENTS

Introduction

MOVING TOWARD A WORLD WITHOUT FOSSIL FUELS

The 101 Solutions

TEN SOLUTIONS FOR INDIVIDUALS

TEN SOLUTIONS FOR CITIZENS ORGANIZATIONS

TEN SOLUTIONS FOR CITIES, TOWNS & COUNTIES

TEN SOLUTIONS FOR BUSINESSES AND ORGANIZATIONS

FIFTEEN SOLUTIONS FOR ENERGY COMPANIES

FIVE SOLUTIONS FOR AUTO COMPANIES

TEN SOLUTIONS FOR STATES & PROVINCES

TEN SOLUTIONS FOR NATIONAL GOVERNMENTS

TEN SOLUTIONS FOR DEVELOPING NATIONS

TEN GLOBAL SOLUTIONS

Foreword

by Ross Gelbspan

AT THE BEGINNING OF THE 21ST CENTURY, America stands virtually alone in its denial of the urgency and magnitude of the accelerating pace of climate change.

It is no accident.

The reason for this denial lies in a relentless campaign of deception and disinformation in the U.S. by the coal and oil industries to persuade the public that the issue is either non-existent or negligible. That campaign for the longest time targeted the science. It then misrepresented the economics. Most recently, with its new champion in the White House, it has attempted to demolish the diplomatic foundations of the climate convention. And it has been extraordinarily successful in maintaining a relentless drumbeat of doubt in the public mind. It has also inhibited the press from the kind of aggressive, consistent coverage that would long ago have promoted the formation of a mass constituency around the issue.

But the very nature of the challenge has created fertile ground for the industry campaign. When a person is confronted with a problem that seems overwhelming — and, at the same time, insoluble — her or his natural reaction is to not want to know about it. To acknowledge a problem which is both inevitable and insoluble leaves one with a deeply disorienting and paralyzing feeling of impotence.

Both the urgency and the magnitude of the climate crisis are, indeed, overwhelming. Unintentionally, we have begun to alter massive systems of the planet with huge amounts of inertia that have kept this Earth hospitable to civilization for the last 10,000 years. We have raised atmospheric carbon levels to their highest point in 420,000 years. We have heated the deep oceans. We have loosed a wave of violent and chaotic weather. We have altered the timing of the seasons. We are living on a very precarious margin of stability and the evidence is everywhere around us.

A recent peer-reviewed study by 11 prominent researchers found that unless the world gets half of its energy from non-carbon sources by the year 2018, the planet will see a quadrupling of atmospheric carbon levels — which would clearly be catastrophic. To put this in perspective, the oceans are warming, the tundra is thawing, the glaciers are melting, infectious diseases are migrating, the timing of the seasons has changed — and all that has resulted from *one degree* of warming. By contrast, the Earth will warm from 4 to 10 degrees later in this century, according to the scientific consensus.

In speaking about the climate crisis, the most difficult — and inevitable — question that Americans ask is: "What can I do about it?"

The answer lies in *Stormy Weather* with its "101 Solutions to Global Climate Change" — and in a 102nd solution for the world as a whole. Guy Dauncey and Patrick Mazza's invaluable work is based on a critical understanding: to be effective, change must involve every segment of society working at every level of action. There is no dichotomy between a "grassroots" and a "top-down" approach. The climate crisis requires actions at every level of society. To the extensive list of remedies catalogued in *Stormy Weather*, I would add one more that combines several mechanisms cited in its pages.

The World Energy Modernization Plan, proposed by a small group of energy company presidents, economists and energy policy specialists combines three global-scale, interactive, macro-level strategies that would, if implemented, reduce global carbon emissions by the 70 percent required by nature. At the same time, it would create millions of jobs all over the world, raising living standards in poor countries without compromising them in the industrial world.

The strategies involve:

- Switching national subsidies away from fossil fuels and to renewable energy technologies. The federal government in the U.S. today spends $20 billion a year subsidizing fossil fuels. If those subsidies were put behind renewable energy sources (with a small portion dedicated to job retraining for displaced coal miners), that would create the incentive for the large oil companies to become aggressive developers of solar, wind and hydrogen energy sources.
- Replacing the ineffectual mechanism of international "emissions trading" in the Kyoto Protocol with a progressively more stringent Fossil Fuel Efficiency Standard. If every country began at its current baseline to improve its Fossil Fuel Efficiency by 5 percent a year (until the 70 percent reduction was achieved), that would create the mass market for renewable technologies which would bring down their price and make them economically competitive with coal and oil. Under such a progressive Fossil Fuel Efficiency Standard, countries would either create the same amount of goods with 5 percent less fossil fuel use each year or, alternatively, produce 5 percent more with the same amount of fossil fuels.
- Creating an Energy Modernization Fund of about $300 billion a year for the transfer of climate-friendly technologies to developing countries. Virtually all poor countries would love to go solar; almost none can afford it. Such a fund could be financed by a small tax on international currency transactions (a Tobin Tax, named after its creator, Dr. James Tobin, a Nobel Prize-winning economist — see Solution #93). Since currency transactions today total $1.5 trillion per day, a tax of a quarter-penny per dollar would net in the order of $300 billion a year for windfarms in India, solar assemblies in El Salvador, vast, hydrogen-producing photovoltaic farms in the Middle East, and fuel cell factories in South Africa.

The fund is not a North-South giveaway program. More accurately, it represents a transfer of speculative, non-productive capital away from the finance sector into intensely productive, wealth-generating, job-creating investments in the industrial sector. It reflects the kind of thinking that gave rise to the Marshall Plan after World War II. Today, instead of a collection of impoverished and dependent allies, the U.S. enjoys a fruitful and prosperous economic relationship with the nations of Europe. A plan of this scope and scale — regardless of the details — would have a similarly enriching effect on the developing economies of the world.

Regardless of what policies are adopted, the very act of truly addressing the climate crisis on the scale it requires would bring home to people all over the fact that we are living in a finite planet with limits. Ultimately, a crash program to rewire the globe with clean energy would yield far more than a fuel switch. It would generate a whole new ethic of sustainability that would permeate our institutions and dynamics in ways we cannot begin to imagine.

Any macro-level plan, of course, requires the participation of the leaders of the world's governments, finance centers and corporations.

For all the rest of us, the wealth of solutions included in *Stormy Weather* eliminates forever any excuse to remain passive in the face of what is perhaps the most profound challenge ever faced by humanity.

In all her anguished expressions, nature is calling on us to save the world. *Stormy Weather* is telling us how.

Ross Gelbspan is author of *The Heat Is On: The Climate Crisis, The Cover-up, The Prescription* (Perseus Books, 1998), and maintains the website: www.heatisonline.org.

Author's Preface

ON FEBRUARY 28, 2001 a 6.8 magnitude earthquake rocked Seattle. For years, Seattle's planners, builders, engineers, councilors and citizens had prepared for just such an earthquake. They knew they were in the danger zone, so they became earthquake-smart and took the necessary precautions. Because of their preparations, Seattle escaped with only a few minor injuries and some damaged buildings.

Today, the whole Earth is in the danger zone — not from earthquakes, but from our warming climate. Our purpose in this book is to help everyone become climate-smart, and eliminate the excuse that we don't know what to do. We can either remain in denial, and continue until our addiction to fossil fuels ruins us; or we can learn from Seattle — accept that we are in the danger zone, and use our creativity to lift our civilization to a higher level where it can operate in harmony with nature, not against nature.

The alarming thing about writing this book has been the information I received every day about the impacts of climate change. I feel sick to the stomach at the thought that within my lifetime, by the year 2040, we could lose the polar bears due to the melting Arctic ice, and see the Amazon rainforest start to die as the air above it dries out. I feel every alarm bell ringing when I realize how little we understand about the relationship between the oceans, the atmosphere, and the climate. We are like teenagers, playing with things we don't understand. We do not know why the Earth's previous interglacial periods collapsed, plunging the world into another ice-age, and yet we are conducting this huge unplanned experiment in our skies. If ever there was a call to grow up as a species and to end our obsession with money, power and fossil fuels, this is it.

The exciting thing about writing the book has been the inspiration of so many people who are making a difference, and the organizations they work for. Some, like Greenpeace, work in the full light of publicity. Others, like the Sustainable Energy and Economy Network (www.seen.org), hide their light under their dedication.

It has been so encouraging to read the daily news about solar and wind energy, hydrogen, and creative policy developments. In the few weeks after finalizing the manuscript, the same few weeks that saw President Bush ditch the US commitment to Kyoto, preferring to accept shame from the world's leaders rather than upset the coal and oil industry barons who funded his election, I learned about:

- The world's largest solar project, to serve 400,000 people in the rural Philippines;
- The world's largest wind plant (300 MW), to be built on the Washington-Oregon border;
- The world's largest fuel cell installation, a 1.2 MW system to be installed at the Juvenile Training School in Middletown, Connecticut; and
- A new citizens' organization called 'Clean Air — Cool Planet' (www.cleanair-coolplanet.org) which works with colleges, churches, towns and businesses in the north-east to help them reduce their emissions.

On the downside, I also learned about new efforts by the nuclear industry to promote itself as the solution to climate change, and the efforts of the US auto-industry to shut down California's Zero Emission Vehicles rules despite a study from the Scripps Institute of Oceanography that shows a clear and definitive link between the warming of the world's oceans and the rising CO_2 emissions.

The technology and the solutions are ready: but are we?

Before getting to the answers to that question, there are a few brief things to note:

- For easy access, every website listed in the book is available on-line at the book's website, www.earthfuture.com/stormyweather.
- We have done our best to be accurate, but if you find any errors, miscalculations or corrections, please let us know via the book's website.
- This book needs to be read all over the world, and soon. If you are interested in cooperating on worldwide editions, as co-author or publisher, please let us know.
- We would also like to see the book as an educational TV series; maybe even as a dramatic musical on climate change that can reach people's hearts.

I would like to thank the many people who have given their time to help create the book, while absolving them from any responsibility for the text, and any mistakes we have failed to catch. Top of the list is Patrick Mazza, my Seattle-based co-author, who carefully checked every section, and provided a stream of useful suggestions, almost all of which found their way into the text.

I would also like to thank Ben Addlestone, Michael Armstrong, Steve Baden, Yves Bajard, Jack Barbash, Sally Bingham, Angela Bischoff, Åke Bjørke, Matthew Bramley, Jack Brautigam, David Brook, Rinaldo Brutoco, Mark Burch, Paul Burnett, Wil Burns, Arthur Caldicott, Ralph Cavanagh, Steve Clemmer, Rob Clay, Caspar Davis, Karl Davies, Derek Dexheimer, Anne Ducey, Jason Edworthy, James Elkins, Dennis Elliot, Tracy Ewen, Jeanette Fitzsimons, Brian Fleay, Ned Ford, Forest, Peter Fraenkel, Sidney Freeman, Gary Gallon, Sarah van Gelder, Tooker Gomberg, Tom Gray, Michael Grillot, Tom Hackney, James Hansen, Danny Harvey, Rick Heede, Johannes Heister, Cheeying Ho, Mark Holland, Doug Howell, Susan Innis, Marc Jaccard, Phil Jessup, Mark Jimenez, Florentin Krause, Arturo Kunstmann, Diane LeFebvre, Robert Leopold, Rebecca Livermore, Todd Litman, Lee Eng Lock, Doug Lockhart, Victoria Long, Amory Lovins, Ernie Lowe, Lawrence Mansueti, Shawn Marshall, Bo Martin, Pete & Pam Martin, Paul Maycock, Matthew McCullogh, Alan Miller, Trevor Murdock, Chris O'Brien, Susan Ode, Don Osborne, Andrew Pape-Salmon, Mac Post, Bill Prindle, Bill Rees, Chris Rolfe, Michael W. Roschlau, Rhys Roth, Al Rycroft, Paul Salveson, Joe Savage, Gerry Scott, Jean Shaffer, Peter Shepherd, Andrew Simms, Maren Souders, Dave Spittlehouse, Rosalie Stewart, Frederick Stoss, Eric Taylor, Jenny Taylor, Joe Thwaites, John Todd, Bruce Torrie, Ted Trainer, Peter Ullman, Judy Walker, Petra Weiler, Alan Weisman, Jerry Woodfill, Masakazu Yoshimori, Bryan Young, and all who kindly contributed their photos and graphics.

I would like to thank my editor, Maggie Paquet, and my publishers, Chris and Judith Plant at New Society Publishers (www.newsociety.com), who have been so encouraging all the way through. Finally, I would like to thank my friend Roger Colwill, and my ever supportive friend and partner, Carolyn Herriot, whose organic garden blooms so beautifully (www.earthfuture.com/gardenpath).

Guy Dauncey,
Victoria, June 2001

Introduction

MOVING TOWARD A WORLD WITHOUT FOSSIL FUELS

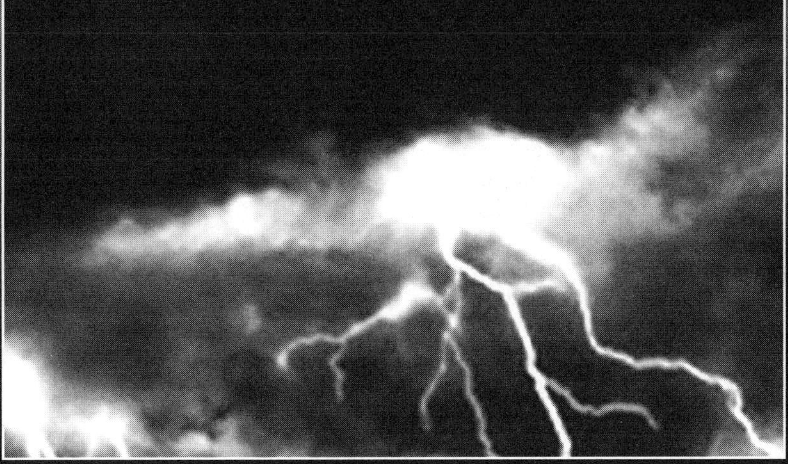

Time to Get Started

The evolution of life on Earth occurs over such a long time period that it is beyond comprehension for most people. The formation of the atmosphere, oceans, forests, and carbon sinks happened long before our time, long before the mammals or reptiles appeared. How then are we to understand what we are doing to our atmosphere today? We have no instinct to guide us, only the knowledge we have gained from science.

For the past 250 years, we have been engaged in a breathtaking process of discovery, exploration, and development -- but now we are beginning to appreciate that what is good for humanity is not always good for nature or Earth's ecosystems. The Arctic ice is melting; the coral reefs are dying.

We need to make some urgent decisions before the greenhouse gases we are releasing accumulate much further. The consequences, if we don't get them under control, are truly awful. We urgently need to change from carbon fuels to solar and hydrogen, and quadruple the efficiency of all our energy systems; we need to overhaul our methods of forestry and farming so that we store nature's carbon, instead of wasting it; we need to control our methane emissions and abandon the use of a number of industrial chemicals. For a species that has managed to visit the moon and build a space station, these things should not really be so difficult. Besides, the world's oil supply is about to start declining, giving us an extra incentive to transform our energy system to one based on solar energy and hydrogen.

This universe is huge, and full of stars like our Sun. "The stars in the cosmos are more numerous than all the grains of sand on all the beaches of the Earth," said Carl Sagan. It is reasonable to assume that life has evolved on some of the planets surrounding some of these stars, since we share the same basic chemistry throughout the universe. On a 50:50 basis, human-equivalents (let's call them "human-es") may be more advanced on half these planets and less advanced on the other half. Maybe all human-es go through this teething problem, as they wean themselves off their mother planet's stored fossil fuels and learn to generate their own energy. "Oh yes," the more advanced human-es might say, "the terrible twos, the fossil fuel blues -- those were difficult times indeed!"

Will we make this all-important shift? Or will the fossil fuel companies continue to holler and demand to be fed, sabotaging humanity's best efforts to wean itself? In the cosmic picture, nature may not care. You win some; you lose some. For all we know, the cosmos may be littered with planets whose human-es blew thousands of years of hard-earned development because they failed to make this critical adjustment. Only Earth's future historians will know. Our task is to get on with it before it is too late.

> **"Humanity is conducting an unintended, uncontrolled, globally pervasive experiment whose ultimate consequences could be second only to a global nuclear war. The Earth's atmosphere is being changed at an unprecedented rate by pollutants resulting from human activities, inefficient and wasteful fossil fuel use, and the effects of rapid population growth in many regions. These changes represent a major threat to international security and are already having harmful consequences over many parts of the globe."**
> **— Statement by WMO, UNEP, and Environment Canada at The Changing Atmosphere: Implications for Global Security Conference, Toronto, June 1988.**

Dan Guravich, Polar Bears Alive, www.polarbearsalive.org, with appreciation.

By 2040, at the present rate of warming, the Arctic summer ice will have melted away to nothing.[1] Without the ice, the polar bears will starve to death.

We can have polar bears, or we can have fossil fuels.

We cannot have both.

Our Story[1]

- 15 billion years ago: the mystery begins. The universe is born.
- 4.6 billion years ago: A star becomes a supernova, giving rise to the Sun and its planets.
- 4.4 - 4.1 billion years ago: Earth forms a thin rocky crust. Volcanoes expel hot magma to the surface, steam condenses into rain, rivers flow into great seas.
- 4 billion years ago: The rich chemical brew produces bacteria, the first living cells.
- 3.9 billion years ago: Cells invent photosynthesis. Blue-green algae store the sun's energy, taking hydrogen from the sea and carbon from the atmosphere, releasing oxygen to the air.
- 2 billion years ago: Oxygen-loving cells emerge; oxygen levels rise to near present-day levels.
- 700 million years ago: Some organisms begin living in colonies, communicating with chemical messages.
- 600 million years ago: Light-sensitive eyespots evolve into eyesight. The first animals (invertebrates) evolve in the oceans.
- 460 million years ago: By now, the ozone layer has formed, making it safe to leave the water. Worms, mollusks, and crustaceans venture out, along with algae, fungi, and insects.
- 395 million years ago: The first amphibians leave the water.
- 345-280 million years ago: Carboniferous Period. Age of amphibians. Great swamps, forests of ferns and early conifers (gymnosperms) appear and start to store carbon. Reptiles appear and evolve rapidly.
- 280-225 million years ago: Permian Period. First dinosaurs emerge; glaciations and marine extinctions; reptiles spread and amphibians decline. The world consists of one large continent called Pangaea. The temperature is 8° C warmer than today.
- 225-195 million years ago: Triassic Period; first dinosaurs and primitive mammals appear; forests of gymnosperms (early conifers) and ferns.
- 195-135 million years ago: Jurassic Period; zenith of dinosaurs, flying reptiles, birds and small mammals appear; first flowering plants (angiosperms) appear; continents drift apart.
- 135-65 million years ago: Extinction of dinosaurs (Cretaceous Period), likely caused by intense atmospheric disturbance when a large asteroid hits Earth near the Yucatan peninsula. Flowering plants (angiosperms), marsupials, and insectivorous mammals become abundant. The age of mammals begins.
- 65-55 million years ago: Something (possibly methane hydrate releases or volcanoes) causes the atmospheric CO_2 level to leap to 3,000-3500 ppm, 8-10 times the current level. Large-scale burial of the surplus carbon through swamps; formation of coal.
- 7-26 million years ago: Miocene Epoch; whales, apes, grazing mammals. Spread of grasslands as forests contract.
- 7 million years ago: Pliocene Epoch begins; large carnivores, earliest hominids (manlike primates) appear. Temperature 2.5° C warmer than today.
- 4 million years ago: Hominids leave the forest and start to spread around the world.
- 1.6 million years ago: Pleistocene Epoch of ice ages begins. Approximately 5° C colder than today.
- 400,000 years ago: Early humans start using fire as a source of energy.[2]
- 395,000 years ago: New ice age begins.
- 335,000 years ago: Ice age ends. Brief warm period.
- 310,000 years ago: Next ice age begins.
- 240,000 years ago: Ice age ends. Brief warm period.
- 230,000 years ago: Next ice age begins.

NASA

View of Earth from space

- 135,000 years ago: Ice age ends. Eemian interglacial warm period begins.[3]
- 125,000 years ago: Half of Greenland's ice-cap melts, raising sea levels by 4-5 meters.
- 122,000 years ago: Sudden cooling event in North Atlantic, switching off ocean current, maybe leading to end of warm period.
- 100,000 years ago: Modern humans leave Africa, develop language, religion, and art.
- 11,500 years ago: Ice age ends. Holocene interglacial warm period begins.
- 10,000 years ago: Humans develop agriculture and begin to shape our environment.
- 3,000 years ago: We develop writing, cities, and classical religions.
- 500 years ago: We learn how to print and share knowledge more widely.
- 350 years ago: We begin serious exploration of the material world (the Scientific Revolution).
- 250 years ago: We begin burning fossil fuels (the Industrial Revolution).
- 177 years ago (1824): Jean Fourier suggests CO_2 emissions from burning fossil fuels might accumulate in Earth's atmosphere, enhancing the planet's greenhouse effect.
- 142 years ago (1859): First oil well drilled at Oil Creek, Pennsylvania.
- 141 years ago (1860): Étienne Lenoir develops the first internal combustion engine, in Belgium.
- 105 years ago (1896): Svante Arrhenius confirms connection between CO_2 and Earth's temperature.
- 46 years ago (1955): Charles Keeling confirms that CO_2 levels in the atmosphere are rising.
- 35 years ago (1966): We see our Earth from space for the very first time.[4]
- 9 years ago (1992): We decide to protect our planet's environment (Rio Conference).
- 4 years ago (1997): We decide to reduce our greenhouse gas emissions (Kyoto Protocol).
- 1 year ago (2000): We fail to reach an agreement on how to reduce our emissions (The Hague).
- 2001: First clear evidence of the greenhouse effect seen by satellite, from space.

Resources

Cosmic Walk: www.forests.org/ric/deep-eco/cosmic.htm

Greenhouse Timeline: www.newscientist.com/global/globaltimeline.jsp

NASA's Earth Observatory: www.earthobservatory.nasa.gov

Planet Atmospheres: www.grida.no/climate/vital/01.htm

Greenhouse: The 200-Year Story of Global Warming by Gale E. Christianson (Greystone, 1999)

Planetary Weaning

Around 400,000 years ago, early humans discovered how to use fire, and realized that wood contained energy that could be used to cook and keep warm. We had embarked on the first energy revolution. Trees capture the energy of the sun through photosynthesis, and turn it into wood. About 10,000 years ago, we learned how to use fire to break up the green stone called malachite and make copper. Later, we learned how to make bronze (3,800 BC), iron (3,000 BC), and steel (1,000 BC). Using metal tools instead of stone, we were able to cut down trees and use them as fuel.

By the Middle Ages, parts of Europe were running short of wood. Houses, ships, watermills, windmills, and bridges were all made from wood; the glass industry used wood in its furnaces; the iron industry used charcoal. France was once almost completely forested. By 1300, forests covered only 32 million of her 134 million acres. In 40 days, a single iron furnace using charcoal could level a forest for a radius of a kilometer.[1]

In the 13th century, as a result of the firewood crisis, people started using coal - the second energy revolution had begun. In 1257, Queen Eleanor of England was driven from her castle in Nottingham by the foul fumes from the sea-coal being burned in the industrial city below. We have been burning coal ever since. In 1998, the world burned five billion tons, 23% of the world's total energy supply, and released 2.2 billion tons of carbon into the atmosphere, 36% of the world's carbon emissions from fossil fuels. The world's coal supply is not running out. There is enough for another 200 years, but we can only afford to use it if we find a cost-effective way to separate the CO_2 emissions and sequester them safely (see Solutions #53, #54).

Then there was oil, which comes from the stored sunlight of mostly marine organisms that lived millions of years ago. Like coal, it releases carbon when burned, forming CO_2. We first used oil in 1861. By 2000, we were using 75 million barrels a day, releasing 43% of the world's carbon emissions from fossil fuels. Natural gas accounts for the remaining 21%.

At this point in the story, we realized that the carbon emissions released by fossil fuels were accumulating in our atmosphere. Their presence has increased from 275 parts per million to 370 ppm, and is getting us into erious trouble. Just as coal replaced firewood in the 13th century, however, a third energy revolution is ready to replace fossil fuels (and it's not nuclear power — see p. 42). In case the climate crisis is not enough to persuade us to change, however, mother nature has another surprise in store: the world's oil supply is about to peak and then start running out.

"Our children and grandchildren are going to be mad at us for burning all this oil. It took the Earth 500 million years to create the stuff we're burning in 200 years. Renewable energy sources are where we need to be headed."
— Jack Edwards, Professor of Geology, University of Colorado

Resources

Are We Running Out of Oil? poster: http://geopubs.wr.usgs.gov/open-file/of00-320

Oil as a Finite Resource: www.wri.org/wri/climate/finitoil

The Coming Global Oil Crisis: www.oilcrisis.com

US Energy Information Administration: www.eia.doe.gov

World Energy Prospects to 2020, International Energy Agency: www.iea.org/g8/world/oilsup.htm

How Much Oil is Left?

The general consensus, formed over 50 years and from scores of studies, is that Earth's total oil supply is between 2,000 and 2,800 billion barrels. The halfway marks are 1,000 and 1400 billion barrels, respectively. By the end of 2000, the world had consumed 899 billion barrels. Each year we are consuming 28 billion barrels (rising by 2% a year), but discovering only 6 billion barrels. (A barrel is 42 gallons). If the global supply is 2,000 billion barrels, we'll reach the halfway mark in 2004; if it is 2,800 billion barrels, we'll reach it in 2018. From that moment on, global demand will outstrip supply, causing prices to skyrocket as everyone competes for what's left. The atmosphere's rising temperature is telling us not to burn any of the remaining oil, but hydrogen derived from non-fossil fuel sources is not yet ready as a fuel for transportation, so the reality is that we're going to go on burning it for a while yet.

The key to understanding the coming oil crisis is not knowing when the oil supply will run out, but when its production will peak: after that date, consumption will outstrip production. In the 1950s, the Shell geologist King Hubbert argued that the big oil fields are always discovered first; when they run out, the industry turns to the smaller, more difficult fields. There comes a point when oil discovery peaks, followed some years later by a production peak. Hubbert used his method to successfully predict that US oil production would peak in 1970. On a global level, discovery peaked in the 1960s. Since then, we have been finding less each year. Geologist Colin Campbell, who has worked for major oil companies for 45 years, has used the same method to predict that the world's production will peak in 2005. "One indisputable fact stands out," Campbell told the British House of Commons in 1999, "discovery peaked 30 years ago. It takes no feat of intellect to conclude that we now face the corresponding peak of production."

In 1998, the International Energy Agency told the G8 energy ministers that the non-OPEC peak for the world outside the Middle East would arrive sometime in 2001, and the global peak would come in 2015. By 2020, they are projecting a 20% shortfall of supply to demand. Somewhere in this period, we're going to face the crunch, with a natural gas shortfall not far behind (see p. 43).

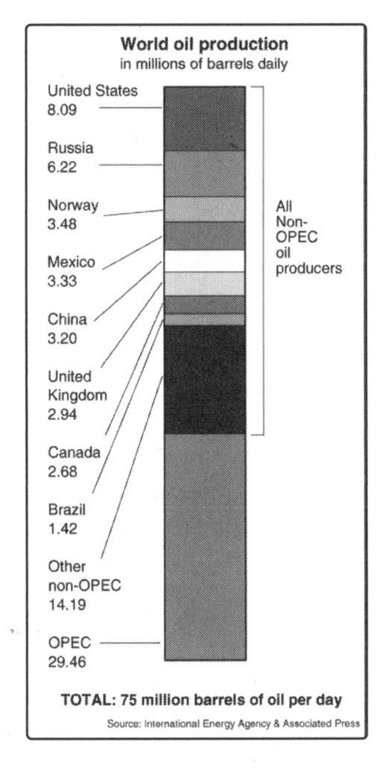

"What about the new discoveries in Kazakhstan, in the Caspian Sea?" you might ask. Yes, it's a big field: between 25 and 60 billion barrels, of which 10%-50% may be recoverable. In a world that uses 28 billion barrels a year, however, this is just 1 to 12 months supply. It doesn't do much to solve the problem.

This book is about finding solutions to the world's climate crisis. If you were looking for an additional reason to stop using fossil fuels, the oil crisis is arriving just in time. We will soon look back with nostalgia on the days when oil was $30 a barrel. From a planetary perspective, the timing could not be better. The single biggest factor that will encourage a shift to renewable energy, better than all the policies in the world, is higher oil prices.

And now - on with the climate change crisis.

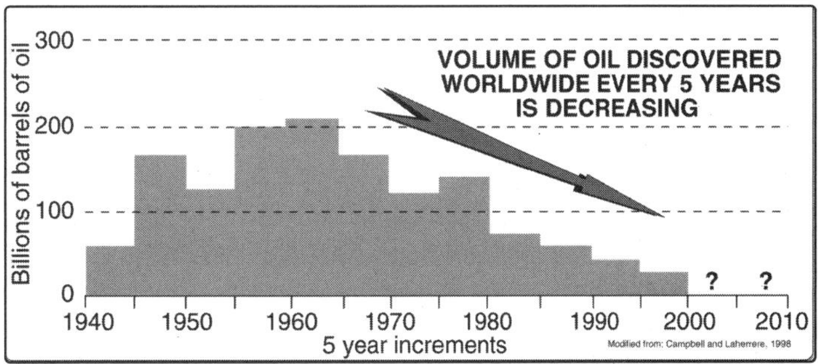

The Greenhouse Effect

The sun pours its energy into the entire solar system, which by nature is rather cold. If you stood on a planet such as Mars, which has almost no atmosphere and no greenhouse effect, the sun would keep you warm during the day at 98 F°, but at night, when the temperature fell to -189 F°, you would discover how cold the cosmos can be. Bye-bye life, hello deep-frozen astronaut. If you stood on Venus, however, where the atmosphere traps so much of the sun's heat that the average temperature is 860° F, you'd learn what "hot" meant. Bye-bye life, hello astronaut fricassée.

The difference is caused by an atmosphere that contains gases that trap the sun's heat, giving us the greenhouse effect. On Earth, as far as humans are concerned, the greenhouse effect happens to have got it "just right" for the past 10,000 years. With an atmosphere that's 77% nitrogen, 21% oxygen, 0.9% argon, 1% water vapor, and 0.3% trace gases (CO_2 275 parts per million, methane 700 parts per billion, nitrous oxide 275 parts per billion), we have been able to enjoy a pleasant average temperature of 59° F (15° C), just right for gardening and growing forests. It's the miniscule presence of water vapor and trace gases that makes all the difference. They create the *good* greenhouse effect. If we had no greenhouse effect, Earth's average temperature would be 0° F (-18° C), which is bad for tomatoes and trees.

It was not always so pleasant. Water vapor, CO_2, ozone, methane, and nitrous oxide all trap the Earth's heat as it is reflected back into the coldness of space, but the gas that makes the biggest difference is CO_2. Over the past 400,000 years, there have been four occasions when CO_2 levels fell below 200 ppm, causing the global average surface temperature to fall by approximately 5° C and plunging the world into an ice age. In all those years, CO_2 levels never rose above 298 ppm, and Earth's temperature was never more than 3° C above today's. By 2000, CO_2 levels had reached 370 ppm, the highest in 20 million years.[1] It is desperately important that we understand how the greenhouse effect works, and reduce our emissions to less worrying levels.

When the sun's rays hit the Earth, 30% of the incoming visible light is reflected back into space, 25% by clouds and 5% by ice; the remaining 70% is captured by the greenhouse gases warming the earth and especially the oceans. Plants, algae, and ocean phytoplankton capture the sun's energy through photosynthesis, and use it to convert carbon dioxide into carbon as part of nature's carbon cycle, a process we call carbon sequestration. The carbon cycle determines how much of our excess CO_2 is absorbed and how much remains in the atmosphere to trap more heat.

"On the island where I live, it is possible to throw a stone from one side to the other. Our fears about sea level rise are very real. Our Cabinet has been exploring the possibility of buying land in a nearby country in case we become refugees of climate change."
—Teleke Lauti, Minister for the Environment, Tuvalu.

Resources

Carbon Cycle (1980-1989):
www.grida.no/climate/vital/13.htm

Carbon Cycle (1992-1997):
http://cdiac.esd.ornl.gov/pns/graphics/globcarb.gif

Carbon in Live Vegetation:
http://cdiac.esd.ornl.gov/ftp/ndp017/table.html

Greenhouse Effect: www.grida.no/climate/vital/03.htm

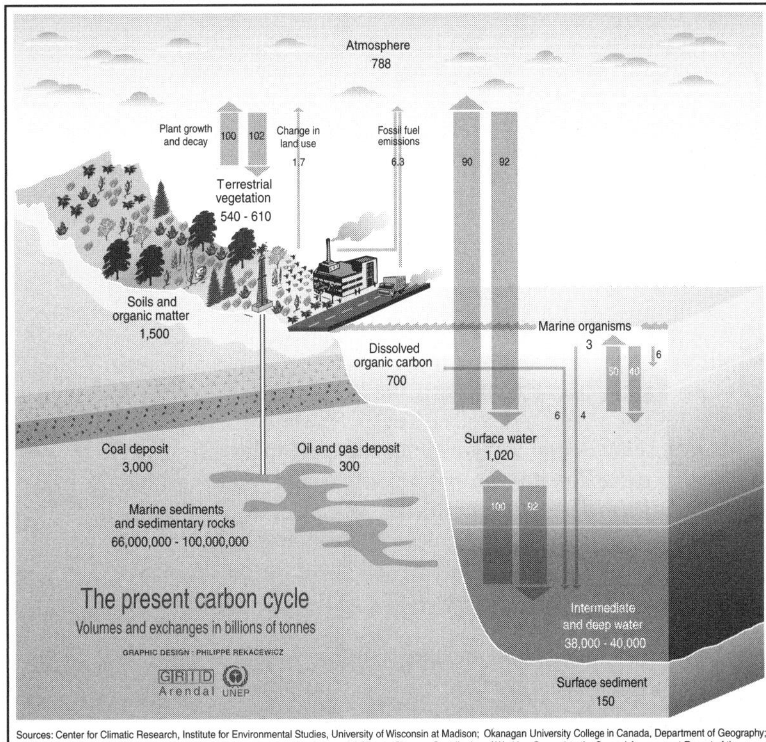

Sources: Center for Climatic Research, Institute for Environmental Studies, University of Wisconsin at Madison; Okanagan University College in Canada, Department of Geography; World Watch, November-December 1998; Climate Change 1995, The Science of Climate Change, Contribution of Working Group 1 to the Second Assessment Report of the Intergovernmental Panel on Climate Change, UNEP and WMO, Cambridge University Press, 1996. Updated with data from the IPCC WG1 Third Assessment Report, Chapter 3, and the Carbon Dioxide Information Analysis Center, Global Carbon Cycle 1992-1997, with atmospheric carbon updated for 2001, with permission from Åke Bjørke at GRIDA.

The First Sink: The Oceans

In general, half of nature's photosynthesis is carried out in the oceans by phytoplankton,[2] and half on the land by bacteria, plants, and trees. Every year the oceans absorb around 92.4 billion tonnes of carbon and release 90 billion tonnes, storing 2.4 billion tonnes as dissolved inorganic carbon in the deep ocean (see diagram). Over the millennia, the oceans have socked away almost 40,000 billion tonnes of carbon, where, hopefully, it will remain. A quarter of this is stored as frozen methane hydrates, which have the potential to thaw as the ocean warms, triggering a very nasty turn of events (see p. 19). There are also concerns that in a warming ocean, diatoms might prevail over the algae that absorb carbon, stalling the ocean's ability to absorb CO_2.

The Second Sink: The Soil

Soil is not just "dirt". It is a mass of minerals, moisture, bacteria, and thousands of other micro-organisms that absorb carbon from plants and trees as they die. Every year, the world's soils absorb 50 billion tonnes of carbon from dying vegetation and release 50 billion tonnes through decomposition. Forest destruction and farming weaken the soil, however, causing 1.5 billion tonnes a year to be lost to the atmosphere. In western Canada, when an oldgrowth Douglas fir forest is clearcut, it can take 150 years for the soil and forest to recover the carbon they had before the trees were felled, which is why it is so important to change our methods of forestry.[3] Over the millennia, the world's soils have accumulated 1,500 billion tonnes of carbon, of which 500-800 billion tonnes is locked up in the world's peatlands, including 500 billion tonnes in the Arctic tundra, where rising temperatures are unlocking it as the snow cover disappears,[4] opening another troublesome scenario (see p. 18).

The Third Sink: Forests and Vegetation

The world's forests and vegetation store around 550 billion tonnes of carbon, 40% in the tropical forests. Every year, forests lose 50 billion tonnes of carbon to the soil and 50 billion tonnes to the atmosphere through respiration, but they absorb 101.5 billion tonnes from the atmosphere, reducing its load by 1.5 billion tonnes. Yes, trees matter. In our warming world, however, scientists fear that by 2050, the tropical forests will cease storing CO_2 and start releasing it, adding another factor to the scary prospect of a runaway greenhouse effect.[5]

In the balanced natural carbon cycle, approximately half of the atmospheric CO_2 is exchanging with the forests and soil, and half with the global oceans, resulting in no net atmospheric increase. In the disturbed carbon cycle that we are creating, approximately a quarter is going into the soil and vegetation, a quarter is going into the oceans, and half is accumulating in the atmosphere, trapping the sun's heat, which is why we are having this little problem.

"We have a mighty task before us. The Earth needs our assistance."
— Laurens van der Post

The Greenhouse Gases

Before the industrial age, we didn't have a problem with greenhouse gases. If anything, according to the pattern of the past 420,000 years, we were due for another ice age. Then we started adding greenhouse gases to the atmosphere. If we take carbon dioxide, methane, and nitrous oxide, the three main gases after water vapor, and express them as carbon equivalents, we are adding 10 billion tonnes of carbon-equivalent to the atmosphere every year that nature would not have put there. At the present trend, this will rise to 10.6 billion tonnes per year by 2010.

The best way to measure the impact of the various gases is by their "radiative forcing". This is a measure of the extent to which a gas alters the balance of incoming and outgoing energy in the atmosphere. The sun's natural radiation is around 240 watts per square meter (240 W/m2); the greenhouse gases are increasing this by about 1% (2.78 W/m2). Most aerosols (such as dust) have a negative radiative forcing, though there is still much uncertainty about the data. The various gases also have different global warming potentials (GWP). GWP is the standard used to compare the radiative forcing of each greenhouse gas over 100 years, using CO_2 as a baseline, which is given a GWP of 1 (see chart opposite).

Carbon Dioxide is released by the fossil fuels we burn to generate electricity, drive the world's 530 million vehicles, and heat our homes and buildings. It is also released by the flaring of natural gas and oil, and by the production of cement. (Cement is made from limestone, the crushed shells of ancient carboniferous sea-creatures that surrender their carbon when we turn them into patio tiles.) It is released when forests and savannah grasslands are burned down, and when forests are clearcut, which destroys the soil they depend on. It is released by plowing and disking when farmers don't farm sustainably, and lose the carbon in their soil. In 1999, the worlds carbon emissions from fossil fuels came to 6.144 billion tonnes; global forest and net biomass loss added another 1-2 billion tonnes. Cement production may add 100 million tonnes, and there are no clear figures for soil loss. The present concentration is probably the highest it has been for 20 million years; the rate of increase is the highest for at least 20,000 years.

To convert carbon emissions into CO_2 emissions, multiply by 3.667 to include the addition of oxygen. Most global data is expressed as carbon, while most national and local data is expressed as carbon dioxide — just to keep us on our toes.

Resources

Canada's greenhouse gas emissions:
www2.ec.gc.ca/pdb/ghg/English/eDocs.html

Climate forcings in the industrial era:
www.giss.nasa.gov/research/intro/hansen.05

CO_2 in the atmosphere - the Mauna Loa Curve:
www.grida.no/climate/vital/06.htm

CO_2 Mauna Loa, Hawaii:
http://cdiac.esd.ornl.gov/ftp/maunaloa-co2/maunaloa.co2

Current greenhouse gas concentrations:
http://cdiac.esd.ornl.gov/pns/current_ghg.html

Greenhouse gas emissions in 35 developed nations:
www.grida.no/db/maps/collection/climate6

Methane in the atmosphere:
www.epa.gov/outreach/ghginfo/topic1.htm

US greenhouse gas emissions:
www.eia.doe.gov/oiaf/1605/ggrpt

World CO_2 emissions from fossil fuels:
www.eia.doe.gov/iea/tableh1.html

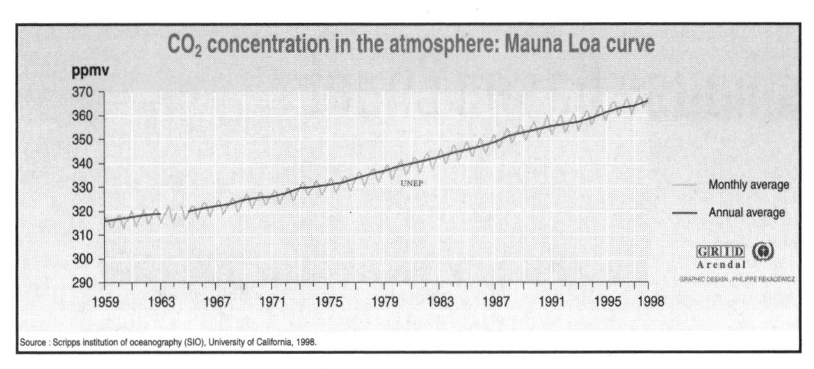

Methane comes from a variety of sources. It escapes from landfills. Cows release it during digestion, and liquid slurry ponds release it when animal wastes break down without oxygen. It is released when natural gas escapes (natural gas is 85%-95% methane), and it is released by coal mines, oil wells, sewage treatment plants, and flooded rice fields. It was recently discovered that methane is produced by rivers that have been dammed to generate hydroelectricity, when biomass in and entering the reservoir breaks down without oxygen. The present atmospheric concentration is the highest for at least 420,000 years.

Nitrous Oxide is released by the use of nitrogen-based chemical fertilizers in farming and by transportation, nitric acid production, poor manure management, and various other sources.

The CFCs, HCFCs and HFCs. Remember the CFCs (chlorofluorocarbons), the chemicals we used as coolants in fridges and air-conditioning systems and as propellants in spray cans, that are blowing holes in the ozone layer? Turns out they are potent global warming gases, too. They are being phased out, but their replacements, the HCFCs and HFCs (hydrochlorofluorocarbons and hydrofluorocarbons) are also potent greenhouse gases. (HCFCs also attack the ozone layer, but much less so.) See Solution #95 for more on this muddle. Life gets complex here, because the ozone that CFCs destroy is a natural greenhouse gas, counterbalancing some of their radiative forcing.

Other Industrial Chemicals. There's perfluorocarbons; used in the manufacture of aluminum and semi-conductors; sulfur hexafluoride, used in the production of magnesium, as a dielectric in electrical transmission and distribution and in the semi-conductor industry; and trifluoromethyl sulfur pentafluoride, that nobody knows much about.[1]

Tropospheric Ozone is also a greenhouse gas. Its accumulation in the lower atmosphere (troposphere), up to seven miles high, is caused by pollutants from burning fossil fuels. Ozone in the upper atmosphere (seven to 30 miles up) is a good thing.

Aerosols. These are very fine dust particles or liquid droplets that come from fossil fuel burning, forest and biomass burning, and industrial pollution. With the exception of black carbon, they have a negative radiative forcing, shielding Earth from the impact of the main greenhouse gases. The data on the next page is from the 2001 IPCC report, but there is not a good understanding of how the different aerosols work. James Hansen, Director of NASA's Goddard Institute for Space Studies, thinks that black carbon, or soot, may contribute 15%-30% to the overall radiative forcing.[2] If this is so, it changes the responsibility of the various greenhouse gases: CO_2 – 41%; Black Carbon – 22%; Methane – 13%; Ozone – 10%; CFCs etc – 10%; Nitrous Oxide – 4%. 90% of black carbon comes from burning biomass and fossil fuels.

Resources:

- Carbon Dioxide Information Analysis Center: http://cdiac.esd.ornl.gov
- CO_2 emissions from industrial processes: www.grida.no/climate/vital/09.htm
- CO_2 emissions from land-use changes: www.grida.no/climate/vital/10.htm
- Global atmospheric concentration of CO_2: www.grida.no/climate/vital/07.htm
- Global greenhouse gases: www.grida.no/climate/vital/05.htm
- Global GHG emissions (real time): www.grida.no/db/maps/collection/climate5
- HFCs, PFCs, and SF6: www.epa.gov/globalwarming/emissions/national/xfcs-sf6.html
- Industrial greenhouse gases: www.epa.gov/outreach/ghginfo/topic8.htm
- Radiative forcing: www.grida.no/climate/vital/04.htm
- Surface/atmosphere methane exchange: www.giss.nasa.gov/data/ch4fung

Greenhouse Gases Chart

n/a = not applicable n/k = not known	Pre-Industrial Concentration (1860)	Concentration in 2000[1]	Average atmospheric lifetime[2]	Growth Rate % per yr.
Water Vapor	Variable (1-3%)	Variable (1-3%)	Few days	0.2%[6]
Carbon Dioxide CO_2	288 ppm	370 ppm Rising by 1.5 ppm per year	50 - 200 years	0.45%
Methane CH_4[9]	750 ppb	1750 ppb	12 years	0.6%
Nitrous Oxide N_2O	285 ppb	312 ppb	120 years	0.25%
CFCs (eg CFC-12)	0	533 ppt	102 years	1%
HCFCs (eg HCFC-22)	0	142 ppt	12 years	4.2%
HFCs (eg HFC-23)	0	12ppt[11] (+0.55ppt pa)	1 - 264 years	5.1%
Perfluorocarbons (PFCs, eg CF_4)	0	79 ppt	3,200 – to 50,000 years	1.4%
Sulfur hexafluoride SF_6	0	4.7 ppt	3,200 years	6.3%
Trifluoromethyl sulfur pentafluoride SF_5CF_3	0	0.12 ppt	3,500 years	n/k
Tropospheric ozone O_3	25 ppb	25/26 ppb	Weeks	n/k[13]
TOTAL				
Aerosols (net effect)	>0	Variable	Days/weeks Very variable	n/k[15]

"There is no known geologic precedent for the transfer of carbon from the Earth's crust to atmospheric carbon dioxide, in quantities comparable to the burning of fossil fuels, without simultaneous changes in other parts of the carbon cycle and climate system. (...) The present level of scientific uncertainty does not justify inaction in the mitigation of human-induced climate change and/or the adaptation to it."
— American Geophysical Union, policy statement, 1998

Anthropogenic Sources (from human activities)	Global Warming Potential (GWP) over 100 years[3]	Current Radiative Forcing Watts per sq. meter[4]	Current Share of Overall Climate Change[5]
All of the below[7]	0[8]	n/a	n/a
Fossil fuel combustion (75%) Poor forest management (n/k) Deforestation (24%) Cement production (0.6%) Poor soil management (n/k)	1	+1.46	52.5%
Fossil fuel extraction (20%) Dams, reservoirs (20%) Livestock digestion (18%) Rice paddies (17%) Landfills (10%) Animal manure/slurry (7%) Carbon monoxide emissions[10]	23 (62 over 20 years)	+0.48	17.3%
Poor soil management (70%) Transportation (14%) Industrial processes (7%)	296	+0.15	5.4%
Liquid coolants, foams	8,100		
Liquid coolants	1,500		
Liquid coolants CFC & HCFC substitutes	12,000		
Aluminum manufacture (59%) Solvents and other (26%) Plasma etching (15%)	5,700	+0.34[12]	12.2%
Magnesium production Dielectric fluid	22,200		
Unknown	~18,000		
Indirectly, via industrial pollutants[14]	n/a	+0.35	12.5%
		2.78	**100%**
Fossil fuel combustion Biomass burning	n/a	-0.6[16]	-21.6%

The Enhanced Greenhouse Effect

So what do these gases do when you add them to the atmosphere? That's what the world's climatologists and oceanographers have been looking at.

One of their projects has been to drill 3,600 meters into the ice at Vostok station in East Antarctica, retrieving frozen samples of ice that date back 420,000 years.[1] When they melt the ice and analyze the air trapped in it, they can produce readings for temperature, carbon dioxide, methane, and dust throughout the period. The record confirms that there have been four major ice ages, each lasting 65,000 to 100,000 years. During each ice age, the temperature was 6-8° C lower than it has been for the past 11,000 years, and levels of CO_2 and methane were equally low. The connection between temperature and CO_2 and methane is almost mathematical.

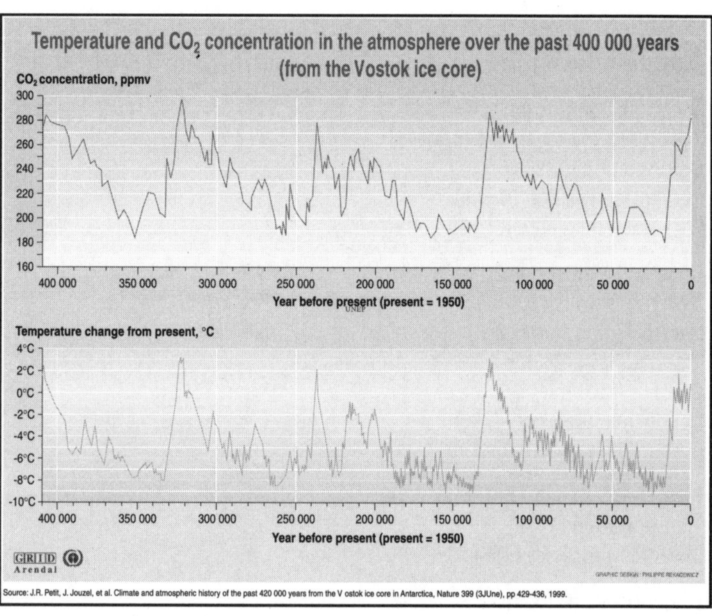

Source: J.R. Petit, J. Jouzel, et al. Climate and atmospheric history of the past 420 000 years from the V ostok ice core in Antarctica, Nature 399 (3JUne), pp 429-436, 1999.

Each of the ice ages was followed by a relatively sudden "termination," when the temperature rose to 5° C above the recent norm, accompanied by a rise in CO_2 and methane and a fall in the amount of dust in the atmosphere. (Dust prevents solar radiative forcing, creating a shield against the sun's rays.) CO_2 rose from 200 parts per million to 280-300 (today's level is 370), while methane rose from 300-400 parts per billion to 550-800 (today's level is 1750). The warm periods lasted 6,000-10,000 years, then the world plunged back into the next ice age, accompanied by falling CO_2 and methane.

The scientists do not yet understand why the climate stabilized 11,000 years ago instead of peaking and freezing. The Vostok research shows that the recent stability (the Holocene) has been the longest in the past 420,000 years. This has been the *exception*, not the rule; without the warmer temperature, there would have been no farming, no food surpluses, no cities, no civilizations. We mess with this stuff at our peril.

The Vostok scientists believe that each warming period starts with a change in the nature of the Earth's orbit, bringing an increase in solar radiation. The warming is amplified by simultaneous increases in CO_2 and methane caused by the massive increase in flooding as the ice melts, and by a decrease in dust. As the temperature rises, the melting snow and ice lose their albedo effect, allowing the earth and oceans to absorb more heat. Albedo is the name given to the reflection of the sun's light off a surface. Snow has a very high albedo, reflecting 75%-95% of the light; tundra and forest have a low albedo (15%-20%) and absorb a lot of heat. Water's albedo can be as low as 5%. The climate is also influenced by changes in the albedo of the clouds — cirrus clouds have low albedo (15%-20%) while stratus clouds have high albedo (70%) — and by particles of dust in the atmosphere from volcanoes and industrial pollution, which have a cooling effect.

"The scientific consensus presented in this comprehensive report about human-induced climate change should sound alarm bells in every national capital and in every local community."
— Klaus Topfer, United Nations Environment Program, commenting on the IPCC's Third Assessment Report, January 2001.

So what is happening in the greenhouse today? The concentration of CO_2 in the atmosphere is the highest that it has been for 20 million years.[2] If the increase continues, in 100 years time it will be the highest since the Eocene Epoch, 50 million years ago, when there were no ice caps at all, London (England) was a tropical swamp and there were crocodiles on today's Arctic islands.

There is no doubt that the Earth is warming. The evidence of warming has been confirmed by research that looked at 15 different records of past climate, including tree rings in Colorado, ice cores from Tibet, old English shipping records, ancient Chinese writings, and mud from the bottom of the Sargasso Sea.[3] From 1750 to 2000, the temperature in the Northern Hemisphere rose by 2° F (1.1° C). An analysis of more than 600 deep boreholes shows that temperatures rose by 1.25° F (0.7° C) from 1750 to 1990.[4] During the 1990s, temperatures rose six to seven times faster, by 0.72° F (0.4° C). Since 1990, the rate of warming has increased to 7° F (4° C) per century. Scientists have concluded that there is only a 1-in-20 chance that the current string of high temperatures is a result of natural phenomena; the evidence points overwhelmingly to human activities as the cause.[5] British scientists at the Hadley Centre for Climate Research have predicted that the global temperature could increase by as much as 14° F (8° C), as the increased warmth causes forests and oceans to stop absorbing CO_2 and begin releasing it.[6]

Climate science is like detective work, piecing together clues about the past, studying the present, and making informed guesses about the future. Climate modeling is very complex and there are many different feedback loops to consider, including the effects of changing cloud cover, albedo, solar radiation, ocean temperatures, carbon sinks, sulfate aerosols, ozone, and so on. The UN Intergovernmental Panel on Climate Change (IPCC) is an elite body of 2,500 global scientists who work together to establish a consensus on what the science is telling us. In their 2001 report, they said that "there is now stronger evidence for a human influence," and that man-made greenhouse gases "have contributed substantially to the observed warming over the last 50 years." Their estimate is that the global temperature will rise by 2.5° F to 10.4° F (1.4° to 5.8° C) by 2100, replacing their previous estimate of 1.8° F to 6.3° F (1.0° to 3.5° C). Of the ten warmest years on record since 1860, eight have occurred since 1990.

Resources

Albedo for water, snow, forest, clouds: www.stvincent.ac.uk/Resources/Weather/albedo.html

Climate Projections for the New Millennium: www.environmentaldefense.org/programs/GRAP/y3k

Cooling factors: www.grida.no/climate/vital/14.htm

Global average surface temperature, 1860-2000: www.grida.no/climate/vital/17.htm

Global Warming - Focus on the Future: www.enviroweb.org/edf

Greenhouse Lessons from the Geologic Record: http://gcrio.ciesin.org/CONSEQUENCES/winter96/geoclimate.html

IPCC Publications: www.ipcc.ch/pub/pub.htm

IPCC Report *Climate Change 2001: The Scientific Basis:* www.meto.gov.uk/sec5/CR_div/ipcc/wg1

New Scientist: www.newscientist.com/global

Precipitation changes: www.grida.no/climate/vital/18.htm

Sea-level rise over the past century: www.grida.no/climate/vital/19.htm

State of the Cryosphere (snow and ice): http://nsidc.org/NASA/SOTC

Temperature and Atmospheric CO2 over 420,000 years: www.grida.no/climate/vital/02.htm

UNEP Climate Change Info Kit: www.unep.ch/conventions/info/ccinfokit/infokit-1999.htm

US Global Change Research Program: www.usgcrp.gov

The Climate Impacts

> "I've never seen anything like this before. What surprises me is how much [of the ocean warming] is in the deep water."
> — Peter Rhines, Oceanographer, University of Seattle

What will it all mean? All over the world, people have been studying how the enhanced greenhouse effect will affect life on Earth.

Prediction: The arctic ice will melt.
Reality: Temperatures in the Arctic have risen by more than 4° F in the past 30 years, twice as much in as the rest of the world. Some areas are up to 7° F warmer. Arctic ice is normally up to 3 meters thick, but when Russian and US submarines compared their data, they found that it had lost 40% of its depth since 1970. At this rate, the summer ice will be gone by 2040, and the polar bears will starve to death since they will be unable to hunt on the summer sea ice. In 2000, a scientific cruise ship found open water at the North Pole: the last time this happened may have been 50 million years ago. The Greenland ice sheet has thinned by a meter on its southern and eastern edges. At the other end of the world, the Antarctic ice has been melting slowly for the past 10,000 years, but the speed of melting may be increasing. On the Antarctic Peninsula, the temperature has risen by 4.5° F over the past 50 years; between 1973 and 1993, the sea ice decreased by 20%. The IPCC scientists do not foresee any loss of the grounded ice (e.g. the west Antarctic ice sheet) during the 21st century.

Prediction: The glaciers will melt.
Reality: In Glacier National Park, the number of glaciers has fallen from 150 to 50 since 1850. In Africa, Mount Kenya's largest glacier has lost 92% of its mass, and Mt Kilimanjaro has lost 75% of its ice. In Argentina, the Upsala Glacier is retreating by 60 meters a year. In New Zealand, the Tasman Glacier has retreated by 1.7 miles since 1971. Scientists predict that by 2050, a quarter of the world's mountain glacier masses will have disappeared.

Prediction: The oceans will warm.
Reality: Between the mid-1950s and the mid-1990s, the overall world ocean warmed by 0.06° C to a depth of 3,000 meters. In the top 300 meters, the warming was 0.31° C, or 0.3 watts per square meter.[1] In the northern hemisphere, the ocean surface is warming by 1° F per decade, similar to the amount that the entire planet has seen over the past 100 years.[2]

Prediction: Sea levels will rise; coastal areas will flood.
Reality: As the ocean warms, it expands. The 2001 IPCC forecast is for a rise of 3.5 to 35 inches (0.09 to 0.88 meters) by 2100, and a 7-13 meter rise over the next 500 years. If global warming is halted within the century, thermal expansion will still raise the oceans by 0.5 to 4 meters as the heat reaches the ocean deeps. In 2000, floods kept two-thirds of Bangladesh under water for two months, setting back many of their social and economic gains. This much sea level rise will affect far more than Polynesia and Micronesia. Thirty of the world's largest cities are close to sea level, including London, New York, and Shanghai.

Prediction: Spring will come earlier.
Reality: In Europe, an extensive study of gardens has found that spring is arriving six days earlier and fall is arriving almost five days later.[3] In England, frogs are spawning 9-10 days earlier;[4] in the US, spring migrants are being recorded earlier as they reach Michigan's Upper Peninsula.[5]

Prediction: There will be increased precipitation and floods.
Reality: As the temperature warms, more water evaporates from the oceans, causing more downpours and more snow. In 1998, Sydney (Australia) broke all records with 12 inches of rain in two days; Texas had two rainstorms that dumped 10 and 20 inches, causing $1 billion in damage and 31 deaths; the Black Hills of South Dakota had 8.5 feet of snow in five days. In 1999, Mount Baker (WA) had 95 feet of snow between November 1998 and June 1999, a world record.[6] The NOAA data show that in stronger storms, the amount of rainfall per storm has increased by about 10% in recent decades.[7] The clearing of lands and filling of wetlands makes the flooding worse.

Prediction: More heat waves and droughts.
Reality: In 1998, Tibet had its warmest June on record; Christchurch (NZ) had its warmest February; Edmonton (Canada) had its warmest summer; Little Rock, Arkansas had its hottest May; Cairo (Egypt) had its warmest August. From April to June 1998, Florida, Texas, and Louisiana had their driest period in 104 years; in Texas, temperatures stayed over 100° F for 15 straight days. In 1999, New York had its hottest and driest July on record; the eastern US had its driest growing season on record, with agricultural disaster areas being declared in 15 states. Since 1980, droughts are visiting the US, Europe, Africa, and Asia more frequently. There will also be much colder periods in some places.

Prediction: More forest fires and insect damage.
Reality: In 1998, Florida had its worst wildfires in 50 years, and Mexico had its worst fire season ever. In Canada, the area of forest consumed by fire each year has risen steadily since the 1970s; the Canadian Forest Service is predicting a 50% increase by 2050.[8] Scientists in Alaska have reported that 20 million hectares of forest are suffering from an unprecedented attack by spruce budworms as a result of warmer weather.

Prediction: Ecosystems will be disrupted.
Reality: From Bermuda to the Great Barrier Reef, coral reefs are dying from the warmer ocean temperatures — these are the rainforests of the ocean, where 65% of the world's fish species dwell. In the North Pacific, salmon are starving as their food disappears, and migrating north to escape waters that are warmer than 7° C.[9] If the salmon go, so will the orca whales, eagles, and bears that depend on them for food. Alpine plants are retreating up the mountainsides in Austria. Butterfly ranges are shifting north in Europe and California. Adelie penguin populations in the Antarctic have fallen by 33% since 1975 as the winter sea ice melts. The World Wide Fund for Nature has warned that a third of the world's habitat could disappear or change beyond recognition by 2100, including Newfoundland, Ontario, British Columbia, Quebec, Alberta, Manitoba, and the Yukon where more than half the area is at risk.[10]

Prediction: Tropical diseases will spread.
Reality: Malaria has been transmitted by mosquitoes in the northeastern US states and in Toronto; the West Nile virus has arrived in New York; in Mexico and Columbia, dengue fever has spread to higher elevations — 4,000 feet above its normal range.

Prediction: Increase in weather-related disasters.
Reality: From Mozambique to Honduras, and from Canada to China, weather disasters have been proliferating. In 1998, violent weather caused $89 billion worth of damage, killing 32,000 people and displacing 300 million. Insurance companies paid out $91.8 billion in losses from weather-related natural disasters from 1990-1998, four times more than weather-related claims in the 1980s.[11] In China, the 1999 flooding of the Yangtze River region killed 3,500 people, destroyed 5 million homes, and dislocated 200 million lives. In 2000, natural disasters rose by 100 to a record 850; storms accounted for 73% of the insured losses, floods for 23%.[12]

And this is only the tip of the melting iceberg.

Resources

Biodiversity impacts: www.pacinst.org/wildlife.html

Climate Change Impacts on the US: www.gcrio.org/nationalassessment

Climate Hot Map: www.climatehotmap.org

Country-by-country warming rates: www.tyndall.uea.ac.uk/resources.htm

IPCC Report, *Climate Change 2001 - Impacts, Adaptation and Vulnerability:* www.usgcrp.gov/ipcc

Operational Significant Event Imagery: www.osei.noaa.gov

Waterworld: www.pbs.org/wgbh/nova/warnings/waterworld

The Runaway Greenhouse Effect

"If you think mitigated climate change is expensive, try unmitigated climate change." — Dr. Richard Gammon, University of Washington, speaking on the steps of the US Congress, June 28, 1999.

The greenhouse effect is not a static process. It only seems that way because the past 10,000 years have been so stable. But this is *not* the way the world's climate works. If you think the enhanced greenhouse effect is troublesome, wait until the runaway greenhouse effect kicks in. As the saying goes, "you ain't seen nothin' yet."

Every previous interglacial period in the past 1.8 million years (the Pleistocene Epoch) ended with the temperature climbing to 2-3° C above the pre-industrial level and then crashing down to the next ice age. Prior to the Pleistocene, Earth's temperature was permanently higher. During the Cretaceous Period, 65 million-135 million years ago, it was 6-8° C higher - and there were monkeys and alligators on Ellesmere Island near the North Pole. London was a tropical mangrove swamp.

What caused the interglacial temperatures to crash during the Pleistocene? Scientists don't know. But they do know that climate transitions can occur suddenly in as little as one or two decades.[1] The scary thing is that we may be about to find out what causes these changes, if we don't act quickly to stabilize the climate. We're grown-up planetary citizens now, not squabbling nations. We've got to accept this responsibility before it is too late.

Far more dangerous than the enhanced greenhouse effect, the runaway greenhouse effect includes feedback from the various climate change impacts, which intensifies things. There are seven acknowledged climate crunchers, any or all of which might trigger a runaway effect, or play a contributing role:[2]

The Tundra Thaws

The Siberian and Canadian permafrost stores billions of tonnes of frozen CO_2 and methane close to the surface — a third of the Earth's stored soil carbon. When the tundra warms by 3.6° F (2° C), the gases escape into the atmosphere.[3] The Arctic temperature has already increased by 4-7° F, so the process has begun.[4] Methane is 62 times more potent a greenhouse gas than CO_2 over 20 years, accelerating the rate of warming as the Arctic thaws.

The Ice Sheets Melt

As long as the Arctic summer ice exists, its albedo reflects 80% of the sun's light back into space. When it goes, the sun's light and heat will penetrate the water, adding to ocean warming. The same applies to Greenland's ice cap. By 125,000 years ago, at the very end of the last interglacial period, three quarters of Greenland's entire ice cap had melted away, raising global sea levels by 5 meters.[5] The possibility also exists that the west Antarctic ice sheet might collapse, creating a 5-6 meter rise in sea levels.[6]

Estuaries and Farmland Flood

When there is serious flooding from rising sea levels, the water drowns huge areas of vegetation, which rots anaerobically and releases large quantities of methane, further intensifying the greenhouse effect.

The Methane Hydrates Melt

Methane hydrates are huge blobs of pressurized frozen methane that are locked under the ocean floor along the world's continental shelves. They contain ten trillion tonnes of methane; twice as much as the world's entire fossil fuel reserves. As the ocean warms, the deposits could begin to break free, releasing an enormous burst of greenhouse gases.[7] Such an explosion may have happened 15,000 years ago, triggering the end of the last ice age;[8] one may also have happened 55 million years ago, sending CO_2 levels to 700-2,000 ppm and raising the global temperature by 4-8° C.[9]

The Phytoplankton Stop Working

The world's oceans absorb 25% of the CO_2 from human activities. Plankton thrive on the annual upwelling from the cool, deep, nutrient-rich waters below, which is triggered by the temperature difference between the surface

Resources

Sudden climate transitions during the Quaternary: www.esd.ornl.gov/projects/qen/transit.html

The Great Climate Flip-Flop: www.williamcalvin.com/atlantic

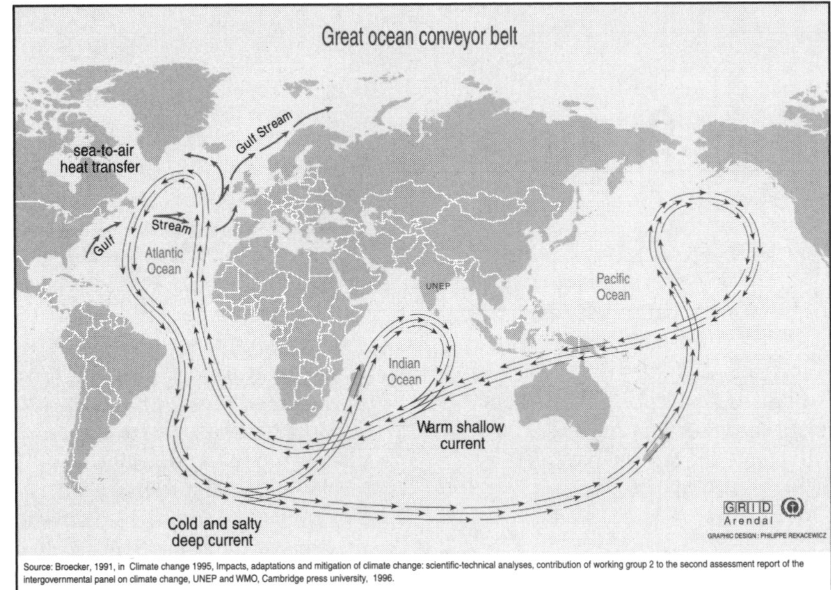

Source: Broecker, 1991, in Climate change 1995, Impacts, adaptations and mitigation of climate change: scientific-technical analyses, contribution of working group 2 to the second assessment report of the intergovernmental panel on climate change, UNEP and WMO, Cambridge press university, 1996.

and the deep ocean. If the warming ocean waters start to circulate less efficiently, fewer inorganic nutrients will be brought up from the bottom, cutting off a key food supply for the plankton, reducing their ability to absorb CO_2. This happens during El Nino events, when it is responsible for the collapse of many Pacific fisheries. Without the plankton, the ocean food chain would collapse. The lack of carbon-absorbing plankton would accelerate CO_2 accumulation in the atmosphere, causing ocean circulation to slow more. It would also scupper the whales, which eat the krill that eat the plankton.

The Forests Die Back

As the world warms, the world's forests will begin to feel the stress. British scientists fear that by 2050, the world's forests and terrestrial vegetation will cease being a carbon sink and become a carbon source. If we continue on our current trajectory, the entire Amazon rainforest will begin to dry out around 2040 due to changes in ocean temperature that alter the normal storm tracks, drying the soil and causing massive fires.[10] These are terrible possibilities to express in so few brief words. After 2050, scientists estimate that as it dies, vegetation will release two billion tonnes of carbon a year instead of absorbing 2-3 billion tonnes, sending CO_2 levels soaring. This is all the more alarming because the projections come from the UK Hadley Center's climate model, which many scientists believe to be the world's best at projecting ocean changes and their consequences.

The Ocean Conveyor Belt Stops

Northern Europe is the same latitude as Labrador, but it is warmed by the Gulf Stream, part of the global thermohaline current that pushes warm water around the world. The current starts off Iceland, propelled into motion by heavier salt water that falls to the bottom of the ocean as the sea-ice forms. By the time the cold water has been south to the tropics and returned, it is warm enough to create the warm westerlies that Europe enjoys. So far, so good. The Gulf Stream has been flowing for 10,000 years, but geological evidence suggests that the current can switch off at very short notice. The trigger is an increase in fresh water from flooding rivers and melting ice in the Arctic, which dilutes the salt water and stops it from sinking. If the current switches off, temperatures in northern Europe could fall by 10° C within ten years, and London would experience the winter cold that now grips Irkutsk in Siberia.[11] This could happen when CO_2 concentrations reach 750 ppm, which will happen within 100 years if we don't get things under control.[12] Since 1980, the salinity of water flowing south has dropped by 0.01 grams of salt per kg of seawater, after being constant for at least 100 years.[13] One of the deep water currents flowing south from the Greenland Sea has already stopped and gone into reverse.[14 15] In 2000, scientists discovered that west Greenland's melting glaciers were sending a Nile's worth of fresh water into the ocean in this very area.

This is no picnic. Any or all of these events could flip the world into a new ice age or lock it into a permanently sweltering climate. We have a choice. We can wait to find out what will happen, or start reducing our emissions now, before it is too late.

The Counter-Arguments

One of the difficulties facing scientists who are trying to alert the world to the dangers of climate change is the presence of a handful of "naysayers," scientists who dismiss the accumulating evidence and say there is no proof that climate change is happening. For our purposes here, we will ignore the fact the naysayers are often financed by right-wing think-tanks and fossil fuel industry trade associations (see p. 50), and focus on the arguments.

Argument #1: The climate models are not good enough to be trusted.

The world's climate is incredibly complex, and climate modelers have to account for a huge number of factors. At the NASA Geophysical Fluid Dynamics Laboratory in Princeton, NJ, they use a Cray C90 Supercomputer with the power of 256 top-notch Pentium personal computers, able to perform 15 billion operations a second. In the early 1990s, the computers were predicting a global temperature rise higher than was showing up, which cast doubt on the accuracy of the models. In 1994, however, when the cooling effect of sulfate smogs from industrial areas was factored in, the numbers came out right. The models are not perfect, and there are complexities concerning ocean currents, clouds, and biotic feedback from forests and tundra that have yet to be properly understood, but they are generally accurate to within 10% when it comes to predictions of wind, rain, temperature, radiation, and cloud cover. We have to work with what we have, and improve on them as we learn more.

Argument #2: The temperature increase is caused by natural variation or solar radiation.

One of the commonest arguments by the greenhouse skeptics has been that Earth's surface warming is due to natural variation or solar radiation. Scientists think there is only a 1-in-20 chance that the increase could be due to natural variation. The solar radiation striking the Earth fluctuates according to various cycles, including an 11-year sunspot cycle. Over the past 100 years, there has been a 0.1% increase in radiation. Danish scientists found a northern hemisphere correlation between the solar cycle and temperature changes between 1861 and 1980, which might explain 80% of the 0.2° C increase between 1900 and 1980. After 1980, the correlation ceases and there is no relationship to the subsequent 0.4° C increase.[1] The consensus of the IPCC scientists is that from 1900 to 1950, 0.3 Wm2 of increased radiative forcing (10% of the present forcing) came from an increase in the sun's radiance, but almost none since 1950.[2] There is an ongoing debate about the sun's magnetic field increasing the cosmic rays that hit the Earth, and the impact of solar UV radiation on ozone at different levels of the atmosphere.[3]

> **"This is not some slow, controlled change we're talking about. It's fast, it's unpredictable, and it's unprecedented during human civilization."**
> **— Adam Markham,**
> **World Wide Fund for Nature.**

Resources

A Scientific Critique of the Greenhouse Skeptics, in *The Heat Is On* by Ross Gelbspan (Addison Wesley)

Global Warming is Here: The Scientific Evidence:
www.climatesolutions.org/global_warming_is_here

Global Warming:
The History of International Scientific Consensus:
www.edf.org/pubs/FactSheets/d_GWFact.html

"Greenhouse Wars," *New Scientist:*
www.newscientist.com/global

The Science - Frequently Asked Questions:
www.epa.gov/globalwarming/faq

Argument #3: Cloud cover will reduce global warming.
Warmer temperatures increase evaporation, putting more moisture into the atmosphere and creating more clouds. Water vapor is a greenhouse gas, so this amplifies the greenhouse effect. Not so, say the nay-sayers. An increase in low stratus clouds will result in greater rainfall, removing the water vapor from the atmosphere; it will also reflect more heat back into space, reducing the amount of radiation that reaches the ground. But nothing is simple in the world of clouds. Cirrus clouds (those high, wispy, romantic clouds) allow heat to pass through them, so if the water vapor forms as cirrus, it will have the opposite effect. One of the frustrations of debating the skeptics is that their work is usually published by right-wing think tanks, not in journals which require peer reviewing. There is no consensus yet on the effects of clouds and water vapor, so this particular argument is not proven, one way or the other.

Argument #4: The satellite readings of the atmosphere do not show any warming.
Over the past 20 years, the Earth's surface temperature has risen by 0.25 to 0.40° C. Since 1979, however, satellites readings have been saying that the atmosphere 5 miles above the surface was cooling by 0.05° C per decade, not warming, causing the skeptics to challenge the models. The satellites are positioned in orbit 450 miles above the Earth and they read the temperature by measuring microwave emissions from different angles, subtracting the result from readings taken directly below. In 1998, one reason for the apparent cooling emerged: the satellites had been losing altitude as a result of atmospheric drag, which skewed the readings. The corrected data shows a rise of 0.05° C per decade, less than the 0.15° C per decade increase on the ground, but closer.[4] The debate has revealed the need for more comprehensive data collection about ozone, water vapor, clouds, and aerosols, as well as temperature and wind.[5]

Argument #5: There is too much uncertainty in the science to merit taking any action.
Response: There will always be uncertainty; the data could equally well be *underestimating* the reality. As a metaphor, imagine you are in an airplane and you have just taken off. The captain comes over the intercom and says the fuel indicator has broken, so he doesn't know if there's enough fuel to reach your destination. He puts it to a vote among the passengers. Do you want to go on or turn back? How would *you* vote? The greenhouse skeptics are saying, "Maybe the meter will fix itself. We don't know if this is for real yet, so let's go on."

"In light of new evidence and taking into account the remaining uncertainties, most of the observed warming over the last 50 years is likely to have been due to the increase in greenhouse gas concentrations." — IPCC Third Assessment Report, 2001, written and reviewed by 960 of the world's scientists and approved by the IPCC member governments.

An 80% Reduction by 2025

So what's to be done? The Kyoto Protocol calls for a 5.2% reduction in emissions below 1990 levels by 2012. Back in 1990, the IPCC scientists said we needed an immediate 60% reduction in CO_2 emissions below 1990 levels in order to stabilize their concentration in the atmosphere. Jerry Mahlman, director of the Geophysical Fluid Dynamics Laboratory at Princeton, NJ, says, "it might take another 30 Kyotos over the next century to cut global warming down to size."[1]

We have seen what the consequences are likely to be if we don't get our greenhouse gas emissions under control. The atmospheric lifetimes of most gases are long (CO_2 100+ years, N_2O 120 years, CF_4 50,000 years), so the gases that are already up there and the gases we continue to release will have impacts on the world of our children, grandchildren, and great-grandchildren for years to come.

If it were not for the stubborn resistance of the US government and the intense lobbying by the fossil fuel interests on the Saudi and Kuwaiti delegates,[2] the countries that met to negotiate the Kyoto Protocol in 1997 would probably have signed onto a far bigger reduction, perhaps 12%-15%. The British government has committed itself to a 20% reduction by 2012, and several other European nations think the same way. The French government has committed to produce 21% of its electricity from renewable sources by 2010. As we were going to print, we heard that Holland is considering an 80% reduction in greenhouse gases by 2040.

The chief reason for the US government's resistance is that many senators and congressmen owe their election success to campaign donations by the coal, oil, and auto corporations. These corporations have stuck their heads in the sand and want the good times to keep rolling. The same corporations have funded bodies such as the Global Climate Coalition (see p. 50) that have spent large sums of money persuading the public and politicians that there is no scientific agreement about climate change, and no need for reductions. The public has been made to believe that addressing climate change will cost an economically crippling sum of money, and the whole mindset around making a transition out of fossil fuels has become negative and intransigent. The self-confidence of the American people about tackling challenges and realizing dreams has vanished and been replaced by a rather whiny insistence that nothing should change. The fact that the world's oil supply is also about to start declining is ignored, because it interferes with this attitude.

As soon as you step out of the negativity, however, and start meeting people who are involved with solar and wind energy, hydrogen, organic farming, sustainable forestry, electric vehicles, the greening of cities, green architecture, and so on, the excitement is palpable. They argue that we achieved a successful energy revolution 100 years ago, when oil, electricity, gas, and automobiles became dominant within a 20-year period from 1890-1910, and that a new energy revolution will be good news all round, for jobs, investment, the economy, and the environment. They point to the revolution in computers, which has fundamentally changed the world economy in just 20 years, as an indication of how fast economies can change when we set our minds to it. So let's get going with the clean energy revolution!

> **"There is broad agreement within the scientific community that amplification of the Earth's natural greenhouse effect by the buildup of various gases introduced by human activity has the potential to produce dramatic changes in climate. Only by taking action now can we ensure that future generations will not be put at risk."**
>
> **— Statement by 49 Nobel Prize winners and 700 members of the National Academy of Sciences, 1990**

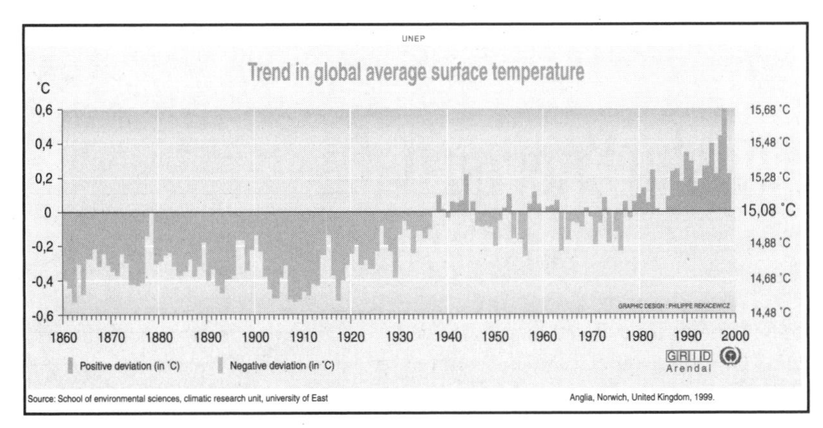

What is Needed?

With a 60% reduction, greenhouse gas concentrations would still be high. We need an 80% reduction below the 1990 level, and we need it by 2025. This has been our goal as we wrote this book.

- For the industrial gases (12% of the problem), this means 100% elimination, which can be achieved by legislation since alternatives already exist (see #95).
- For nitrous oxide (5% of the problem), it means phasing out fossil fuels and shifting from nitrogen fertilizers to organic farming and better soil management. In Austria, 8% of all farms had been certified as organic by 1997.
- For methane (17% of the problem), it means phasing out fossil fuels, ending methane escapes from landfills, and shifting to sustainable livestock management.
- For tropospheric ozone (12% of the problem), it means phasing out fossil fuels, eliminating the pollutants that cause the ozone to form.
- For black carbon aerosols (percentage uncertain), it means phasing out fossil fuels.
- For CO2 emissions from the destruction of the world's forests (13% of the problem), it means a global shift to eco-certified forestry, and an end to illegal logging.
- Finally, for CO2 emissions from burning fossil fuels (39% of the problem), it means phasing out fossil fuels.

Overall, phasing out fossil fuels will eliminate 57% of the problem;[3] shifting to sustainable farming practices will eliminate 8%.

In 2000, the scientist James Hansen, sensing how difficult it was to obtain political support in the US for reductions in CO_2, turned his attention to the other gases, suggesting that an emphasis on reducing methane and black carbon would yield quicker results because of their shorter lifetimes, while buying the world time to reduce CO_2 emissions.[4] He also noted that since aerosol pollutants such as sulfuric oxides shield the world from climate change, and since these result from burning fossil fuels, reducing fossil fuel emissions would also reduce the aerosols, canceling out much of the benefit. Sulfuric oxides are already being phased out by legislation and by new pollution control technologies, because they cause acid rain and are a major health threat - a trend that is unlikely to be reversed. His paper was misread by some as saying that CO_2 reduction doesn't matter, which was not his intent at all. The vast array of solutions described in these pages show that reducing CO_2 emissions may be much easier than people think, but only if we work together to create a strategy that combines sustainable technologies, government policy and tax initiatives (including the elimination of fossil fuel subsidies), the establishment of strong citizens action groups, and the creation of a global compact as part of a new global treaty (see Solution #92).

So back to the 80% reduction in CO_2 emissions. Is it possible? Before we look, we need a brief diversion into the often confusing world of energy statistics. Take a deep breath!

An 80% Reduction: What's Needed?

	Share in 2000	Reduction	By 2025
CO_2 - fossil fuels	39%	80%	7.6%
CO_2 - deforestation	13%	80%	2%
Methane	17%	80%	4.8%
Nitrous oxide	5%	80%	1%
Tropospheric ozone	12%	80%	2%
Industrial gases	12%	100%	0%
Black carbon	Uncertain	80%	0%

Those Pesky Energy Data

To make life fun, energy is measured in at least five different ways, including joules, calories, tonnes of oil equivalent, Btus, and watts. The units do speak to each other, but it's a mighty complicated procedure. The watts are the easiest, since they come in a big friendly family named after their great-great-grandpapa, the Scottish inventor and engineer James Watt (1736-1819).
A watt is the transfer of one joule of energy in one second:

- Watt (Weeny watt): 1 watt is one Christmas lightbulb.
- Kilowatt (Kiddy watt): 1,000 watts (KW)
- Megawatt (Mama, Middle watt): 1000 kilowatts (MW)
- Gigawatt (Grandpapa, the Gigantic watt): 1,000 megawatts, or a million kilowatts (GW)
- Terawatt (Tyrannosaurus watt): 1,000 gigawatts, or a billion kilowatts (TW)

There are two sides to the family. There are the Watts, who are a measure of capacity, and their cousins the Watt-hours, who are a measure of delivered energy. If a 500 MW coal-fired power plant produces power for 11 hours a day, 365 days a year, it will produce 2 million megawatt-hours in a year. A geothermal plant might produce power 23 hours a day, a nuclear plant 17 hours a day, a hydro plant 10 hours a day, a wind plant 6-12 hours a day. A solar plant (or PV panel) might average 6 hours a day, depending on the weather.

It was Napoleon who introduced the metric system, to bring order to the chaos and cheating that was rampant when every town in Europe had its own version of weights and measures. The metric system is easy, but the British and Americans didn't like Napoleon, so they stuck to the Imperial system, which measures energy in British thermal units, or Btus. A Btu is the heat needed to raise the temperature of a pound of water by 1° F. The British and Canadians have given up on their pounds, miles, and gallons, along with their hundredweights, furlongs, and perches, but the Americans hang onto the old system. So while Europe, Canada, and the rest of the world measure large quantities of energy in exajoules (= joule x 10^{18}), Americans measure large quantities in quadrillion British thermal units, or "quads." A joule is the energy generated when a force of one Newton moves an object by one meter. One kilowatt-hour is 3412 Btus; one quad (a quadrillion Btus) is 293 terawatt-hours.

To complicate things further, oil is measured in barrels and tonnes (1 tonne of oil = 7 to 11 barrels, depending on the precise fuel), natural gas in cubic feet, and coal in short tons. That makes it hard to compare fuels, so the fossil fuel corporations have taken to describing energy in "Million tonnes of oil equivalent" (Mtoe). One tonne of oil equivalent is 41.868 gigajoules, which is the net heat content of 1 tonne of crude oil. 1 Mtoe = 4.1868 x 10(4) TJ = 10(7) Gcal = 3.968 x 10(7) MBtu = 11630 GWh. — Don't you love this stuff?

Sorting Out The Numbers

Amazing "kilowatt-hours into anything" calculator: www.wattsonschools.com/calculator.htm

Common Conversion Factors: www.eia.doe.gov/oiaf/1605/ggrpt/appendixf.html

Energy A-Z: www.eia.doe.gov/fueloverview.html

Energy Glossary: www.energy.ca.gov/glossary

Energy Information Administration: www.eia.doe.gov

International Energy Agency: www.iea.org

Mtoe conversions: www.iea.org/stats/files/mtoe.htm

Oil-gas energy conversion calculator: www.bp.com/worldenergy/ccalculator

World electricity generation: www.eia.doe.gov/emeu/iea/table63.html

World Energy Overview: www.eia.doe.gov/iea/overview.html

World Energy Review (BP): www.bp.com/worldenergy

More Muddling Math

In 1998, the world used 378 quads of primary energy, which is 111,000 TWh (terawatt-hours), 399 exajoules, or 9,546 Mtoe. 39.8% of this energy came from oil, 23.2% from coal, 22.4% from natural gas, 7% from hydro, 6.4% from nuclear, and a tiny 1.3% from solar, wind, biomass, and geothermal power combined. That's what we have to change. Overall, the USA used 19% of the energy, Russia 11%, and China 9%. The USA used 26% of the world's oil production, 26% of the natural gas, and 21% of the coal. China used 26% of the world's coal.

Americans use American gallons, but Canadians use British gallons, which are 20% larger. This makes the new hybrid gas-electric Hondas more fuel-efficient in Canada (81 mpg) than in the USA (70 mpg). Hey, it all adds to the fun. A Canadian metric tonne (1,000 kilos) is also 10% larger than an American ton (2,000 pounds). After all, Canada is 6% larger (USA: 3.61 million square miles; Canada: 3.83 million square miles), so it makes sense.

At least electricity is always measured in watts. In 1998, the world consumed 12,637 terawatt hours of electricity, which it derived from 3.1 terawatts of installed capacity. To obtain this, it used 144 quads of energy (42,206 TWh), of which 30,000 TWh was lost through inefficiency in the generating process. Approximately 38% of the world's primary energy was used to generate electricity. 62% came from burning fossil fuels, 19% from hydro, 17% from nuclear, and 1.5% from solar, wind, biomass and geothermal.

In general, we have used American measurements for American energy and greenhouse gas issues, and metric measurements for Canadian, European, and global data. In 1998, the US produced 1,495 million metric tonnes of carbon from burning and flaring fossil fuels, 24% of the world's total of 6,124 tonnes. To express this as carbon dioxide, we multiply by 3.667 to factor in the oxygen, which gives 5,482 billion tonnes of CO_2. Divide by the US population of 280 million people and it comes to 20 tonnes of CO_2 per person. Canada's 31 million people produced 16.4 tonnes of CO_2 per person, because they get more of their electricity from hydro. There's an enormous gap between the USA and Canada and a country like the Philippines, whose people produce 0.7 tonnes of CO_2 per person. Everyone is going to suffer, however, if we don't get our emissions under control.

So back to the big question: Can we reduce our greenhouse gas emissions by 80% by 2025?

"A child born in a wealthy country is likely to consume, waste, and pollute more in his lifetime than 50 children born in developing nations. Our energy-burning lifestyles are pushing our planet to the point of no return. It is dawning on us at last that the life of our world is as vulnerable as the children we raise."

— George Carey,
Archbishop of Canterbury, UK

Sustainable Energy for a Sustainable Planet

In 1998, the world produced 382 quadrillion Btus of primary energy, 85% of which came from fossil fuels (petroleum 39.8%, coal 23.2%, natural gas 22.4%) and the rest from hydroelectric (7%), nuclear (6.4%), and biomass, geothermal, solar and wind (0.65%). Can we achieve an 80% shift to renewable energy by 2025? That's the big question. The figures that follow are back-of-the-hand calculations, but they give a sense of what's needed.

We will start with the USA's level of energy consumption and work from there. In 1998, the USA consumed 94.8 quadrillion Btus of energy, 0.34 quads for every million people. The first assumption we make is that we can save 50% of this energy by being more efficient with our vehicles, appliances, homes, industrial equipment, and energy generation. The technologies exist to yield a 50% reduction, and with the right policies and incentives, the savings could be realized. This reduces the energy needed to 0.17 quads per million people.

Our next assumption is that we all need to live more sustainably, making lifestyle changes in the way we travel, the way we design our communities, the rate at which we recycle and re-use materials, the volume of "stuff" we buy, and so on. We have become hooked on cheap energy and cheap raw materials. The rest of the world sees Americans and Canadians as hugely wasteful and over-indulgent, so we make no apologies about saying we need a cultural diet, reducing the energy needed by another 50% to 0.085 quads per million people. We'll call this a sustainable energy level — which it would be if the energy was generated sustainably.

By 2025, the world's population will be around 7.5 billion. If everyone were to enjoy a sustainable standard of living, the world's primary energy need would be 637 quadrillion Btus. In reality, many developing nations will need 50 or 100 years to lift themselves out of poverty and achieve this kind of sustainability, so we'll arbitrarily reduce the figure to 500 quads. The IIASA/World Energy Council's scenario for an ecologically driven future suggests 454 quads, so we may be in the right ballpark.[1] 80% of 500 is 400 quads, so we'll make that our goal.

The Energy Math for a Sustainable World

1998 global primary energy production	382 quadrillion Btus (quads)
1998 primary energy from fossil fuels	326 quads
1998 primary energy from renewables	2.5 quads
1998 US consumption	94.8 quads
US energy per million people	0.34 quads per million
50% efficiency gain	0.17 quads per million
50% sustainable lifestyles gain	0.085 quads per million
World population in 2025	7.5 billion people
7.5 billion people, 0.085 quads per million	637 quads
Reduction for slower development in some countries: 137 quads	400 quads
Total primary energy needed for a sustainable, steady state world in 2025	500 quads
80% of primary energy to come from renewable sources	400 quads
Land needed to generate 400 quads of solar PV	330,000 square miles
Total Earth land-area	57.5 million square miles[6]
Global wind energy potential	341 quads
Global geothermal potential	468 quads
Global tidal energy potential	72 quads
Global biomass potential	254 quads

Energy potentials: World Energy Assessment: www.undp.org/seed/eap/activities/wea

Green Mountain Power Corporation

The 6-MW Green Mountain power plant in Vermont consists of eleven 550-kW wind turbines.

A quadrillion Btus (1 quad) is 293 terawatt-hours, so 400 quads is 117,000 terawatt-hours. Assuming that a typical renewable energy power plant produces energy for 8 hours a day, we will need a primary renewable energy capacity of 40 terawatts (40 million MW) compared to today's renewable total (including existing hydro) of 0.8 TW. This is 5 kW of solar PV per person in 2025, or 16 - 20 kW per household.[2] 38% of the energy will be needed to generate electricity and 62% will be available as ethanol and to manufacture hydrogen for use in cars, trucks, ships, airplanes, space rockets, factories, and heavy equipment. If we learn how to strip the CO_2 off fossil fuels and sequester it away safely and affordably, we may be able to continue using fossil fuels as a source of energy and a feedstock for hydrogen, but for our purposes here, we will set this possibility to one side, along with nuclear and cold fusion energy (see pp. 43 and 46). So the question is: Can we generate 117,000 TWh of renewable energy by 2025?

Solar: In 1997, the USA used 3,570 terawatt hours of electricity, which could be supplied by 10,000 square miles of solar PV in a desert area (see p. 30). We need 33 times as much, which could be met from 330,000 square miles of solar PV.[3] The desert nations of North Africa and the Middle East from Iran to Mali have 7 million square miles of solar-irradiated, mostly desert land, which is 21 times more than we need. Solar PV can also be generated in the 24-hour Arctic summer, and on every rooftop.

Other Renewables: The world's realistically achievable land-based wind turbine energy is put at 53,000 terawatt hours per year; off-shore wind could probably double this to 100,000 TWh, 85% of what's needed.[4] Achievable geothermal potential is put at 470 quads (117% of the needed supply); tidal energy potential offers a further 74 quads.[5] These three energy types alone could provide 220% of our needs, not counting solar, biomass, and micro-hydro. The missing pieces are not technology, or even cost — they are vision, policy, activism, and global cooperation.

Building 40TW of renewable energy capacity by 2025, however, means building an average 1.6 TW per year. The world's entire non-hydro renewable energy capacity in 1998 was only 0.143 TW, so this will require an enormous expansion. Over the next few pages, we will consider how it might be done.

"Globally, emissions may have to be reduced, the scientists are telling us, by as much as 60% or 70%, with developed countries likely to have to make even bigger cuts if we're going to allow the developing world to have their share of growing industrial prosperity...The Kyoto Protocol is only the first rather modest step. Much, much deeper emission reductions will be needed in future. The political implications are mind-blowing."
— Michael Meacher, UK Environment Minister, November 2000

Building a More Efficient World

Tim Ellison

National Association of Home Builders' 21st Century solar townhouse with PV roofing, Bowie, Maryland.

Before we begin, let us be totally honest with you. Our conclusion that we need an 80% reduction in emissions by 2025 is far more progressive than most other proposals. Out of a database compiled for the Intergovernmental Panel on Climate Change (IPCC) of 400 scenarios of global greenhouse gas emissions for the next 100 years, the fastest scenario suggested a 75% reduction by 2060.[1] Our proposal says that progress needs to be made almost ten times faster than the World Energy Council's ecologically driven scenario, which projects a 17% reduction in emissions below the 1990 level by 2050, and a 66% reduction by 2100.[2]

Many of the models and scenarios are poorly informed about the true potentials of renewable energy, however. They are also poorly informed about the way corporations can reduce emissions if they put their minds to it, the way smart policies can accelerate change, and the impact that non-profit societies can have on market behavior. In Colorado, a non-profit society called the Land and Water Fund used grassroots marketing to achieve a five-fold increase in the local production of premium-priced wind energy (see #47). This book is full of examples that demonstrate how a rapid reduction in greenhouse gas emissions can be achieved. The success of these examples invites the question: *What if every family, school, college, church, city, business, state, province, and nation were to do the same?*

The quest for an 80% reduction in greenhouse gas emissions begins with the need for a 50% improvement in efficiency. Under a business-as-usual scenario, the energy intensity of the economy improves by 0.7% per year. This delivers a 14% improvement in efficiency over a 25-year period. When economies are growing and increasing their use of energy by 2% to 4% per year, however, a 0.7% per year gain is nowhere near enough. In 1997, Germany's Wuppertal Institute and Colorado's Rocky Mountain Institute called for a 4% annual improvement, yielding a 60% reduction over 25 years.[3] In the book *Natural Capitalism*, the authors suggest that a ten-fold increase in resource efficiency is possible.[4] We need to achieve a 5% to 8% annual improvement, which could be done if the best policies were implemented everywhere, as they would if there was a war. For examples of 50% to 90% savings in the efficiency of appliances, buildings, vehicles, and industrial equipment, see Solutions #3, #4, #5, #24, #32, #45, #46, #65, and #72.

> **"Within one generation, nations can achieve a ten-fold increase in the efficiency with which they use energy, natural resources, and other materials."**
>
> **— The Carnoules Declaration, 1994[5]**

 Follow the comic adventures of Zed, last of his species, in www.GristMagazine.com, gloom and doom with a sense of humor.

The problem with achieving greater efficiency is that everyone thinks it a grand idea but nobody does anything about it. Because of this laziness, schools spend more money on energy than they do on books and computers.[6] There are barriers that cause families, businesses, schools, colleges, cities, and governments to stop caring, but the right policies and incentives can yield the needed savings, as players as diverse as Seattle City Light, San Francisco, New York, Dow Chemical, and the New Zealand government have shown. Even China has brought in tax incentives for the construction of energy-efficient buildings.[7] When Dow Chemical's Louisiana Division got serious about efficiency 20 years ago, they achieved savings that brought an average return on investment of 240% on 575 different projects, and a profit to the shareholders worth $110 million a year.

What about cars and trucks? There are vehicles on the road today that use a half to a quarter the fuel that the average vehicle uses. With mandatory standards, the overall efficiency of the vehicles we drive could easily be raised. It is politics and attitudes that stand in the way, not technology.

The second big assumption we are making is that our lifestyles must become far more sustainable. This means recycling more, eating more locally grown food and less meat (meat is heavy on the emissions, see Solutions # 8 and #36), cycling and walking more, taking the bus or car-sharing instead of driving alone, planting more trees, and generally living more simply, not leaving such a heavy material footprint on the Earth.

This involves profound cultural change. A healthy culture does not stand still, however. It evolves. Consider the huge changes that have taken place over the past 20 years on issues such as child abuse, smoking, and recycling. Now imagine the changes that will take place when there is a mass realization that we are seriously upsetting the climate, and that our everyday decisions make all the difference.

Yes, this involves difficult choices, but they are wise ones, for they may lessen the painful impacts that our planet's climate will impose on us through floods, storms, mudslides, droughts, and heat waves. We can choose to ignore the signs, but we would be extremely foolish to do so when the solutions are so close at hand.

Policies That Accelerate Efficiency

Buildings and equipment: 29% of global energy-related CO_2 emissions

- Raised building codes, mandatory on sale of a building
- Mandatory efficiency labeling on all buildings and appliances
- Tax incentives for efficiency investments
- Free energy audits and information
- Rebates on efficient appliances
- Subsidized home improvement loans
- Energy efficient mortgages

Industry: 45% of energy-related CO_2 emissions

- Reward employees for efficiency improvements
- Corporate efficiency goals and programs
- Detailed energy monitoring
- 7-year paybacks for efficiency investments
- Materials recycling goals

Transportation: 21% of energy-related CO_2 emissions

- Mandatory fuel efficiency standards and labeling
- Tax incentives for the purchase of efficient vehicles
- Investments in public transportation
- Policies to encourage cycling
- Planning for smart communities

Governments

- Public Benefits Charges on fuel bills to finance efficiency improvements
- Energy efficiency and conservation legislation
- A cabinet-level chief for energy efficiency

Solar Energy

"**It is my plan that every household in Japan be solar-powered, and little by little, it is happening." — Yasuo Kishi, Managing Director, Sanyo Industries**

If 400 quadrillion Btus (117,000 TWh) is our goal for a sustainable world in 2025 and solar energy provides 20%, we will need to produce 80 quads by solar means. This comes to 23,400 TWh. If we assume that solar will produce energy for an average 5 hours a day, this will require 13 TW of capacity. What will it take?

The sun is our number one source of energy. It is over 93 million miles away from Earth, yet we still feel its heat. Every year it radiates 220 million terawatt-hours of energy onto the Earth's surface, 2,000 times more than the world's consumption of primary energy today (111,000 TWh).

Solar energy is pollution-free and, once an installation has been paid for, it is also cash-free. At today's solar efficiency, the entire US electricity demand could be met from 10,000 square miles of solar PV (100 miles by 100 miles), or 8% of Arizona.[1] Assuming half the solar gain, Canada would need 3,000 square miles, or 1.25% of Manitoba. Greenpeace has calculated that solar panels fitted to all suitable buildings could supply two-thirds of Britain's electricity needs. We have calculated that solar panels fitted to all rooftops in the US could meet 27% of today's US electricity demand.[2]

In 1971, the world's production of photovoltaic (PV) cells was 0.1 MW. By 1999 it had reached 201 MW and was increasing by 30% a year. Cumulative capacity had reached 1,000 MW, compared to the world's total electrical capacity of 3 million MW. If solar capacity is to reach 13 TW by 2025, production will need to expand by an average 52% a year.

80% RENEWABLE ENERGY BY 2025: WHAT'S NEEDED?

	Share of 400 quads	Hours per day	Quads in 2000	Quads in 2025	TWh in 2025	GW in 2000	TW in 2025[5]
Wind	40%	8	0.15	160	47,000	15	16
Solar	20%	5	0.007	80	23,500	1	13
Geothermal	15%	23	0.6	60	17,500	20	2
Biomass	15%	10	41	60	17,500	410	5
Tidal/wave	5%	10	0.003	20	6,000	0.26	2
Hydro	5%	10	8.8	20	6,000	678	2

THE PRICE OF SOLAR ENERGY

	2001	2005 (500 MW/year)
PV module	$3.25 per watt	$0.80 per watt
Balance of System	$2.00 per watt	$0.70 per watt
Total per watt	$5.25	$1.50
Total cost for 2kW	$10,500	$3,000
A: 2000 hours/year (e.g., AZ) Cost over 20 years @ 3% interest	$13,920 17 cents/kWh	$4,080 5 cents/kWh
B: Same at 8% interest	$21,120 26 cents/kWh	$6,000 7.5 cents/kWh
A, B @ 1,000 hours/ year (e.g., Seattle)	**A:** 34 cents/kWh **B:** 52 cents/kWh	**A:** 10 cents/kWh **B:** 15 cents/kWh

"I'd put my money on the sun and solar energy. What a source of power! I hope we don't have to wait 'til oil and coal run out before we tackle that."
— Thomas Edison (1847–1931)

Guy Dauncey

Soltek's first solar home in Victoria, Canada.

The biggest barrier to a worldwide solar take-off is still the price. In 1975, installed photovoltaic panels cost $75 per watt. Today, they cost $5.25 per watt.[3] Once installed, solar energy is free, so the best way to compare costs is over 20 years. On this basis, solar costs 17–50 cents per kWh, compared to 5–12 cents for electricity in the USA (but 19–70 cents for peak electricity in California). In Japan, where electricity costs 17–20 cents per kWh, solar is already becoming competitive.

What about the energy needed to manufacture solar PV cells? Siemens Solar has calculated that the energy payback of solar cells is 1–3 years, and that the cells will generate 9–17 times more energy than is required to produce them.[4] When PV is installed as solar-shingles there is no need for an aluminum frame, which is a large part of the energy cost.

Seven Steps to Solar Take-Off

With each year that passes, the price of solar energy falls slightly — but not enough. What will it take to make the price low enough that every home and business owner will want to pick up the phone and order a solar roof, paying for it with an easy solar mortgage? The following seven steps could make solar competitive within five years, if they were taken together:

Step One: Work to expand and aggregate the world's future demand. KPMG did a study in 1998 that showed if a solar factory could increase its production to 500 MW a year, the price would fall from $5.25 to $1.50 per watt, enabling PV to produce energy for 5–15 cents/kWh, depending on the region.[6] Such a factory would cost $660 million. BP (which owns BP Solar) made a profit of over $8 billion in 2000, so it's not as if there is a shortage of cash.

In 2001, the largest plant was First Solar's 100 MW plant in Toledo, Ohio. The Worldwatch Institute has estimated that PV will become competitive when annual production reaches 500 MW, which will happen by 2003 at the current 30% growth rate.[7] The solution is to form a Global Solar Compact between governments, businesses, cities, utilities, and non-profit societies around the world, making a joint commitment to buy enough solar energy to justify investments in 500 MW/year factories (see #92).

Step Two: Phase in an eco-tax on fossil-fuel-derived electricity to pay for the damage caused by acid rain, lung diseases, and climate change (see #48), increasing the price from 8–10 cents/kWh, recycling all of the revenue as subsidies for efficiency measures and renewable energy.[8]

Step Three: Pass federal legislation guaranteeing net metering, entitling anyone who produces solar or other forms of renewable energy to sell their surplus to the grid.

Step Four: Legislate a guaranteed, subsidized transition price for the sale of solar energy (e.g., 15 cents/kWh). Germany pays 45 cents per kWh as an incentive.

Step Five: Establish a legal framework that facilitates 25-year solar mortgages and provide a federal subsidy to underwrite a 3% interest charge, similar to California's.

Step Six: Establish a tax credit for all solar investments and additional incentives. In New York, a 1kW solar rooftop system costs as little as $3 per month (see #43), thanks to supportive policies.

Step Seven: Pass legislation:

- requiring utilities to produce 2% of their of energy from solar by 2010 (175 TWh, needing 32 GW, and a 75% pa growth rate);
- requiring all buildings up to eight storeys high to produce solar hot water, as Israel does;[9]
- requiring all new buildings to include a 2kW rooftop solar system, starting in five years time;
- requiring all existing buildings to be fitted with a 2kW solar system by 2015.

As soon as solar energy hits its take-off point, every roof, window, and wall will become a potential source of both income and energy. With annual production reaching 500 to 1,000 GW of PV per year, the price will fall still lower and North Africa and Middle East nations will start generating solar energy to manufacture hydrogen for sale to Europe. For Solar Resources, see #6, #43, and #82.

Wind Energy

The winds that can free us from climate confusion blow all the time. All we need do is build the turbines.[1] That's exactly what they are doing in Europe, where concern about climate change has persuaded the British and German governments to reduce their greenhouse gas emissions to 21% below the 1990 level, and persuaded France to generate 20% of her electricity from renewables, all by 2010.

The market for wind turbines in Europe is growing by 40% a year, with no sign of slowing up. The secret is not technology or price — it's policy. In Denmark, until recently, electric utilities were required to pay wind producers 85% of the consumer price of electricity, and turbine owners received a 9 cents/kWh subsidy, guaranteeing a good return on investment. By the end of 2001, wind energy will be producing 15% of Denmark's electricity, and the cooperatives and entrepreneurs involved will have built the Danish wind industry to the point where it is exporting 55% of the world's wind turbines and employing 17,000 people. In 2000, Danish wind turbines generated 4,500 gigawatt hours of electricity, the equivalent of 1.4 million tons of coal being delivered to a thermal power plant in a train stretching all the way down the I-5 from San Francisco to Los Angeles (366 miles).

Germany followed Denmark's lead and introduced a feed law that obliges utilities to buy renewable energy at 90% of the retail price. By 2000, the north German state of Schleswig-Holstein was producing 19% of its electricity from wind, with a goal of 50% by 2005. In 2000, out of the world's installed wind capacity of 17,000 MW, Germany had 6,100 MW, Denmark 2,140 MW, Spain 2,250 MW. The USA had 2,500 MW, and Canada 137 MW. Europe's target for wind energy in 2010 is 60,000 MW, 150,000 MW by 2020.[2]

Globally, we need 40 TW or 117,000 TWh of non-fossil fuel primary energy by 2025 for a worldwide sustainable lifestyle. If 45% comes from the wind, that is 18 TW (53,000 TWh) — more than a thousand times greater than today's 17 GW. This could be achieved by an average growth rate of just over 45% a year for 25 years.

The future may lie with large, offshore turbines. At Blythè, off Britain's coast at Northumberland, two large 2 MW turbines with 33 meter-wide blades spin slowly at 23 rpm, generating enough power for 6,000 homes. Britain has enough off-shore wind potential to meet three times her current electricity needs.

Resources

(see also Solutions #15, #20, #44, #64, #75)

American Wind Energy Association: www.awea.org

Canadian Wind Energy Association: www.canwea.ca

Danish Wind Turbine Manufacturers Association: www.windpower.dk

Europe's ocean and land wind energy maps: http://130.226.52.108/oceanmap.htm http://130.226.52.108/landmap.htm

European Wind Energy Association: www.ewea.org

Renewable Energy Policy Project: www.repp.org

US Wind Resources Map: www.eren.doe.gov/windpoweringamerica/images/wherewind800.jpg

US Wind Resource Database: www.nrel.gov/wind/database.html

Wind Energy Incentives: www.eia.doe.gov/cneaf/solar.renewables/rea_issues/windart.html

Windpowering America: www.eren.doe.gov/windpoweringamerica

Wind Energy Made in Germany: www.dewi.de

Wind Energy Potential in the USA: www.nrel.gov/wind/potential.html

Wind on the Web: www.igc.org/energy/wind.html

Wind Resources Atlas of the US: www.homepower.com/windmap.htm

World Offshore Wind Map: www.esru.strath.ac.uk/projects/EandE/98-9/offshore/frame.htm

Jim Green

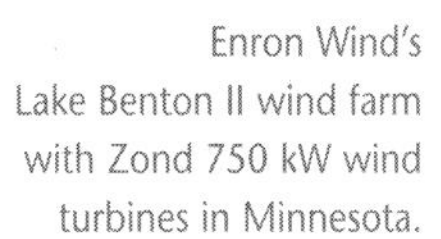

Enron Wind's Lake Benton II wind farm with Zond 750 kW wind turbines in Minnesota.

What is the World's Wind Power Potential?
In 2000, the World Energy Assessment team estimated that 27% of the world's land-surface has winds that blow at Class 3 or higher,[3] of which 4% can be used for wind farms, giving a total of 19,000 TWh.[4] Grubb and Meyer estimated 50,000 TWh.[5] Both looked only at land-based wind. Off-shore potential may double this, giving a range of 38,000 to 100,000 TWh. The Gaviotas pioneers in Colombia developed a wind turbine that can generate energy in a very slight breeze (see #84), so this may rise further. Building 18 TW of capacity will require a combination of the best policies, grassroots marketing, and an end to the power of the fossil-fuel utilities, which use their influence to prevent renewable energy from sharing the grid. Denmark's cooperative wind pioneers only succeeded because they organized to overcome the power of Denmark's utilities (see #20).

If 45% of the world's non-fossil-fuel primary energy comes from wind, a system of energy storage will be needed to provide base-load stability for the grid. Possibilities include high-speed flywheels (see #49), industrial-scale batteries, and compressed air storage (see #88). 60% of the energy will be needed to create hydrogen for use in industry, and in the world's vehicles.

What is the US and Canadian Potential?
In 2000, the US had 2,500 MW of wind power, 0.3% of its electricity capacity. A 1992 study showed that the lower 48 states have the wind potential to produce 10,871 TWh of electricity, three times more than was generated in the entire US in 1998. North Dakota, Texas, and Kansas could provide almost 100% of US electricity needs.[6] Another study suggested that the wind corridors in the Kansas and Nebraska wheatlands could provide twice as much electricity as America needs.[7] Now that the cost has fallen to 3-6 cents/kWh, this potential is beginning to be realized. In 2001, 1,946 MW of wind turbines were planned for construction in the US, some by farmers in Minnesota and Iowa who have realized they can earn $1,500 to $2,000 a year for each large-scale turbine on their land, and carry on farming beneath them.[8]

Canada is still dominated by large, centralized coal, nuclear, and hydro power producers who discourage diversity. In 2000, Canada had 137 MW of installed wind energy, and no sense of what the potential might be.

What about bird kill by wind turbines? Ten years ago, some wind turbines killed birds, including hawks and eagles. Today's wind farms follow careful siting guidelines, and turbines have been redesigned with much larger blades that turn more slowly, minimizing bird deaths.

> **"Wind energy...could kindle a truce in civilization's century-old fossil-fuel driven warfare with nature, and among nations."**
> **— David Case, reporter for TomPaine.com**

US Wind Power Potential[9]

(Land-based)

US electricity generation, 1998	3,600 TWh	% of 3,600 TWh
North Dakota	1,200 TWh	33%
Texas	1,190 TWh	33%
Kansas	1,070 TWh	30%
South Dakota	1,030 TWh	29%
Total of the above	4,500 TWh	125%
Total for lower 48 states	10,871 TWh	300%

Geothermal Energy

Dick Moore, USGS

A drillhole on Aguade Pau Volcano in the Island of Sao Miguel, Azores.

In 1892, Boise, Idaho became the first town in the world to heat its buildings with geothermal energy, using hot water from underground springs. Today, geothermal energy is used in 27 countries and generates more than 20,000 MW of heat and power. Globally, 39 countries could obtain 100% of their electricity from geothermal energy, including Bolivia, Peru, Costa Rica, Guatemala, Honduras, Nicaragua, Iceland, Fiji, Indonesia, the Philippines, Ethiopia, Kenya, Uganda, and Tanzania. Together, they hold 17% of the world's population.

There are five kinds of geothermal energy: hot water, hot dry rocks, magma, compressed hot water aquifers, and ground-source heat. The main kind that people talk about is the hot water variety — heat from hot, briny water deep below the Earth's crust that is brought to the surface as steam or hot water, and used to generate electricity in a steam turbine, then returned to the wells. It is also used to provide space heating for buildings (not to be confused with ground-source heat from *surface* soil or water, used to heat buildings — see below). At Geysers, near Santa Rosa in northern California, Calpine operates the world's largest geothermal power plant with 1030 MW of capacity, enough to power San Francisco. California produces 2,800 MW of electricity from geothermal energy, 34% of the world's total. In the US western states, geothermal energy has the potential to produce 10% of the region's electricity by 2020, including 20% to 30% of California's, a goal that the Department of Energy is encouraging through its "GeoPowering the West" initiative. Cost-wise, geothermal is cheaper than coal-fired power. A study by the Geothermal Energy Association estimates the global potential of this type of geothermal energy to be 80,000 MW (671 TWh), enough to meet 8.3% of the world's electricity demand.

The second, third, and fourth kinds of geothermal are hot rocks, magma, and compressed hot water aquifers. These have yet to be developed commercially, but in Western Queensland, a research program is underway to explore the possibilities in the Charleville area, which has sufficient hot rocks 3-5 kilometers underground at 200° C to meet Australia's entire energy needs for hundreds of years, if the program is successful. The World Energy Assessment estimates that geothermal reserves expected to become economical within 10 to 20 years total some 500 exajoules,[1] or 138,500 TWh, 118% of the non-fossil fuel energy needed for 2025. The main sources are in North and South America, Russia, sub-Saharan Africa, and central and south Asia. We are assuming that geothermal energy has the ability to generate 17,550 TWh by 2025, 15% of our goal.

The fifth type of geothermal energy is ground-source heat, which can be used in any new building and gives an average payback in 6 years (see #6 & #34). In two commercial/institutional cases that were examined in a Canadian study, ground-source heat pumps achieved CO_2 reductions of 15% to 77%.[2] As far as we know, ground-source heat's global potential has not been studied.

Geothermal energy does involve a few environmental issues. Some of the best sources are in remote locations, including wilderness areas where development may not be appropriate. Roads, pipes, and generators need to be sited sensitively, and some geothermal production produces toxic emissions from within the earth that need to be controlled.

Resources

Australian hot rocks program:
www.petrol.unsw.edu.au/research/medr.html

GeoPowering the West:
www.eren.doe.gov/geopoweringthewest

Geothermal Energy Association: www.geotherm.org

Geothermal links:
www.millennium-debate.org/linksgeothermal.htm

Geothermal Resources Council: www.geothermal.org

World Geothermal Power Sites:
www.demon.co.uk/geosci/world.html

Hydroelectric Energy

> "Major energy transitions — from wood to coal or coal to oil — take time to gather momentum. But once economic and political resistance is overcome and the new technologies prove themselves, things can unfold rapidly."
> — Chris Flavin, Worldwatch Institute

Humans have been using the flow of rivers to generate energy for millennia. In 31 AD, Tu Shih, Prefect of Nanyang in China, invented a system of water-powered bellows to generate the heat needed to cast iron agricultural implements.[3] In 1086, Britain's Domesday Book listed more than 5,000 water mills that were being used to power agricultural machinery. In 1883, the first hydroelectric plant opened in Portrush, Ireland, powering an electric tram.

In 1998, the world produced 2,584 TWh of hydroelectricity, 19% of the world's electricity consumption and 7% of the world's primary energy production, with capacity for another 506 TWh under construction. Almost all the energy came from 45,000 large dams, 21% from Western Europe (mostly France and Norway), 15% from the USA, 10% from Canada, and 9% from China. Countries like China and India are eager to expand production, but there is growing concern about the ecological costs, and the social costs to the millions of people who will lose their homes, farmlands, and ancestral villages to the new dams. To address these concerns, the Low Impact Hydropower Institute has started certifying hydro facilities that meet environmental protection criteria.

When it comes to climate change, there is new evidence that hydroelectric reservoirs produce methane, a powerful greenhouse gas. In 2000, the World Commission on Dams reported that organic matter washed into reservoirs breaks down as methane, a process that continues for the life of the reservoir. Methane is also released from the rotting forests and vegetation in the bottom of shallow reservoirs. In Brazil, it has been estimated that the Balbina reservoir, which is only four meters deep, will produce 3 million tonnes of carbon a year over its first 20 years, almost ten times more than the equivalent-sized coal-fired power plant. Different reservoirs produce different quantities of methane. Methane emissions from reservoirs studied in Brazil varied by a factor of 500, so not all hydroelectric dams need be written off as a source of renewable energy. Detailed assessment is needed for every project.

In addition to large hydro, there is also micro-hydro (up to 10 MW) and pico-hydro (up to 5 kW), which harvest the energy from small rivers and streams, often using run-of-the-river technology instead of large dams. 6% (37 GW) of the world's hydroelectricity comes from small hydro, mostly in China, which is estimated to increase to 55 GW by 2010. In Britain, there used to be 70 small hydro installations in the catchment area of the river Dyfi in west Wales, making many villages self-sufficient in power, but only 15 operate today. Microhydro can be generated for 4-10 cents/kWh.

Globally, the world has a potential for 40,500 TWh of hydropower, of which 14,320 TWh is thought to be technically feasible.[4] We are assuming that 5% (6,000 TWh) of the 117,000 TWh needed by 2025 can be provided by hydro, including 3,000 TWh that exists or is under construction, 2,400 of new hydro, and 600 TWh of microhydro.

Resources

Canadian Hydropower Association: www.canhydropower.org

IEA Hydropower Agreement: www.ieahydro.org

International Hydropower Association: www.hydropower-dams.com/iha

International Network on Small Hydro Power: www.digiserve.com/inshp

Low Impact Hydropower Institute: www.lowimpacthydro.org

Microhydro: www.geocities.com/wim_klunne/hydro

National Hydropower Association (US): www.hydro.org

Pico-Hydro: www.eee.ntu.ac.uk/research/microhydro/picosite

Universal Electric Power: www.uepholdings.com

US Hydropower Program: www.inel.gov/national/hydropower

World Commission on Dams: www.dams.org

Tidal and Wave Energy

"It has been estimated that if less than 0.1% of the renewable energy available within the oceans could be converted into electricity, it would satisfy the present world demand for energy more than five times over."
— Marine Foresight Panel, UK

The world's tides, and their offspring the waves, run 24 hours a day, 365 days a year, and offer enormous potential to harvest what is, in effect, lunar energy. As long as the moon remains where it is, tidal energy will continue. Energy from the tides has been used since the Middle Ages, when millers used to catch the incoming tide in millponds to drive their water-wheels. In the family of renewable energies, tidal energy is the baby sister who has just started dating. Once she makes the right commercial marriages, her potential will be enormous. Water has four times the energy intensity of a good wind site, so a water-driven turbine can generate more energy for the same area of pressure.

There are five main types of tidal technology: estuary impoundments, ocean impoundments, tidal fence turbines, tidal current turbines, and wave technologies. And there's ocean thermal energy.

Estuary Impoundment
At St. Malo in northern France, a tidal impoundment or barrage was built at the mouth of the Rance river in 1966, and has been reliably generating 240 MW ever since. There is also a 23 MW pilot project in the Bay of Fundy, Nova Scotia, where the tidal flux is so high that there is a 30,000 MW potential. The Derby Tidal Energy Project in northwest Australia is exploring the potential to generate energy from the high tidal range in that region's many inlets and bays. Estuary impoundment is not at all popular with ecologists, however, since it disrupts estuarine ecology.

Ocean Impoundment
Tidal Electric, a Connecticut company, is developing plans to create open ocean impoundments, like miniature walled cities filled with water, with turbines in the city gates. The structures are built just below the low-tide line, and so avoid many of the environmental problems associated with estuary impoundment. Pilot projects are underway or being considered in southeast Alaska, southern Chile, Wales, Gujarat (India), and NW Mexico.

Tidal Fence Turbines
Blue Energy, a British Columbia company, has developed an underwater vertical axis Davis turbine for use in areas where the tidal flow is 3 knots or more. A tidal fence is built between two areas of land, with a series of turbines contained in large caged concrete structures which turn slowly at 25 rpm, allowing fish to swim through. A gap is left in the fence for seals, whales, and larger marine mammals. The technology has been successfully prototyped, and a 500 kW pilot is being built on the British Columbia coast.

Resources

Archimedes Wave Swing:
http://easyweb.easynet.co.uk/~friendly/wave.html

Blue Energy, Canada: www.bluenergy.com

Energy from the Oceans:
www.zebu.uoregon.edu/1998/ph162/l15.html

European Wave Energy Atlas: www.ineti.pt/ite/weratlas

Marine Current Turbines, UK: www.marineturbines.com

Ocean Power Delivery, Scotland: www.oceanpd.com

Ocean Thermal Energy Conversion (Hawaii):
www.nelha.org/otec.html

Offshore Renewable Energy Centres:
www.esru.strath.ac.uk/projects/EandE/98-9/offshore

Sea Solar Power International: www.seasolarpower.com

The Rance, St. Malo, France:
www.muohio.edu/~phy2cwis/slides/tidal.html

Tidal Electric: www.tidalelectric.com

Wave energy in Denmark: www.waveenergy.dk

Wavegen, UK (includes global wave density map):
www.wavegen.co.uk

Wavegen

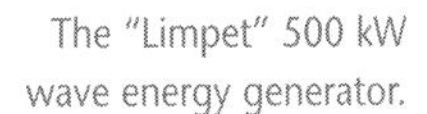

The "Limpet" 500 kW wave energy generator.

Tidal Current Turbines
Marine Current Turbines in Britain is developing plans for large stand-alone underwater turbines, ranging from 600 kW to 1 MW, to be grouped in clusters under the sea in areas with strong tidal currents. The first 300 kW turbine will be built in 2001. Estimated price: $10 per watt, 10 cents/kWh.

Wave Technologies
Wavegen, a Scottish company, has constructed a 500 kW shoreline system called the "Limpet" on Islay, an island off Scotland's west coast, and secured a 15-year purchase agreement with a Scottish utility that is buying the energy for 9 cents/kWh. The Limpet involves an inclined concrete tube with an opening below the water level. The waves cause the water in the tube to oscillate, which compresses and decompresses the air to drive a Wells turbine. Future designs may produce electricity for 4-7 cents/kWh. Various other wave technologies are being developed, including "Salter's Duck," which was killed off prematurely by the British nuclear industry in the mid-1980s. The duck involves a floating canister that drives a generator as it bobs up and down on the water.

Wavegen has also developed the "Osprey" (Ocean Swell Powered Renewable Energy), designed to operate in 15 meters of water within half a mile of the shore, that generates up to 2MW of power. Wavegen has signed a 15-year purchase agreement with the government of Ireland. By adding a 1.5 MW wind turbine to the Osprey, Wavegen created the "WOSP" (Wind and Ocean Swell Power), with an output of 3.5 MW, which is awaiting its first commercial application. Teamwork Techniek of Holland is developing the Archimedes Wave Swing, with a 2 MW system being tested in Portugal in 2001. They estimate a global potential of 48,000 MW along 1000 km of coastline, selling at 9-15 cents/kWh.

Ocean Thermal Energy Conversion (OTEC)
There is often a large temperature difference in the world's tropical oceans between the warm surface waters and the colder deep waters. Where the difference is more than 20° C, energy can be generated using an evaporator to drive a turbine (plus other benefits). The Natural Energy Laboratory of Hawaii runs an experimental OTEC plant. The global potential is enormous; the World Energy Assessment suggests 7,200 EJ (2 million TWh), 18 times more than the world's primary energy production in 1998, but the technology is a long way from being commercially viable.

Global Potential
It is difficult to assess the overall global potential of tidal and wave energy since there are too many unknowns around the potential for scaled-up commercial development. The World Energy Assessment suggests a realizable potential of 79 EJ (22,000 TWh) for tidal energy and 65 EJ (18,000 TWh) for wave energy, totaling 40,000 TWh, which is 34% of the 117,000 TWh we are seeking for a sustainable world in 2025. The World Energy Council, by contrast, estimates that 2TW (6-10,000 TWh) could be harvested from the world's oceans. We are assuming the development of 6,000 TWh by 2025, 5% of our goal.

Biomass Energy

"We're on the threshold of an energy use revolution. If we're smart, we'll stay one step ahead of the market and provide the clean energy decade with the tools it needs: solar panels, biomass, wind power, and fuel cells."
—Gary Locke, Washington Governor

Now we come to biomass, which we have been using ever since we learned to make fire, 400,000 years ago. Plants and trees use photosynthesis to capture the sun's energy, storing it as carbohydrates. We burn it, and the released energy keeps us warm. Nature is so amazing.

Fast-forward to the present. In China, firewood (biomass) accounts for 19% of the country's primary energy supply. In India, it is 42%. In the world's poorest countries, it is 90%. The US gets 2% of its electricity (60 TWh) and 1% of the fuel used in its cars and trucks (1.5 billion gallons) from biomass, the Department of Energy is proposing a 20- to 30-fold increase in its use by 2050. The potential exists for America to get 16% to 20% of its electricity[1] and 50% of its vehicle fuels[2] from biomass. Minnesotans get 7% to 10% of their transportation fuel from ethanol, which is liquid biomass. So what is biomass? Put simply, it is any kind of biological material. Biomass can be burned to produce electricity, digested by bacteria to produce biogas and biodiesel, or processed to produce ethanol.

Resources

American Coalition for Ethanol: www.ethanol.org

Biodiesel: www.VeggieVan.org/biodies.html

Bioenergy: www.eren.doe.gov/RE/bioenergy.html

Bioenergy Information Network: http://bioenergy.ornl.gov

Canadian Renewable Fuels Association: www.greenfuels.org

Carbohydrate Economy Clearing House: www.carbohydrateeconomy.org

Governors' Ethanol Coalition: www.ethanol-gec.org

How Biomass Energy Works: www.ucsusa.org/energy/brief.biomass.html

IEA's Bioenergy: www.ieabioenergy.com

Iogen Corporation: www.iogen.ca

Millennium Environment Debate: www.millennium-debate.org/biomass.htm

NREL's Bioenergy Initiative: www.eren.doe.gov/bioenergy_initiative

Renewable Fuels Association: www.ethanolrfa.org

Electricity

In Spain, a power plant in Cordoba uses olive oil wastes ("orujillo") to generate 32 MW of electricity; Spain produces enough wastes for 500 MW. In Finland, one of the world's largest biofuel plants burns milled peat, bark, and wood chips mixed with coal to produce 265 MW of electricity, 100 MW of steam, and 60 MW of district heating for a nearby city. Sweden gets 19% of its energy from bark, straw, rice husks, and wood chips, and plans to get 40% by 2020. In many tropical countries, sugar cane wastes ("bagasse") make an excellent fuel. Biomass absorbs CO_2 before releasing it, so its use reduces CO_2 emissions by 90% compared to fossil fuels. (The 10% comes from fuel needed to harvest and transport the biomass.) The US gets 10,000 MW of energy from biomass fuels.

Biogas

Farmers from China to the US pump animal manure into methane digesters, solving their water pollution and odor problems while producing energy and fertilizer. In New York state, dairy farmer Bob Aman generates 70 kW from the leavings of his 600-cow herd, selling the power he doesn't need back to the grid. Over 100 years, methane is 23 times more potent a greenhouse gas than CO_2, so Bob is also reducing his farm's climate impact. In St. Louis, Missouri, the Anheuser-Busch Brewery treats its brewing wastes with bacteria and uses the resulting methane to fuel its boilers.

Warren Gretz

A biomass gasifier that uses resudue from the nearby Paia sugarcane mill, Hawaii.

Ethanol

The immediate opportunity lies with ethanol, an alcohol fuel that is currently made from corn starch, using 6% of the US crop. When it is made from corn, it produces greenhouse gases from the fossil fuels that are needed to farm, process, and transport it, but 38% less CO_2 than oil. When it is made from grasses and agricultural wastes ("cellulosic wastes"), it can eliminate greenhouse gases altogether. In Canada, the Iogen Corporation is operating an ethanol-from-cellulose plant in partnership with Petro-Canada that produces 1 million gallons of cellulosic ethanol a year, one of the first of its kind in the world.

In 1980, the US produced 250 million gallons of ethanol, rising to 1.5 billion gallons by 2000. Ethanol's potential has been blocked by the oil companies, however. In the early 1920s, cars needed an additive to boost octane and eliminate engine knocking. With its high octane rating, ethanol was the perfect choice, but it would have displaced 10% of the gasoline, so the oil industry chose to add lead instead. By 1970, lead's harmfulness had caused it to be banned. Once again, the oil industry could have chosen ethanol. Instead, they reformulated their gas to include more benzene, toluene, and xylene. By 1990, the government had to limit these because of their toxicity, and Congress required the oil companies to add 2% oxygen to their gas in big cities. This can be achieved by adding 6% ethanol. About 20% of gasoline now does contain ethanol, but for the remaining 80%, the oil companies chose MTBE, an oil-derived product that causes groundwater pollution when it leaks. Thanks to MTBE, communities from California to New England now have polluted groundwater; by 2001 it was being banned all over the USA. For the fourth time, ethanol sits waiting. The oil companies complain that ethanol is subsidized, but the subsidy is 15 times less than the oil companies' own subsidies[3] (see #72). In Brazil, 24% of fuel is made from sugar-cane ethanol. The US Department of Energy believes that 50% of US transportation fuel could come from ethanol by 2030. Canada produces enough plant wastes to replace 50% to 75% of its gasoline consumption.[4]

Biomass can also be used to manufacture hydrogen. The World Energy Assessment suggests that the world has the technical potential to produce 94-282 quads (27,500-82,500 TWh) of primary energy from biomass during the 21st century. The world currently produces 11% of its primary energy (12,000 TWh) from biomass energy sources.[5] We are assuming that biomass could produce 17,500 TWh, 15% of the energy needed by 2025.

Beyond Energy

It is important to think beyond the use of biomass to generate energy. For the past 250 years, we have been riding the hydrocarbon cycle. As intelligent beings, we should be building a carbohydrate, not a hydrocarbon economy. Today's ethanol refineries could become tomorrow's biorefineries, converting biomass into chemicals, plastics, clothing, paper, film, and fuel. Today's paper could be made from hemp, kenaf, and rice-wastes, keeping the carbon in the forest.

Hydrogen

An integrated solar/hydrogen house and car. (www.zeenergy.net)

Jan Johanson

H2. It's the simplest, lightest element in the universe, 14 times lighter than air. Combine two atoms, each having one proton and one electron, and you have it. Over 90% of the matter in the universe is made from it; it may be the source of all elements. By nuclear fusion inside stars, it makes helium, sending us solar radiation. You'll find it in water (H_2O), in all living matter, even in dead matter such as fossil fuels (CH_4), but you'll never find it on its own because it attaches to other elements so readily. When it meets oxygen in its pure state, it burns. When it mixes with oxygen in an electrochemical process, it produces electricity.

Hydrogen was discovered in 1766 by Henry Cavendish. In 1839, the Welsh physicist Sir William Grove invented the fuel cell by combining hydrogen with oxygen to produce electricity. As the hydrogen enters the fuel cell, it meets a thin platinum membrane, causing the hydrogen protons and electrons to separate. The protons pass through the membrane, meet oxygen, and produce water. The electrons, which can't pass through, are channeled along an external route to drive an electric motor.

During the 19th and 20th centuries, various people grasped the vision of a hydrogen economy in which the energies of the sun and wind would be used to split water and create hydrogen, creating energy without pollution. Since the 1960s, NASA has been using hydrogen in its space flights, but it was too expensive for general application. Then in 1990, the California Air Resources Board decided that by 2003, 10% of all new vehicles in California (100,000 vehicles a year) must be zero-emission vehicles. Overnight, investment dollars poured into fuel cell companies, such as Ballard Power in Vancouver, British Columbia. Today, every major car manufacturer — from Ford and GM to Toyota, Honda, Hyundai, VW, Opel, and BMW — is planning to produce a hydrogen fuel cell vehicle. DaimlerChrysler will be producing 100,000 a year by 2003; a Deutsche Shell scenario suggests that 50% of all new vehicles could be fuel-celled by 2020, 20% of the total fleet.[1]

And it's not just cars. There's an equally big rush to manufacture domestic fuel cells for use in homes and offices, where they can generate reliable electricity for 7-8c/kWh.[2] A Texas developer is building 500 residential homes with built-in fuel-cell units costing $6,000 each; in time, that could fall to $3,000. When you use the excess heat to preheat water, their efficiency reaches 70% to 85%. The New York City Police have installed a 200 kW system; Ballard Power is manufacturing 250 kW units for use in businesses and neighborhoods.

The Hydrogen Business

Avista Corp: www.avistacorp.com

Ballard Power: www.ballard.com

DCH Technology: www.dcht.com

FuelCell Energy: www.fce.com

General Hydrogen: www.generalhydrogen.com

Global Thermoelectric: www.globalte.com

H Power: www.hpower.com

IdaTech: www.northwestpower.com

IMPCo: www.impco.ws

Melis Energy: www.melisenergy.com

Metallic Power: www.metallicPower.com

Plug Power: www.plugpower.com

Solar Hydrogen Energy Corp: www.solar-h.com

Stuart Energy: www.stuartenergy.com

Xantrex: www.xantrex.com

ZeTek: www.zevco.co.uk

> **"We are at the peak of the oil age but the beginning of the hydrogen age. Anything else is an interim solution. The transition will be very messy, and will take many technological paths... but the future will be hydrogen fuel cells."**
> **— Herman Kuipers, Royal Dutch Shell**

Where Will the Hydrogen Come From?
Hydrogen is everywhere. It's in your coffee (water) and it's in the pages of this book (biomass). It's in all biomass and all fossil fuels. It can also be produced by algae if you tweak them a little, which Melis Energy is doing (see #54). There's lots of water in the world, so the best solution is to use renewable energy to produce hydrogen by splitting water (electrolysis), which Stuart Energy is doing. By 2015, we need to see major solar, hydroelectric, and off-shore wind stations producing hydrogen, along with hydrogen being produced on farms from biomass-wastes and algae. Iceland is already exploring these potentials (see #51).

For now, the cheapest way of producing hydrogen is by stripping it from a fossil fuel such as methanol, natural gas, or gasoline. This can be done by means of a hydrogen reformer on board a vehicle, at the service station, or in a central plant. The process releases CO_2, but the amount released varies widely between the different methods. There are big differences between the various methods. The wrong approach will waste the potential of the hydrogen revolution and do almost nothing to slow climate change.[3]

Sources of Hydrogen
Here's how much CO_2 the different methods will release, based on a life-cycle assessment over 575 miles (1000 km):
For contrast: Mercedes-Benz A class internal combustion engine: 248 kg CO_2

A: **Water**, via electrolysis, using electricity from natural gas turbines: 237 kg CO_2 (5% less)

B: **Gasoline**, reformed into hydrogen on board a vehicle: 193 kg CO_2 (22% less)

C: **Methanol**, reformed on board a vehicle: 162 kg CO_2 (35% less)

D: **Natural gas**, reformed in a service station: 80 kg CO_2 (68% less)

E: **Natural gas**, reformed in a central up-stream plant: 70 kg CO_2 (72% less)

F: **Water**, using electricity from renewably generated energy: 0 kg CO_2 (100% less)[4]

G: **Biomass** wastes, sustainably harvested: 0 kg CO_2 (100% less)

H: **Algae**: 0 kg CO_2 (100% less)

Another possibility is that natural gas producers will strip the hydrogen off the gas at the wellhead, pump the CO_2 underground for permanent storage, and ship the hydrogen in retrofitted pipelines.[5] Hydrogen can be shipped or stored as a compressed gas (11.3% efficient), a liquid (8% efficient), or a solid — solid oxides, metal hydrides (3% efficient), sodium powerballs, or graphite nanotubes.[6] If it is stored as a solid, cars could recharge with hydrogen cartridges. By eliminating the need for an on-board reformer and adopting ultra efficient "hypercar" designs (ultralight carbon fiber design, ultra-low drag, hybrid-electric motors[7]), fuel-cell cars could achieve 80 to 100 mpg. If the hydrogen came from D or E above, this could reduce the CO_2 emissions by 90%.[8] If from F, G, or H, it would reduce them by 100%.

Fuel cells will also be installed in homes and offices where they will generate electricity using hydrogen reformed on-site from natural gas, enabling the price to fall through mass production. People living in the buildings could drive fuel-cell cars, using hydrogen generated by the reformers. Plugged into the grid, their vehicles could generate 20-40kW while parked as a source of extra income. If 5-10% of the cars in the US adopted this approach, and electricity use was twice as efficient, they would generate enough electricity for the entire US electrical grid.[9]

Resources

(see also Solutions #51, #54, & #56)

Great fuel cell links: www.me3.org/issues/fuelcells

History of Hydrogen: www.borderlands.com/journal/h2.htm

Hydrogen basics: www.eren.doe.gov/hydrogen/basics.html

Hydrogen Information Network: www.eren.doe.gov/hydrogen

Hydrogen Now!: www.HydrogenNow.org

Hyweb: www.hydrogen.org

Strategy for Hydrogen Transition (RMI): www.rmi.org/sitepages/pid175.asp

Natural Gas and Nuclear Energy

All over the world, people are switching to natural gas. The switch is generally supported because gas contains less carbon than coal or oil per unit of energy burned. Surely, people say, this is a good thing.

Throughout this book, however, you will find scant praise for gas. The reasoning is solid. As well as CO_2 emissions, the production, processing, and distribution of gas also produce "fugitive" methane emissions. Natural gas is 75% to 90% raw methane and escapes are inevitable. The industry has worked hard to eliminate escapes and repair leaky pipelines, but the US data still shows a considerable quantity of methane emissions (see chart below, cols. 4 &7). Compared to countries like Russia, where older gas pipelines are notoriously leaky, the North American natural gas industry is squeaky clean; but that does not change the figures. Coal and oil production also cause methane escapes, but nowhere near as much.

Methane has an average atmospheric life of around 12 years, after which it breaks down into other gases, chiefly CO_2. Over 100 years, it is 23 times more powerful than CO_2. Over 20 years, it is 62 times more powerful.[1] When you include the fugitive methane emissions, gas begins to lose its advantage. Using CO_2 and methane emissions data from the US Energy Information Administration (see endnotes for details of the calculations), it becomes apparent that over 100 years, gas is 38% cleaner than coal, but only 7% cleaner than oil (Col. 6). Over 20 years, which is what counts if we are trying to reduce emissions as quickly as possible, natural gas is 30% better than coal, but 9% *worse* than oil (Col. 9).

In 1998, the US obtained 21% of its primary energy (310 quads) from natural gas. 21% was consumed in people's homes, 14% in commercial businesses, 45% by industry, 15% by electric utilities, and 3% in vehicles. If all these uses were converting from coal, the 43% saving in greenhouse gas emissions would be significant. The reality, however, is that most people are either converting from oil, or creating new demand, causing a net *increase* in greenhouse gas emissions. In the US, 96% of new electricity plants are planning to use gas — but coal-fired plants are not being closed down. The gas is being used to meet new demand, not to offset existing demand.

"Ignoring climate change will be the most costly of all possible choices, for us and our children."
— Peter Ewins,
British Meteorological Office

US Energy Climate Impacts[2]				Total over 100 years			Total over 20 years		
	1	2	3	4	5	6	7	8	9
	Quadrillion Btus consumed in 1999	Carbon in million metric tons (MMTC)	Carbon MMTC per quadrillion Btus	Methane MMTCE* per quad	Carbon + Methane MMTCE per quad	Total MMTCE per quad	Methane MMTCE per quad	Carbon + Methane MMTCE per quad	Total MMTCE per quad
Natural gas	22	310	14.1	38	348	15.8	101	411	18.7
Oil	38	636	16.7	6	642	16.9	17	653	17.2
Coal	22	539	24.5	18	557	25.3	49	588	26.7

*MMTCE = Million Metric Tonnes of Carbon Equivalent.

This is not the only problem with natural gas. First, it is not like oil. It cannot be shipped around the world in tankers without being converted into liquid natural gas (LNG) by refrigerating it to -160° C, requiring expensive new plants, which also need energy.

Second, the world's proven reserves of natural gas in 2000 were 5,146 tcf (trillion cubic feet). At the current rate of consumption (83 tcf a year), this will last 62 years. At the rate forecast for 2020 (167 tcf/year), it will last 31 years. Unless the world shifts to LNG technology, however, global reserves are irrelevant. What matters is North American reserves, which can be shipped by pipe. The US, Canadian, and Mexican natural gas reserves in 2000 were 258 tcf. The US consumes 22 tcf per year (rising to 35 by 2020)[3], Canada 3 tcf/year, and Mexico 1.3 tcf/year. At this rate, North America's reserves will be exhausted by 2010. Is it any wonder the price keeps rising? There are rigs searching for gas everywhere, but production is not keeping up with consumption. With 96% of the USA's new electricity planned to come from natural gas, producers will find it hard to obtain even half the gas they need.[4] The price will continue to rise, and future supplies will be a major concern.

Third, natural gas generates public concerns about pipeline explosions, local air pollution from gas-fired plants, and the devastation of wilderness areas where gas-exploration is underway. In Russia, the industry is investing billions to ship gas from the Yamal Peninsula north of the Ural Mountains on the specific understanding that global warming will melt the Arctic icecap and make it possible for the gas to be extracted. This really is the devil's bargain. Gas is not a solution to global climate change unless it is stripped of its CO_2 at the wellhead and reformulated into hydrogen, with the CO_2 being stored safely away underground.

What about Nuclear Energy?

The world's nuclear plants generate 17% of the world's electricity, but the industry has not sold a plant in North America since 1976. The nuclear salesmen see climate change as their last hope. Nuclear energy does not create CO_2 emissions, except during construction, so on the surface of things, it might seem like a good option. Dig deeper, however, and the problems appear. There are still safety concerns awaiting the inevitable human error. No one has found a solution to the problem of long-term nuclear waste storage; if the Pharaohs of Egypt had developed nuclear power, we would be guarding their wastes today, and for another 96,000 years. Investors won't touch nuclear energy because of huge unknowns around the cost of decommissioning, so any development would have to be taxpayer-financed. In the US, the nuclear industry has already received $145 billion in subsidies, compared to $5 billion for solar and wind energy. In Canada, it has received Can$16.6 billion. Finally, if nuclear energy was to displace coal, a similar investment in efficiency could displace twice as much CO_2. Whichever way you look at them, new nuclear plants don't appear to make sense. (See Campaign for Nuclear Phase-out: www.cnp.ca)

Resources [Gas]

Natural Gas (Energy Source): www.naturalgas.com

Natural Gas Information & Education: www.naturalgas.org

Natural Gas Data: www.eia.doe.gov

Resources [Nuclear]

Atomic Energy of Canada Ltd.: www.aecl.ca

Greenpeace nuclear campaign: www.greenpeace.org/~nuclear

Nuclear Energy Institute: www.nei.org

Nuclear links: www.n-base.org.uk/public/links/links4.htm

Nuclear Safety: www.ucsusa.org

World Information Service on Energy: www.antenna.nl.wise

Carbon Sequestration

Another solution to global warming involves storing the carbon dioxide out of harm's way before or after it is released. The fancy word for this is carbon sequestration from the Latin *sequestrare*, meaning "to place in safe-keeping." There are five ways in which nature or humans can store carbon: in forests, in soils, in oceans, in carbonate rocks, and underground. Some governments and industries love the idea. They want to use carbon sequestration to meet their Kyoto commitments. "Hey," they think, "we can continue burning fossil fuels and sequester our emissions!" As you will see, however, there are some major spooks lurking in the carbon-pile that could come back to haunt us.

Storing Carbon in the Forest

Forests store carbon. That's what makes them forests, along with birds, mosses, mice, eagles, insects, bears, salmon, and wolves. The non-profit organization American Forests estimates that an average acre of newly planted American forest will store 50 tons of carbon over 40 years. You can sign up with a carbon storage company and buy enough trees (or other carbon offsets) to cover the emissions released by your car or business (see #3, #10, #35, #37, #52). The forest industry is becoming very interested. If they cut more trees and plant more trees to replace them, they argue, will this not store more carbon than leaving the forest standing, earning them carbon credits in addition to the income from the lumber?

The answer, in most situations, is no. As well as storing carbon in the trees, forests also store carbon in the soil. In the old-growth coastal Douglas fir rainforests of British Columbia, the forest ecosystem needs 150 years or more to rebuild the carbon it stored before logging.[1] The soil in undisturbed tropical, German, and Siberian forests contains carbon that is over 1,000 years old; centuries would need to pass after industrial logging before the carbon content was restored.[2] Where forests are managed in an ecologically certified manner with little or no soil disturbance, the picture probably changes. If we want to use our old-growth forests to sequester carbon, we need to leave them standing.

So can we store carbon by preventing the world's existing forests from being cut down? Let's hope so; anything that helps us save the forests is worth trying. But global warming will bring more forest fires and more insect damage (see #97 and p. 17). At the present rate of warming, the Amazon rainforest will start dying by 2040 as its soil dries out.[3] What happens if the forest where you are storing your carbon emissions dies or burns down? You've already released the emissions, and now the forest that was supposed to be storing them is re-releasing them. Forests are critically important, but we mustn't fool ourselves that we can use their carbon-storage abilities as a substitute for reducing our own emissions. There is also a slight practical consideration: the US would have to cover 58% of its land with forest every year in order to absorb the CO_2 emissions from just its fossil fuel emissions; Canada would have to cover 5% of its land every year.[4]

"As parliamentarians, we have to stand on platforms around the planet and explain to electors ...why the forest is burning, the cattle are dying, ...why there is surf in the High Street. To explain ...that these are not Acts of God, but Acts of Man."
— Tom Spencer,
Member of the European Parliament

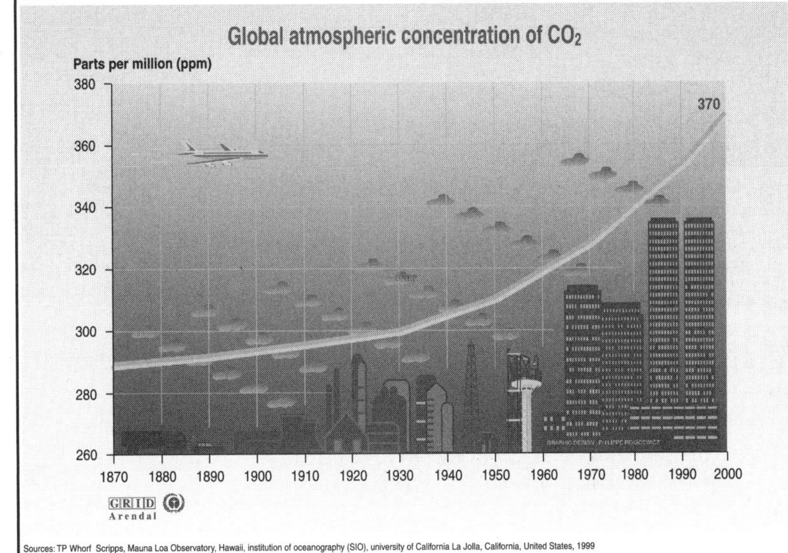

Storing Carbon in the Soil

Soils store carbon. When you disturb the soil by poor farming practices, urban development, or clearcut logging, you disturb its carbon. About 2% of soil's organic matter is very short-lived (1-2 years), and up to 20% has a moderate age (3-100 years). What is astonishing (at least to a non-soil scientist) is that 66%-75% of the organic matter lives from 1-to 4,000 years, and 12%-14% may live for 5,700 to 15,000 years.[5] Disturb that, and you really upset the balance. If we want to maximize the soil carbon, farmers must add compost and manure, adopt zero-tillage instead of plowing, and grow more cover crops. In 1998, the Rodale Institute completed a 15-year research project in which they enriched the soil on some fields while plowing under immature plants on others. The carbon in the experimental plots soared, and yields were just as good as if the crops had been grown with commercial fertilizers.[6] If all of the corn and soybean growing regions of the US were to adopt these practices, 1% -2% of CO_2 emissions from US fossil fuel combustion could be absorbed.[7] (See also #36, #69, #78).

Storing Carbon in the Oceans

Good forestry and farming absorb CO_2 emissions. The other three methods presume that CO_2 will be separated at the power plant or stripped from natural gas at the wellhead and then stored away. This needs energy (releasing more emissions) and costs money, but let's leave these concerns aside for now. The oceans absorb CO_2 naturally, dissolving it in the water as a bicarbonate. The proposal that researchers are pursuing is to pump separated CO_2 to the bottom of the ocean, hoping that it stays there. In ecological terms, we are now in Eco-X-Files territory. Marine biologists are alarmed because CO_2 is an acidic gas that would kill most marine life in the vicinity. Climate activists are alarmed because no-one knows how long it would stay there. What if it returns to the surface (and the atmosphere) in 30 or 100 years? See #53.

Storing Carbon in Rocks and Underground

Nature turns CO_2 into calcium carbonate rocks, so why can't we? That's one line of research that the US government is supporting, including the possibility that algae might help to strip CO_2 from the gas flues at power plants. The final "CO_2 graveyard" being considered involves pumping CO_2 into deep geological formations underground, depleted oil and gas wells, and empty coal-seams. The technology is already being used at enhanced oil recovery sites around the world, and the Norwegians are busy doing it, linked to the vision of a CO_2-free hydrogen economy (see #53). Maybe there's some potential here.

Resources

Carbon Sequestration: http://cdiac2.esd.ornl.gov

CO_2 Ocean Sequestration Field Experiment: www.co2experiment.org

CO_2 Sequestration Technology: www.lanl.gov/CO2

DOE Center for Research on Ocean Carbon Sequestration: http://esd.lbl.gov/DOCS

Ocean Chemistry of Greenhouse Gases: www.mbari.org/ghgases

World Ecosystems - Carbon in Live Vegetation: http://cdiac.esd.ornl.gov/ftp/ndp017/table.html

Wild Cards

What about some of the other ideas that float around? Might they have the answer to global warming? It pays to keep an open mind and remember these immortal words: "Heavier-than-air flying machines are impossible." (Lord Kelvin, President of the Royal Society, 1895) "Who the hell wants to hear actors talk?" (H.M. Warner, Warner Brothers, 1927) "Everything that can be invented has been invented." (Charles Duell, Commissioner, US Office of Patents, 1899)

Anti-Matter

NASA is working on a system of spacecraft propulsion based on anti-matter engines — like the ones used to fuel the starship *Enterprise*. The universe consists of an equal quantity of particles and anti-particles that have the same mass, spin, and lifetime, but the opposite charge. When matter collides with antimatter, they cancel each other and release pure energy. So far, no one has been able to gather enough anti-matter to be useful.[1]

Anti-Gravity

NASA is also trying to understand gravity, the inherent attractive force that exists between all bodies with mass. When scientists cracked the mysteries of electromagnetism in the 19th century, they gave us the miracle of electric power. Could gravity's secrets bring us gravity power? Maybe vehicles could travel without friction, and water could flow upwards and then downwards into an electric turbine? So far, little progress.

Black Light

BlackLight Power was founded by Randell Mills as a spin-off from his work on a grand unified theory. BlackLight power cells generate energy through a patented chemical process that apparently causes the electrons of hydrogen atoms to drop to lower orbits, releasing excess energy as a hot ionized gas that can be converted to electricity and used to create hydrogen from water at 10% of the cost of using fossil fuels. The company is being financed by power utilities, but others are skeptical.

Resources

BlackLight Power Inc: www.blacklightpower.com

Canadian Space Power Initiative: www.space.gc.ca/cspi

Car exhaust into diamonds:
www.newscientist.com/nl/1007/little.html

Chemtrails Conundrum:
www3.bc.sympatico.ca/Willthomas/homepage.html

Geo-engineering (PDF, Chapter 28):
http://stills.nap.edu/html/greenhouse

Infinite Energy magazine: www.mv.com/ipusers/zeropoint

Institute for New Energy: www.padrak.com/ine

NASA's Breakthrough Propulsion Physics:
www.lerc.nasa.gov/WWW/bpp

Space Studies Institute: www.ssi.org/energy.html

SunSat Energy Council: www.sunsat-energy.org

The Gravity Society: www.gravity.org

Zero Point Energy:
www.planetarymysteries.com/hieronimus/zeropoint.html

Cold Fusion

Fusion is the process by which atomic nuclei fuse together to produce a single nucleus (contrasted with nuclear fission, where the nucleus breaks apart.) Fusion has been achieved on an uncontrolled scale in the hydrogen bomb, but no one has yet been able to do so in a controlled manner. In 1989, Stanley Pons and Martin Fleischmann announced that they had produced enough energy to heat a cup of tea, but so far, no one has been able to duplicate their results. To its believers (including Arthur C. Clarke), cold fusion is the answer to all our climate worries.

Zero Point Energy

Quantum theory tells us that empty space contains a huge amount of untapped electromagnetic energy, known as zero point energy (ZPE). According to its believers, there is enough energy in the empty space of a coffee cup to evaporate all the world's oceans. To some, ZPE is the energy of consciousness that pervades the universe. The energy may exist, but no one has been able to demonstrate the theory, and so far, you can't plug a kettle into it.

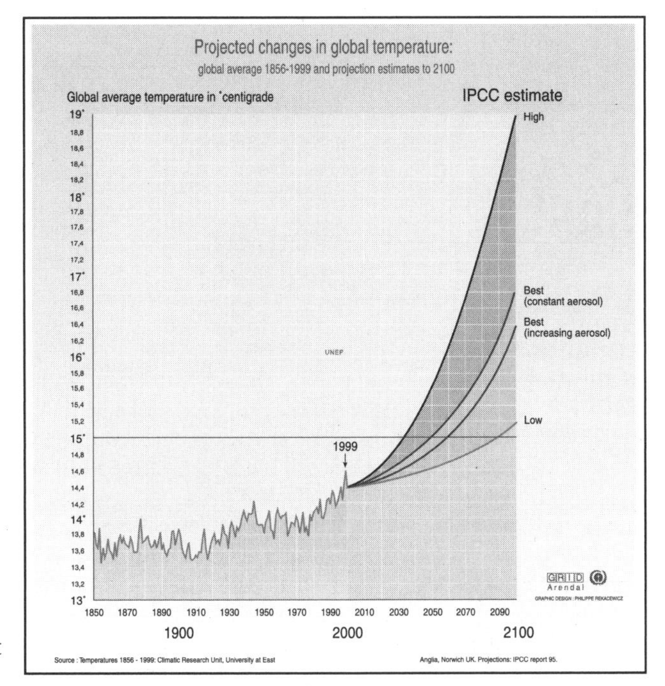

Methane Hydrates

There is an enormous quantity of frozen methane hydrates along the continental shelves of the world's oceans, theoretically enough for 2,000 years of energy-use. The potential for disaster, however, is appalling; just one underwater landslide could set off a huge destabilization event (see p. 19). The Japanese National Oil Company is already drilling for hydrates in a $60 million research program.[2] Even if they extract them safely, the hydrates are a fossil fuel, so they will only make things worse.

Solar Power from Space

The sun shines for 24 hours a day in space, so could solar collectors beam its energy back to Earth? NASA is investigating the possibility of a 22-mile long 1,200 MW "sun tower" in stationary orbit 22,000 miles away, transmitting its energy to Earth by microwave. Japanese researchers suggest sending the energy by laser to atmospheric airships, then beaming it by microwave to a receiving antenna 15 miles in diameter. If the technology can be made to work, the obstacle will be public concerns about large-scale microwave radiation. Another approach is a 22,000 mile long, 1/4 inch superconducting "beanstalk" tether made from carbon-fiber Buckeytubes, named after the visionary engineer R. Buckminster Fuller.

Car Exhaust into Diamonds

The most fascinating of these wild cards is a device the size of a wine bottle called a "microwave emissions converter" that was announced by Australian engineers Elias Siores and Carlos Distefani in October 2000.[3] Attached to the tail-pipe of a vehicle, it microwaves the car exhaust to 5000° C, breaks the molecular bonds of the gas and creates a plasma of free ions. As the mix cools, the ions recombine into less harmful substances, reducing CO_2 and other emissions by 70%. This increases the number of tiny carbon particles being released, so the engineers have created an electro-static filter that lines the exhaust pipe to collect the particles. The particles can be harvested and treated by another process to create industrial-grade diamonds. Watch this space!

Geo-Engineering

Some of this is wacky; some is weird and worrying. First, there's the idea of covering the world's oceans with white styrofoam balls to reflect the sun's light back into space. Yeah, right. Then there's the proposal to position a series of space mirrors between the Earth and the Sun to block the sun's light from reaching us, like permanent rotating eclipses, or to fill the sky with a vast number of aluminized, hydrogen-filled balloons, creating a reflective screen. Hmm. Then there's the idea of seeding the oceans with iron filings to stimulate plankton growth (see p. 9). But most worrying is the proposal to increase the Earth's reflective albedo by injecting quantities of dust, sulfuric acid, or aluminum into the atmosphere.[4] The physicist Edward Teller and others at the Lawrence Livermore National Laboratory have calculated that inserting 10 million tons of dielectric aerosols would increase the Earth's albedo by 1%, eliminating the effect of doubling CO_2 emissions to 560 ppm. The worry is that the experiment appears to be already underway, using aircraft to spray aluminum with their exhaust emissions. There are too many reports and photographs of people seeing strange, grid-like contrails (or "chemtrails") in the skies and reporting strange associated sicknesses for the rumors to be dismissed, even if the interpretations are sometimes paranoid. For details, see Resources on previous page.

What Will It Cost?

Which of these two arguments do you believe?

- *Taking action to reduce greenhouse gas emissions will create an economic Armageddon,*

  or

- *Taking action to reduce greenhouse gas emissions will encourage technological innovation, create new jobs, and benefit everyone.*

The first argument has caused many North American politicians to believe that it is politically impossible to take any serious action to address climate change. A 1998 study by WEFA, funded by the American Petroleum Institute and widely quoted in Congress, assumed that a $265 per ton carbon tax ($971 per ton of CO_2) would be needed to meet the Kyoto requirements. This would increase gas prices by 65 cents a gallon; nearly double the price of natural gas and electricity; cost 2.4 million jobs; lower wages; reduce GDP by 3.2%; increase poverty; harm American competitiveness; reduce investment; reduce average household income by $2,700; increase the cost of food, housing and medical care; and reduce state revenues by $93 billion.[1] Just what you want for Christmas.

Studies such as this are very limited, because they start with worst-case assumptions and limit their methods to a whopping big tax. In 1997, the Canadian government commissioned a study by Standard & Poor's DRI. By including domestic carbon emissions trading, it showed that achieving the Kyoto goals would reduce Canada's GDP by just 2% up to 2010, with a full recovery by 2020.[2]

But why stop there? A US study concluded that the benefits of reduced acid rain, reduced emissions from power plants, and reduced smog were worth $3-$7 per ton of carbon, with a more substantial reduction yielding $12-$18 per ton. In general, the benefits reduced the cost by around 30%.[3] When a US government five-laboratories study examined the role of 200 technological innovations that could reduce emissions by 2010, it found that the cost of $50-$90 billion per year would be offset by a $70-$90 billion per year savings.[4] An International Project for Sustainable Energy Paths study concluded that the US would gain $200 billion by implementing Kyoto, assuming investments in energy efficiency, the expansion of successful policies and programs, domestic CO_2 trading, and redirecting revenues from carbon-related permits and fees into incentives for renewable energy and efficiency investments, plus reduced taxes on non-energy items.[5]

The top-down studies favored by the fossil fuel companies don't factor in these potentials. A 1991 study by the Association to Save Energy found that carbon emissions could be reduced by 80% by 2030, with more than $500 billion in cumulative savings, at a marginal cost of $7 per ton of carbon saved, without including the benefits of carbon-reduction.[6] When the Union of Concerned Scientists and the Tellus Institute compared studies that considered the role of technological innovation and efficiency, they found that there was no net cost at all, rather savings that ranged from $4 to $7 per ton of carbon.[7]

We want to repeat this over and over again: reducing greenhouse gas emissions is going to be exciting, profitable, and create a more sustainable world. The fossil fuel companies are just whining and distorting the evidence, as they try to protect their turf.

The Evolution of Climate Change Reduction Econometric Models

A: Carbon taxes only
B: A + tax and policy initiatives for efficiency & renewables, and domestic emissions trading
C: B + benefits from technological innovation and competitiveness
D: C + benefits from reduced air pollution (acid rain, smog, asthma, mercury pollution)
E: D + costs of climate change damage and mitigation measures
F: E + international emissions trading
G: F + reduction measures for methane, NO_2, black carbon, and industrial greenhouse gases
H: G + carbon credits, global solar compact, tax-shifts, etc, as in Solutions #1-#101.

> "Cutting income taxes (by 10%) while increasing gasoline taxes (by 50 cents-per-gallon) would lead to more economic growth, less traffic congestion, safer roads, and reduced risk of global warming - all without jeopardizing long-term fiscal solvency. This may be the closest thing to a free lunch that economics has to offer."
> — N.Gregory Mankiw, Professor of Economics, Harvard University.[8]

Rising Storms, Rising Damages

But what about the cost of *not* reducing emissions? As temperatures have risen, the insurance industry has seen a doubling in the cost of weather-related disasters every five years.[9] Munich Re, the leading German re-insurance company, has seen a threefold increase in natural catastrophes and a 15-fold increase in costs to the insurance industry since the 1960s.[10]

The message is clear: cleaning up after climate change enhanced weather disasters is going to be extraordinarily expensive. Munich Re has calculated that the world could be paying $300 billion annually by 2050 as a result of more frequent cyclones and hurricanes, the loss of land from rising sea-levels, and damage to agriculture and fish stocks.[11] For most countries, this ranges from 0.2% to 2% of their GDP; for some, including small island states, it exceeds 10%. Andrew Dlugolecki, a director at CGNU, the UK's largest insurance group, has seen claims for weather-related property damage rising by 10% a year and estimates that the costs could exceed the world's GDP by 2065.

Reducing Emissions, Creating Jobs

Global warming has a clear dark side, but tackling global warming has a very promising bright side. Worldwide, 150 million people were unemployed at the end of 1998, plus all those who were holding down unsatisfying part-time jobs. Coal-mining jobs will go, but new energy strategies will create many new jobs to replace them. A 1999 study showed that by using a mixture of policies to reduce emissions to 14% below the 1990 level, the US could save $43 billion a year on energy costs and create 870,000 new jobs by 2010.[12] A similar study showed that a European Union energy tax would create 1.9 million new jobs, including 15-19 jobs for every 1 MW of wind energy.[13]

Investments in energy efficiency produce four times as many jobs as investments in new energy. If 25% of Canada's annual energy expenditure was returned to the economy as energy efficiency investments, it would create 33,000 jobs lasting 20 years. The American Council for an Energy Efficient Economy has estimated that an annual investment of $46 billion over the period 1992-2010 would have resulted in a 24% reduction in carbon emissions and generated 1.1 million new jobs by 2010.[14]

Wherever you look, building a more sustainable society creates more jobs and enables us to spread our wealth around more evenly. Organic agriculture creates more jobs than intensive chemical agriculture; recycling and re-manufacturing create more jobs than disposing of garbage in landfills; sustainable ecoforestry and ecotourism create more jobs than industrial logging. There is an exciting vision here, inviting us to embrace it. Global warming is simply the sign that it is time for us to change.

Resources

A Small Price to Pay: US Action to Curb Global Warming is Feasible and Affordable: www.ucsusa.org

Climate Change Economics and Policy: An RFF Anthology: www.rff.org

Economists Statement on Climate Change, Feb 1997 (2,500 economists, 8 Nobel laureates): www.rprogress.org/pubs/ecstat.html

IPPC Report *Climate Change 2001: Mitigation:* www.rivm.nl/env/int/ipcc

The Carbon Barons

"We're doing great, thank God. The American economy is flourishing. We're using more fossil fuels. We're putting more CO_2 in the air. The coal plants are running at record levels. Business has never been better. We're doing great!"[3]
— Fred Palmer, Western Fuels Association

It is almost impossible to examine solutions to global warming without bumping into the efforts of the fossil fuel industry to sink the Kyoto Protocol, and keep the coal, oil, and gas flowing. The world's fossil fuel corporations have spent millions on PR campaigns, lobbying offensives, and political campaign donations. Their efforts have cast doubt on the science, exaggerated the costs of change, underplayed the impacts, misinformed the public, undermined the Kyoto treaty process, and secured the election of politicians friendly to their cause. It is true that some of the carbon barons have also been investing in solar, wind, and fuel cell technologies (see #51, #54, #56, & #60), and BP has even changed its name to "Beyond Petroleum," but for the industry as a whole, the main agenda is business-as-usual by whatever means it takes.

The Carbon Minstrels

There are 2,000 scientists who sit on the United Nations Intergovernmental Panel on Climate Change, who review the latest peer-reviewed studies before issuing consensus statements. Set against them are a handful of "greenhouse skeptics" who argue that global warming is not happening, or that it is natural variation (see p. 20), and whose work is very rarely peer-reviewed. When the carbon barons pay, the skeptics sing. Their songs are picked up by dozens of right wing think tanks (see list opposite) and broadcast by media outlets around the world. Three of the leading skeptics are Dr. Pat Michaels, who has received $160,000 from Western Fuels, German Coal Mining Association, Edison Electric, and Cyprus Minerals; Dr. Robert Balling, who has received $350,000 from German Coal Mining Association, Cyprus Minerals, British Coal Corporation, and OPEC, and $48,000 from the Kuwait Foundation for the Advancement of Science; and Fred Singer, who has received funding from Exxon, Shell, Unocal, and ARCO.[1]

The Carbon Journalists

Sad to say, many North American journalists don't know the difference between the greenhouse effect and the hole in the ozone. One bunch of scientists says one thing, another says something else, so they report them both equally. There is often more to this than meets the eye, however. The News Corporation, for instance, is owned by the Australian media-baron Rupert Murdoch. Its holdings include Fox TV (Allie McBeal, The X-Files), cable stations, magazines, the *New York Post,* Britain's *Times, Sunday Times, Sun,* and *News of the World*, and 70% of Australia's newspapers. Fox News runs a column called *Junk Science,* written by Steve Milloy, which trashes any talk of global warming. Steve works for the Cato Institute, a right-wing Washington think-tank that fights government regulation and is funded in part by Chevron, Exxon, Shell, the American Petroleum Institute, and Amoco. For its science, it relies on the work of Fred Singer and Pat Michaels (see above) to justify its positions on climate change. Rupert Murdoch sits on the Cato Institute's board of directors, and the whole package supports the interests of the carbon barons, who support them.[2]

The Carbon-Protectors

American Petroleum Institute: www.api.org
Cato Institute: www.cato.org
Center for Energy and Economic Development: www.ceednet.org
Center for the Study of Carbon Dioxide and Global Change: www.co2science.org
Citizens for a Sound Economy: www.cse.org
Coalition for Vehicle Choice: www.vehiclechoice.org
Competitive Enterprise Institute: www.cei.org
Cooler Heads Coalition: www.globalwarming.org
Global Climate Coalition: www.globalclimate.org
Greening Earth Society: www.greeningearthsociety.org
Heartland Institute: www.heartland.org
Heritage Foundation: www.heritage.org
Marshall Institute: www.marshall.org
National Center for Policy Analysis: www.ncpa.org
National Center for Public Policy Research: www.nationalcenter.org
Science and Environmental Policy Project: www.sepp.org
Western Fuels Association: www.globalwarmingcost.org

The Carbon Lobbyists

During the 1990s, the carbon barons financed a lobby group called the Global Climate Coalition (GCC), which spent millions in efforts to confuse the public, including a $13 million ad campaign opposing the Kyoto treaty. Its board of directors included representatives from Chevron, Exxon, Ford, General Motors, Mobil, the Southern Company, and Texaco. Jeremy Leggett, an oil company geologist who was Greenpeace's climate change scientist at the UN negotiating sessions leading up to the 1997 Kyoto treaty, recounts how a close associate of the GCC, Don Pearlman, representing an obscure NGO called the Global Climate Council, gave individual coaching to the Saudi and Kuwaiti delegates in how to obstruct and delay the proceedings, going so far as to pass them handwritten notes during negotiating sessions.[4] The Saudis and Kuwaitis played the same obstructionist game over the IPCC's 2001 report.[5] In 1999 and 2000, BP, Shell, Ford, Daimler-Chrysler, Texaco, The Southern Company, and General Motors all abandoned the GCC, forcing it into a strategic retreat. There are also groups such as the Coalition for Vehicle Choice, representing the major motor manufacturers, which lobbies intensively to stop the corporate average fuel economy (CAFE) standard for vehicles being raised (see #57).

The Carbon Politicians

The carbon barons need friendly politicians in Washington, Ottawa, and elsewhere, and they spend millions to get them. In the 2000 US Presidential campaign, the oil and gas companies spent $30 million dollars (78% to Republicans, 20% to Democrats); transportation companies spent $51 million (72% to Republicans, 27% to Democrats).[6] From 1989-1999, members of the Global Climate Coalition spent $63 million (69% to Republicans, 31% to Democrats). With George Bush and Dick Cheney in the White House, the carbon barons have full control of the system, helped by people such as Alexander Haig, who has promoted oil company investments in China and Turkmenistan; Caspar Weinberger, a former employee of Bechtel with energy contracts in the Middle East; Andrew Card, who was the auto industry's top lobbyist before becoming White House Chief of Staff; and Spencer Abraham, the new Secretary of Energy, who opposed raising the CAFE standard as a Senator and suggested that the Department of Energy should be axed.

The Carbon Bankers

As long as the world keeps expanding its use of fossil fuels, the carbon barons are happy. But what do you do when a country such as China starts to close its coal mines? In 1997, Exxon's chairman Lee Raymond urged China to use more, not less, fossil fuels, and said that nature was to blame for most global warming. Fortunately for the carbon barons, the World Bank is firmly controlled by North Americans and Europeans. Since 1992, it has spent 25 times more on fossil fuel projects than on renewable energy; 90% of the contracts went to corporations from the G-7 nations (see #96). It also funds deforestation projects, and subsidizes oil and gas pipelines. The world has been good to the carbon barons — but for how much longer?

Resources

Carbon War: Global Warming and the End of the Oil Era by Jeremy Leggett (Routledge) 2001

Cool the Planet: www.cool.policy.net

Greenhouse Wars: www.newscientist.com/global

How Industry Blocked Progress at Kyoto: www.corpwatch.org/trac/feature/planet/gods.html

Open Secrets - Money in Politics: www.opensecrets.org

The Heat is Online: www.heatisonline.org

A Global Effort

> **"By deliberately changing the internal image of reality, people can change the world."**
> **— Willis Harman**

We have all enjoyed the benefits of fossil fuels. For better or worse, they have given us an enormous advantage over the millions of other species with which we share this planet, in the vastness of this cold, dark universe. They have enabled us to put power to our intelligence and use it to build roads, fly around the world, soak in hot tubs, and impose our will wherever we choose to.

When we first migrated out of Africa 100,000 years ago, it was as equals among equals. Today, we dominate the Earth, but we are still profoundly ignorant about the ecology of this planet that we profess to "manage." We have been building civilizations for 7,000 years, but most people still do not know that carbon dioxide traps heat in our planet's atmosphere, or that our use of fossil fuels is threatening to end the peculiar climatic stability of the past 11,000 years that has allowed us to farm and build cities. According to the Vostok ice-record (see p. 14), this stability has been without precedent for at least 420,000 years.

This book does not say "Stop! We are doing something wrong." It says "Change course! Switch fuels and adopt sustainable goals!" so we can continue building our civilizations. Before we started burning fossil fuels, the CO_2 in the atmosphere was 265 parts per million (ppm). By 2000, it was 370 ppm and rising by 1.5 points per year, the highest rate of rise for 420,000 years, probably 20 million years. If we can reduce our emissions by 80% by

Where Do North America's Greenhouse Gases Come From?

All data in million metric tonnes of carbon equivalent (MMTCE)

		USA, 1998[1]			CANADA, 1997[2]		
	Source	MMTCE	100 yrs	20 yrs	MMTCE	100 yrs	20 yrs
CO_2	Industry	479	26%	22%	74	40%	32%
	Transport	450	24%	20%	47	25%	20%
	Residential	287	16%	13%	20	11%	9%
	Commercial	239	13%	11%			
CH_4[3]	Landfills	64	3%	8%	6	3%	7%
	Natural gas	37	2%	6%	12	6%	13%
	Coal	19	1%	2%			
	Oil	7	<1%	1%			
	Cattle guts	37	2%	6%	5	3%	6%
	Manure	25	1%	3%	1	<1%	1.5%
N_2O	Farm soils	70	4%	3%	9	5%	4%
HFCs etc.	HFCs etc	40	2%	2%	2	1%	<1%
Misc		91	5%	4%	6	5%	5%
TOTAL		1,845	6.5 per person		188	5.8 per person	

"The forests are dying, the rivers are dying, and we are called to act."
— Dhyani Ywahoo

2025, and 100% by 2035, the CO_2 will stabilize at 393 ppm, then hopefully start declining.

The reason for such critical urgency is that if we are still pumping out greenhouse gases by 2040, the Arctic summer ice will have melted, the permafrost will be pouring out methane and CO_2, the Amazon rainforest will have started to die, and we will be past the point of no return. We need to be really clear about our skills and abilities. We know how to switch fuels, capture landfill gas, and manage farms and forests sustainably. We do *not* know how to fix a broken carbon cycle.

Some scenarios assume that we might stabilize our CO_2 emissions at 450 or 650 ppm. We believe that we can stabilize them below 400 ppm if we use the knowledge we already have, if the carbon barons stop sabotaging our efforts, and if we work together as families, businesses, cities, nations, and the world.

What Will It Take?

- Scientists and engineers, eager to develop renewable energy and other sustainable technologies.
- Entrepreneurs, eager to bring renewable energy technologies to market.
- Investors, eager to finance renewable energy.
- State, provincial, and federal politicians, eager to set targets and legislate policies that promote energy efficiency, sustainable energy use, and greenhouse gas reductions.
- Individuals and families, eager to reduce their personal emissions and get involved with citizens organizations that are working to make the transition to clean energy.
- Schools, colleges, churches, and other organizations, eager to become role models for efficiency and sustainable energy.
- Municipal politicians and staff, eager to make their cities role models for efficiency, sustainable energy, tree-planting, recycling, waste reduction, and greenhouse gas reductions.
- Business men and women, eager to make their businesses role models for natural capitalism.
- Motor corporation leaders, eager to sell fuel-efficient, zero-emission vehicles.
- Energy company leaders, eager to lead the transition out of fossils fuels and become champions of a solar-hydrogen economy.
- Farmers and foresters, eager to manage their farms and forests in ways that protect the soil, store carbon, and reduce greenhouse gas emissions.
- Workers and labor leaders, eager to participate in the transition while supporting and retraining those whose jobs will disappear.
- National leaders, eager to sign a more effective global treaty and create a Global Green Deal.
- Leaders from the developing nations, eager to leapfrog over the fossil fuel age and build sustainable, solar/hydrogen economies.
- Ordinary people, eager to use their personal talents to contribute in whatever way they can, cooperating, avoiding judgments, sharing solutions, and *believing that it is possible.*

The 101 Solutions
TO GLOBAL CLIMATE CHANGE

1

Grasp the Big Picture

"One person can make all the difference in the world. For the first time in recorded human history, we have the fate of the whole planet in our hands."
— Chrissie Hynde

So, here we are, at the beginning of our challenge: to phase out fossil fuels and the greenhouse gases that result from human activities. Who is to do it?

(a) The government?

(b) Big business?

(c) City governments?

(d) You and me?

The answer is: (e) All of the above. We're at the beginning of a major planetary transition and we all have a role to play. Once upon a time, we all lived in the Stone Age. One day soon, we will all live in the Solar Age.

This book offers ideas and solutions that enable everyone to meet the challenge, starting with ourselves. It may seem like a huge responsibility, but if we take it one step at a time, it will be manageable. Already there are households that have reduced their personal greenhouse gas emissions to zero and businesses that have become "climate neutral."

In 1998, the world released 10 billion metric tons of carbon equivalent by burning coal, oil and gas, and other greenhouse gas-producing activities. This includes emissions from industry, transportation, farming, forest destruction, aviation, the generation of electricity, cement production, home heating, and other activities. If we lived the way we did a thousand years ago, that number would be almost zero. In the big picture, we are all collectively responsible.

In North America, our personal use of fossil fuels represents 25% of our national greenhouse gas emissions. This is due to the way we heat and power our homes, the way we travel, and the way we eat. If you include the goods we consume, it increases to around 50%, since almost everything we buy is manufactured and delivered using fossil fuels.

In the United States, this 25% represents an average 5 tons of CO_2 per person per year. In Canada, it's 4 tons, since Canadians use more hydroelectric power. North Americans produce far more emissions than people in other countries. In Holland, they produce 3 tons each from personal use; in Japan, 2.1 tons; in China, 0.55 tons; in Bangladesh, 0.03 tons.

Within the North American figures there are big differences based on how much people consume. People on lower incomes produce lower emissions because they have smaller houses, own fewer cars, and don't fly as often.

Imaginary Families

Family A: Gerry and Martha Hanson.
Small condo, no children, natural gas for water and heat. Cycle, walk, bus for most trips.
Compact Honda 4,300 miles/year.
One annual flight Vancouver to Winnipeg.

Total: 7.7 tons CO_2 = 3.85 tons per person.

Family B: Maria and Henrik van Groot.
2 kids, 2,600 sq ft house, oil-fired heating.
Parents drive to separate jobs in Ford Taurus and Chrysler minivan, 17,000 miles/year.
Annual family vacation flight Seattle to UK.

Total: 54.3 tons CO_2 = 13.57 tons per person.

Family C: Shannon and Gustav Murphy.
2 kids, 2,000 sq. ft passive solar energy-efficient house, energy-efficient appliances, solar water heater. Members of local car-share cooperative, shared use of Toyota Prius, 3,000 miles/year. Use bus, bikes, walk. Try to live simply, buy locally grown organic food. Buying 100% renewable energy from GreenMountain.com.
Annual hiking and cottage vacation. No flights.

Total: 1 ton CO_2 = 0.25 tons per person.

"If you think you are too small to be effective, you have never been in bed with a mosquito."
— Betty Reese

Tim Ogline

Seven ways to reduce your emissions to zero

In the next pages, you will learn how you can reduce your household emissions of CO_2 to zero by following seven basic strategies:

1. Travel more sustainably (see #2).
2. If you must drive, buy a fuel-efficient vehicle (see #3).
3. Invest in more efficient household appliances (see #4).
4. Make your whole home energy efficient (see #5).
5. Use the sun's energy (see #6).
6. Buy green power (see #7).
7. Buy carbon offsets to absorb the remaining CO_2 (see #10).

Take Wider Action

As well as reducing your individual emissions, you can join a local action group and work with others to make a difference. If your local group persuades your city to join the Cities for Climate Protection Campaign (see #21), with the result that the council decides to buy green power for its municipal needs — as Santa Monica and dozens of other cities have done — millions of tonnes of CO_2 would be eliminated. Or if lobbying by your group helped persuade the US Congress to raise the Corporate Average Fuel Efficiency standard for vehicles (see #76), you would have helped to eliminate many more millions of tonnes.

The World is Watching

We need to remember that global warming is not just a North American affair. The whole world is threatened. If you were farming and raising a family in Bangladesh, you would be responsible for almost no emissions, and yet climate change might be threatening to wash away your entire village by river flooding and the rise in sea-level.

By showing that it is possible to reduce our personal use of fossil fuels, we send a powerful signal to others, especially in the developing world. It is up to us to demonstrate that it is possible to enjoy a high quality sustainable life without the use of fossil fuels.

Reducing climate change emissions is not a chore. It is an incredible gift — a gift for the air, for the economy, for your children, and for the Earth as a whole. Earth's polar bears, forests, salmon, and coral reefs have no vote in this. All they can do is hope that we get it right.

Carbon Calculators

Carbon calculators show how much CO_2 we produce from our personal activities, depending on where and how we live:

Canada:
- www.climcalc.net

Britain:
- www.bestfootforward.com/carbonlife.htm
- www.carboncalculator.org

USA:
- www.airhead.org/Calculator
- www.chuck-wright.com/calculator_home.html
- www.infinitepower.org/calculators.htm
- www.pbs.org/wgbh/warming/carbon
- www.safeclimate.net/calculator

US Midwest:
- www.elpc.org/polCalc

"We must make the rescue of the environment the central organizing principle for civilization." — Al Gore

2

Travel More Sustainably

Do you ever pause to watch the traffic and think to yourself: Right now, at this precise moment, a similar rush of traffic is happening on roads and highways all over the world? It's a scary thought. In some places, the traffic continues all through the night.

As a world, we've gone travel crazy. Maybe we're trying to catch up for those thousands of years when we had to go everywhere by foot. It's the cheap fuel that makes it possible — cheap fossil fuel. Every time we turn on an engine, the carbon from millions-of-years-old fossils is released into the atmosphere. Globally, our transportation habits make up 21% of the greenhouse gas emissions from human activities.

The first practical solution is for each of us to stop and think how we travel, and then make the commitment to travel more sustainably. Break the habit of just jumping in the car. Think of it like smoking, on a planetary scale. Say to yourself:

I've got to quit!

We, as a family, have got to quit!

We, as a planet, have got to quit!

Some of the solutions will come in time; a solar-hydrogen economy with fuel cell vehicles is just over the horizon. Other solutions are right in front of us. Part of the trick is to reduce the need for travel by re-organizing our lives. Could you move closer to work, or work from home more often? Could you move to a more compact community where stores, schools, and recreation are closer to home? In older, more compact neighborhoods built before the era of sprawl, destinations are generally closer and local stores are mixed into the community. A comparison of 97 Swedish communities found that when urban density is doubled, gasoline consumption falls by 25%.

When you get a yen to explore, try looking closer to home. Before flying off to some exotic destination, look for the jewels in your own backyard, neighborhood, and green spaces. There is no better way to learn about a place and to find its special nooks and crannies than to walk or bike around it.

> **"If we do not change our direction, we are likely to end up where we are headed."**
> **— Chinese Proverb**

The High Carbon Cost of Flying

- New York-Paris: 3.8 tonnes of CO_2
- Seattle-Beijing: 5 tonnes of CO_2
- Winnipeg-Miami: 2.2 tonnes of CO_2
- To calculate any particular trip, see: www.chooseclimate.org/flying
- To buy carbon offsets for your flight, see: www.TripleE.com/environment
- To calculate how many trees are needed to absorb the CO_2 from a flight, see: www.americanforests.org, or www.treecanada.ca
- One tree absorbs 1 ton of carbon over 40 years. Plant 3 trees to ensure that 1 survives.

Start-up Plan For New Cyclists

Step 1: Buy a new or second-hand bike. Don't forget a helmet; it's the law in many places.

Step 2: Take it to a parking lot on a quiet Sunday morning to try it out. If you rode as a kid, it'll soon come back to you.

Step 3: Visit a local cycling store and ask about bicycle safety classes. Sign up to refresh your road skills.

Step 4: Look at a map to see if you can use quiet residential streets to get to work and to the shops.

Pedestrians Do It Feet First
When is the last time you walked to the corner store? Or tried getting the kids to walk to school? These are great ways to get to know your neighbors — and keep fit. Many subdivisions don't even have sidewalks; that's one of the things we have to fix. See www.scn.org/civic/feetfirst and Solution #28.

Cyclists Do It In The Saddle
Forty percent of all automobile trips in the US are bike-sized: 2 miles or shorter. If you are new to cycling, take it easy as you start (see box at right) and you'll soon discover the joys of bicycling. As a cyclist, your fuel efficiency is equivalent to 1000 miles per gallon — of carbohydrates, not carbon. See www.bikeplan.com and #28.

Motorists Do It With Carbon
Gasoline produces 20 pounds of CO_2 per gallon, so if your car does 20 mpg, it releases a pound of CO_2 every mile you travel. If walking, cycling, carpooling, or riding the bus doesn't work for you, you could buy a more fuel-efficient vehicle (see #3) or join a car share co-operative (see #17).

Flyers Do It With More Carbon
Flying is responsible for 6-10% (see #94) of the global warming problem. In reality, no one talks about it much, perhaps because the climate change negotiators and environmental activists do so much flying to all their conferences. The best solution is: don't fly! Use the train for shorter trips, do more teleconferencing, and take your vacation closer to home. If you absolutely have to fly, calculate your emissions and send a check to buy carbon offsets from an organization such as Climate Partners, American Forests or Tree Canada (see box on p. 56). If you book a flight, car or lodging in Triple E's Travel Cool Program, they will invest your travel dollars in environmental projects and technologies to reduce your carbon emissions.

Bus Passengers Do It Together
If you're not a regular bus user, there's only one way to find out how easy it is. You might be surprised, or despair at how long it takes. The best way to get improved services is to form a transit-users group and lobby for changes. See www.apta.com.

Train Passengers Do It With Style
For some, this a possible option. For most, it'll need a big campaign to get high-speed trains built as the best means of travel for journeys under 550 miles (1,000 km), which is becoming the standard in Europe. See www.narprail.org.

Car Sharers Do It Cooperatively
In Europe, 130,000 people have joined car-sharing groups. In North America, there are car-share organizations in 18 cities, with more being planned. See www.carsharing.net and #17.

Car-Poolers Do It In The Back Seat
If four of you share a ride, that's four times less emissions per person. If your place of work does not run a car-pooling scheme, maybe it's time to think about it. See:

- www.shareride.com
- www.carpool.ca
- http://transit.metrokc.gov/van-car/carpool.html.

Telecommuters Do It In Cyberspace
The best kind of travel is not having to travel at all. In 1990, 3.6 million people telecommuted to work in the US at least occasionally. In 1999, 19.6 million did it. While you're at it, you might also shop closer to home and support your local stores. See www.telecommute.org and www.gilgordon.com.

3

If You Must Drive a Car,

switch to the most fuel-efficient model

Toyota Prius, hybrid vehicle

Guy Dauncey

The average North American motorist drives 15,000 miles a year. Every gallon of gasoline burned releases 20 pounds of carbon dioxide; that's a fact of life. For Canadians, every litre of gas releases 2.5 kg of CO_2. If your car does 20 miles to the gallon, that's a pound of carbon dioxide slipping out of the exhaust pipe for every mile you travel.

Now think about it on a global scale. In 2000, the world had around 530 million passenger cars. Every year, 39 million new cars roll off the assembly lines and 29 million go to the great recycling depot in the sky, creating a net increase of 10 million cars.

If the average car does 25 mpg and travels 10,000 miles a year, the total CO_2 released comes to 2.12 billion tons, which is around 9% of the world's CO^2 emissions from human activities.

An average of 25 mpg is very inefficient, however. If every car did 45 mpg, the emissions would fall by almost a billion tons a year. We could do that today; the cars are already on the roads (see box). The new gas-electric hybrids, such as the Prius (Toyota) and the Insight (Honda), are achieving 50-65 mpg and releasing half the average CO_2, thanks to a design that combines an electric and a gas engine with a regenerative braking system so that every time you apply the brakes the energy is transferred to the car's battery.

Right on the heels of the gas-electric hybrids, another revolution is coming along: hydrogen fuel cell vehicles. Hydrogen is present in water, in fossil fuels, and as a by-product of algae (see #54). When hydrogen is burned, the only emission is water vapor. The only snag is that it takes effort — or energy — to obtain the hydrogen. (See p. 40).

For the first wave of hydrogen fuel cell vehicles, the hydrogen will likely be obtained by splitting it from methanol (a fossil fuel composed of carbon and hydrogen). Since this involves some carbon emissions, it is called "dirty hydrogen." For the second wave, hydrogen will probably be generated from water using renewable energies, or from algae, giving us the ultimate green machine, burning "clean hydrogen" with zero emissions. As car manufacturers switch to ultralight molded plastic composite materials for the car's body, and super-efficient designs, we will arrive at the "hypercar," with a fuel efficiency that reaches 120 mpg or more.

The Best Cars:[1]	City	Hway	Av	$/year	Tons CO_2
Honda Insight (Hybrid)	61	68	65	$338	3.1
Toyota Prius (Hybrid)	52	45	48	$457	4.0
VW New Beetle Diesel (Manual)	42	49	45	$470	4.7
VW Golf/Jetta Diesel	42	49	45	$470	4.7
Suzuki Swift	36	42	39	$554	5.0
Toyota Echo (Manual)	34	41	37	$664	5.2
The Best Minivans:					
Dodge Caravan/Chrysler Voyager	20	24	22	$982	8.8
Chevrolet Venture	19	26	22	$982	8.7
Sport Utility Vehicles					
Toyota RAV 4 2WD	25	31	28	$800	7.0
Suzuki Vitara	25	28	26	$831	7.4
The Greenhouse Monsters					
BMW 540i (A)	15	21	18	$1,383	10.8
Dodge RAM 1500 4WD	12	16	14	$1,564	13.8
Chevrolet Suburban/GMC Yukon XL	14	16	15	$1,460	12.9
Ferrari 550 Maranello	8	13	10	$2,490	18.7

1 Assuming 15,000 miles per year, city 55%, highway 45%, $1.45 gallon

Sources: www.greenercars.com *and* www.fueleconomy.gov

"We can no longer afford to ignore the signs of global warming and the fact that the consumption of gasoline and other fossil fuels is on the rise. Environmentally friendly cars will soon cease to be an option... they will become a necessity."
— Fujio Cho, President, Toyota Motor Corporation, January 2000

Gather the facts:
Step 1: How many miles do you drive per year ? ________ (A)
Step 2: What is the average fuel efficiency of your present vehicle ? _____ (B) (See: www.fueleconomy.gov or http://autosmart.nrcan.gc.ca)
Step 3: How many gallons does your vehicle burn in a year ? ______(A/B)
Step 4: How much CO_2 does your vehicle produce in a year ? (1 gallon = 20 lbs CO_2; 2000 lbs = 1 ton)
Result: ______ tons CO_2 per year.

Reduce your annual mileage
- Buy a bicycle or an electric bike, and use it for shorter trips
- Start car-sharing with friends for trips to work, school, and meetings
- Postpone a driving vacation until you have a more efficient vehicle.

Be a climate-smart driver
- Keep your tires correctly inflated; you'll save 10% of the fuel and CO_2
- Don't accelerate hard when you start
- Slow gradually as you approach a stop; don't rush up to the lights and then brake.

Buy a more fuel-efficient vehicle
- Check the fuel economy web sites when you buy your next car.
- Avoid buying a diesel vehicle, EVEN THOUGH THE VWs ARE VERY FUEL-EFFICIENT, as diesel exhaust contains particulates and is a major threat to public health.
- Go for one of the new gas-electric hybrids: Toyota's Prius or Honda's Insight.

Consider buying an electric vehicle
For short range driving, check out the Sparrow; Ford's Th!nk, Ranger and Explorer; Nissan's Altra and Hypermini; Honda's EV, or Toyota's RAV4.

Join the Sierra Club's Campaign for Fuel Efficiency
20% of carbon dioxide emissions in the US come from cars and light trucks, but the regulations governing fuel economy haven't changed since 1975. The Sierra Club is leading a campaign to raise the standard from 27.5 mpg to 45 mpg for cars, and from 20.5 mpg to 34 mpg for light trucks. We should target 60 mpg for cars and 45 mpg for light trucks, with yearly increments to increase efficiency.

Buy carbon offsets to offset or absorb your emissions
When you know how much CO_2 your vehicle produces, you can buy carbon offsets to reduce global CO_2 emissions by the same amount. Climate Partners (www.ClimatePartners.com) will enable you to buy 5 tonnes of carbon offsets for $50 (for instance), which will be invested in local carbon reduction projects such as carpooling or energy efficiency. In the US, you could pay $15 to American Forests (www.americanforests.org) to plant 15 trees, which will absorb 5 tons over 40 years. You can also buy carbon credits for tree-planting in the tropics through Trees for the Future. (www.treesftf.org)

Resources

Alternative Fuels Datacenter: www.afdc.doe.gov

Buyer's Guide to Cleaner Cars: www.arb.ca.gov/msprog/ccbg/ccbg.htm

Cleancars Campaign: www.cleancarcampaign.org

Earthsmart Cars: www.nrdc.org/earthsmartcars

Electric, Hybrid and Fuel Cell Vehicles: www.evworld.com

Environmental Guide to Cars and Trucks, $8.95. American Council for an Energy Efficient Economy, 1001 Connecticut Avenue, NW, Suite 801, Washington, DC 20036: www.greenercars.com

EPA's Green Vehicle Guide: www.epa.gov/autoemissions

Honda Insight Man: www.insightman.com

Hypercars: www.hypercar.com

Regional Gasoline Costs www.eia.doe.gov/emeu/rtecs/gascosts

Sierra Club Campaign: www.sierraclub.org/globalwarming/cleancars

Tailpipe Tally: http://209.10.107.169/tailpipetally

4

Choose Energy-Efficient Appliances

Your home is like a hungry creature that eats energy. Not only does it have to feed all the appliances you and your family like to use; it also loses energy through its windows, walls, and roof.

Let's say you have just been to a down-to-earth lecture on global climate change and have decided to try to make a difference, starting with your home.

To meet your present needs, we'll assume you use 11,000 kilowatt hours (kWh) of electricity a year to power your various appliances, producing 8.8 tons of CO_2, and that you burn 1,000 gallons of oil to heat your home, producing 11.2 tons of CO_2, for a grand total of 20 tons.

Your goal is to reduce this as low as possible while still keeping warm. For each kWh of electricity saved, your CO_2 will fall by 1.6 pounds. (This will vary depending how your electricity is generated locally.) For every gallon of oil saved, it will fall by 22.3 lbs.

A good way to start is to check out the websites (see box on next page) where you can work out how much CO_2 your household is generating, and how much you could reduce if you made your home more efficient and used energy-efficient appliances. You'll be amazed at the resources that are available. In general, if you buy an appliance with an *Energy Star* (US) or an *Energuide* (Canada) label, you'll be on the right track.

Let's start with your washing machine. If you upgrade to the latest *Energy Star* front-loading design, you can reduce your yearly energy use from 900 kWh to 450 kWh. If you use warm or cold water, you can reduce it to 70 kWh a year.

Next, the lights. Ditch any halogen lamps, since they emit only 5% of their energy as light and the rest as heat. If you invest $60 in 10 sub-compact fluorescent bulbs, assuming they are used for 5 hours a day, each bulb will prevent 292 pounds of CO_2 from going into the atmosphere and save you $10 a year.

Now the air-conditioners. Depending on where you live, our advice is to switch them off and install ceiling fans instead, such as the "Gossamer Wind" (Florida Solar Energy Center www.fsec.ucf.ed). If this is impractical, use the *Energy Star* website to get the right size for each room and find the best equipment, or at least get your system tuned up.

Next, the dryer. One possibility is to unplug it and invest in an old-fashioned clothes line and an indoor clothes rack. It will get the neighbors talking, but it will also save you energy. Alternatively, the latest *Energy Star* model will reduce this demand by half.

Appliances That Don't Cost The Earth

(kWh per year)

	Typical	Best	Best practice
Clothes washer	900	450	70
Dryer	900	400	0
Lights	900	450	300
Air conditioning	1,000	400	100
Water heater*	1,900	1,000	800
Fridge	1,200	520	520
Dishwasher	700	350	300
Waterbed	800	700	0
Various other**	2,950	1,500	1,200
Total	**11,250**	**5,770**	**3,290**
Cost @ 8 cents kWh	$900	$460	$263
CO_2 @ 1.6lbs/kWh	9 tons	4.6 tons	2.6 tons

* Does not include 1,600 kWh for hot water included in the clothes washer and dishwasher.

** e.g., TV, stereo, computers, coffee-maker, etc.

"**Investing in a home on your street could be more profitable than investing in Wall Street.**"
— Home Energy Saver

Your hot water heater may be consuming 3,500 kWh per year. The best will reduce that by half. If you're staying with your existing heater, buy it an insulating jacket and insulate the first 3-6 feet of the hot and cold water pipes using tubular foam, plus any pipes in the crawl-space. We also suggest that you buy water-saving showerheads and faucet aerators, install a time-control on the heater, turn the temperature down to 120°F, and install a solar hot water system (see #6).

And now (drum roll)... the fridge! A typical Florida monster fridge may use as much as 3,000 kWh/yr. Your fridge may be using 1,200 kWh/yr, so we suggest you replace it with an Energy Star model that uses 520 kWh/yr. If you set the temperature to 40°F (each degree colder increases the cost by 2.5%), set the freezer to 0°F, and vacuum the coils every six months, that will save more energy. If your old fridge uses CFC-11 as a coolant, remember that it's a highly potent greenhouse gas and ozone destroyer, so be really careful that the scrap dealer captures the gas and doesn't trash it.

The dishwasher: the worst use 700 kWh/yr; the best as little as 344 kWh/yr. If you use it on a full load on the energy-saving setting without heat to dry, you will save even more.

Even small things such as stereo systems and clock radios may be wasting energy. Your audio products may be using 200 kWh /yr, which could be cut in half with efficient models. The new digital TVs are set to be trouble: they have to be left fully powered all the time and will increase your energy use by 7%.

Finally, if you have a waterbed, we suggest you throw it out: it's using 800 kWh/yr. A pool pump? That's 1,500 kWh/yr. A spa pump and heater? This hungry beast will add 2,250 kWh/yr.

If you do all of these things, you could get your appliance energy needs down to 3,290 kWh/yr and prevent 6 tons of CO_2 from entering the atmosphere — a 30% reduction. Next, we'll look at your house, and how you are going to pay for all this.

Home Efficiency Resources

Where does your electricity come from? (USA):
www.environmentaldefense.org/programs/Energy/green_power/x_calculator.html

Consumers Guide to Home Energy Savings:
http://aceee.org/consumerguide

Home Energy Calculator:
www.pge.com/003_save_energy/003b_bus/003b1d2a_cal

Home Energy Checkup: www.ase.org/checkup/home

Home Energy Saver: http://hes.lbl.gov

Efficient appliances:
www.eren.doe.gov/consumerinfo/energy_savers
www.pbs.org/wgbh/warming/carbon
www.energyoutlet.com

Energy Guide: www.energyguide.com

Energy Star: www.energystar.gov

Greywater Heat Recovery from the bath, shower:
www.vaughncorp.com/gfx.html

Rocky Mountain Institute Home Energy Briefs:
www.rmi.org/sitepages/pid171.asp

Changing Lightbulbs:
www.detroitedison.com/home/energyprog/lightbulb.html

Efficient Light bulbs: www.pnl.gov/cfl

Home Carbon Calculator (Canada):
www.climcalc.net

Energuide (Canada): http://energuide.nrcan.gc.ca

Canada's Office of Energy Efficiency:
www.oee.nrcan.gc.ca

5

Make Your Home More Efficient

ACEEE

Cellulose or fiberglass insulation can be blown into uninsulated walls by an insulation contractor.

In Solution #4, you reduced your CO_2 emissions by 6.4 tons by investing in the most efficient appliances and adopting the best practices. Next, we'll look at the 11 tons per year that result from the gas, oil, or electricity you use to heat your home.

If you visit the Alliance to Save Energy (www.ase.org/consumer) and the Home Energy Saver (http://hes.lbl.gov) websites, you can punch in your zip code and a calculator will tell you how to reduce your heating and cooling needs:

- Upgrade the insulation in your ceiling, walls, attic, crawlspace, and basement.
- Replace the windows with triple-glazed, low-E glass.
- Air-seal and caulk any holes and cracks to prevent energy losses.
- Install programmable thermostats in each room and on the water heater.
- Maintain your furnace.

Depending on how efficient your house already is, you might reduce your energy needs by anywhere from 20% to 60%. If it's 60%, your CO_2 emissions from heating will fall to 4.4 tons. With the 2.6 tons your appliances are producing, you will have reduced your total household CO_2 emissions from 20 to 7 tons.

All this costs money, however. The Home Energy Saver website will help you tally up the cost. Counting renovations and the additional cost of buying energy-efficient appliances, it may be as much as $4,000, so you might want to phase in your appliance replacements over several years.

If you click on the "Financing" section of the Home Energy Saver, you'll find that mortgaging your energy efficiency can end up saving you $26 a month (see box). That's the equivalent of a 16% rate of return on your investment — better than most stocks on Wall Street. In Oregon, residential energy tax credits will further reduce your costs.

The Power Of A Tree

Gainesville, Florida has a strict tree ordinance, while nearby Ocala has a loose one. As a result, Gainesville has twice the tree-cover that Ocala has. Now get this! Gainesville's residents spend on average $126 less per year on electricity than Ocala residents do, representing an annual saving of 1.26 tons of CO_2 .[1] Trees are cool, and they make life cool — so plant trees!

> "[With tax-shifting]...thermal insulation and superwindows ... will have a bigger payout than Microsoft stock. You will be able to make Warren Buffet returns by simple investments in hardware-store technologies."
> — *Natural Capitalism*[2]

Financing Your Home Energy Retrofit

According to the Home Energy Saver, an energy-efficient mortgage can save you $26 a month and reduce your CO_2 emissions by 62.5%:

Typical cost of retrofit: $4,000
Additional mortgage cost: . . . . .$29 per month
Savings on monthly utility bills: .$55 per month
Net savings: . . . .$26 per month

ASE Energy Efficient Financing: www.ase.org/consumer/finance.htm
Energy Efficient Financing Info Center: www.nationalguild.com/residential/hers.html
Energy Efficient Mortgage Guide: www.pueblo.gsa.gov/cic_text/housing/energy_mort/energy-mortgage.htm
Home Energy Saver (financing): http://hes.lbl.gov/HES/makingithappen
Home Performance (Canada): www.homeperformance.com
Profitability of Energy Efficiency Upgrades: http://hes.lbl.gov/hes/profitable.html
Residential Energy Services Network: www.natresnet.org
The Weatherization Pages: www.weatherization.com

If The White House Can Do It...
In the White House, energy-efficient measures and other aspects of the "Greening of the White House" initiative begun under President Clinton save $300,000 a year through more efficient lighting, heating, and air conditioning; high-efficiency refrigerators; the purchase of *Energy Star* rated computers, printers, and fax machines; and replacing inefficient windows.

It's Superwindows!
Superwindows combine two or three invisibly thin coatings that let the light pass through but reflect the heat with a heavy-gas filling, such as krypton, to block the flow of heat and noise, keeping you warm in winter and cool in summer. They cost 10-15% more than double-glazed windows, but they insulate 4.5 times better, equivalent to eight sheets of glass.[3] At the Efficient Windows Collaborative (www.efficientwindows.org), you can click on your state, then your city, and see a detailed list of efficient window options. You will learn about U values, solar heat gain, and visible transmittance, and discover how to reduce your heating bills still further for the price of the new windows. Visit the *Energy Star* site (www.energystar.gov/products/windows) and look for windows by Hurd, Pella, Visionwall, and Viking.

- Single-glazed, clear-glass windows $365 a year
- Double-glazed, low-e coating $230 a year
- Triple-glazed, 2 low-e, vinyl insulation $210 a year

Set Up A Home Energy Account
You are going to need the savings on your utility bills to pay for your investments in efficiency. How can you capture the savings and not lose them in the monthly shopping?

Step 1: Set up a Home Energy Account with your bank or credit union (or in your own bookkeeping system) and make a fixed monthly payment to cover the full cost of all your energy expenditures before the changes, in-cluding gas, electric, oil, wood, and light bulbs.

Step 2: When you take out a loan or extend your mortgage to pay for the investments, arrange to repay the loan or the increased mortgage through this account. The savings on your bills will cover the payments and the account will help you track your energy expenditures.

The *Greenfreeze*
Is there an ozone-friendly, atmosphere-friendly fridge? In the early 1990s, the fridge manufacturers said "It can't be done," and shifted from CFCs to coolants called HCFCs and HFCs. These are less ozone-destroying, but they are powerful greenhouse gases (see #95). So Greenpeace hired a scientist to design a 100% planet-safe fridge, using propane and isobutane as refrigerants. The *Greenfreeze* has become the dominant fridge in northwestern Europe, with 100 different models, and has captured almost 100% of the German market, but it is being kept out of North America by the chemical industry and their manufacturing allies.

See www.greenpeace.org/~climate/greenfreeze and www.acupc.es/homes/montse/gf_eng.html

Are You Investing Wisely?

	Return on Investment
Fluorescent lamps and fixtures	.41%
Duct sealing	.41%
Energy Star Clothes Washer	.7%
Energy Star Thermostat	.30%
Energy Star Refrigerator	.27%
10-point energy efficiency upgrade	.16%
Dow Jones Industrials, 1990–1997	.14%
Money market account	.3.5%
30-year bond	.4.2%

Source: Home Energy Saver

6

Use the Sun's Energy

"I will dip my brush in the sun — and paint a window towards eternity."
— Kjell Pahr-Iversen

Install a Solar Hot Water System
Can you reduce those 7 tons of CO_2 by installing a solar water heater? It's an idea that intrigues many people. Depending on where you live, it will reduce your hot water bill by 33% to 75%; some systems, such as Thermomax, work well in cloudy conditions. Over 1.5 million Americans have solar hot water systems and 94% consider it a wise investment.[1]

How about the cost? An installed system will cost between $2,000 and $4,000. If your household uses 3,000 kWh a year to heat water and the system saves 75%, that's an annual saving of $180, or $15 a month, and is enough to repay the loan. On the other hand, if you only use 1,000 kWh and the system saves 33%, your annual saving will be just $26. For a quote, look under "Solar" in the Yellow Pages and call a local company. If you live in Oregon, you'll get a $1500 tax credit; some utilities (such as the Emerald Public Utility District) also provide rebates and interest-free loans.

On average, for every kWh you save, you prevent 1.6 pounds of CO_2 from entering the atmosphere. If your heater saves 50% of 2,000 kWh a year, that comes to 0.8 tons of CO_2 and reduces your household total to 6.2 tons.

Install Solar Shingles
What about a solar roof? A photovoltaic (PV) roof is not like a hot water system; it needs light, not heat and, as long as the sky is clear, it will produce electricity even in winter, less if it is cloudy. Solar shingles are like regular house tiles but are made from PV cells.

An estimated 200,000 homeowners in the US use PV systems, with more joining them every day. On BP Solar's web page, you can plug in your zip code and see what PV will cost (see box on next page). A 1 kW system might cost $25-$40 a month, but in New York it will only cost you $3 a month and $6 to $9 a month in California, because of state and local subsidies.

A 1 kW system will give you an average 4 kWh of electricity a day (1500 kWh/yr), depending on whether it's cloudy or sunny. A 2 kW system (3,000 kWh a year) will reduce your CO_2 emissions by around 2.4 tons, bringing your household total down to 3.8 tons. Many states allow you to hook into the grid through net metering (see #42), so when your roof generates a surplus, your meter spins backwards and you will receive a credit on your next bill that offsets the cost.

Resources

American Solar Energy Society: www.ases.org
BP Solar Clean Power Estimator: www.bpsolarex.com/calculator
Earth Energy Society of Canada: www.earthenergy.org
Eagle Sun solar hot water roof (new homes only): www.aetsolar.com
Florida Solar Energy Center: www.fsec.ucf.edu/Solar/index.htm
International Ground-Source Heat Pump Association: www.igshpa.okstate.edu
Kyocera's solar electric referral service: www.solarelectricrepair.com
Long Island's Solar Pioneers: www.lipower.org/pioneer.html
Million Solar Roofs: www.eren.doe.gov/millionroofs
My Solar: www.MySolar.com
Net Metering: www.eren.doe.gov/millionroofs/netmeter.html
Solar Hot Water Factsheets: www.eren.doe.gov/erec/factsheets/solrwatr.html and www.seia.org/SolarEnergy/SolarFactSheets/sfheatpower.htm#1
Solar Power Calculator: www.epa.gov/globalwarming/actions/cleanenergy/sol
Solar Radiation Map (US): www.homepower.com/solmap.htm and http://rredc.nrel.gov/solar/pubs/redbook
Thermomax Solar Tubes: www.thermomax.com
The Sunshine Revolution by Harald N. Røstvik, 1992. A truly wonderful book. $39 + postage from Sun Lab, Alexander Kiellandsgt. 2, 4009 Stavanger, Norway.

The Approximate Cost of a 1kW Solar PV Installation[3]

	Best monthly net cost	Tons CO_2 saved per year	Total net cost
Amherst, Mass	$28	1.05	$7,480
Anchorage	$38	0.57	$8,200
Austin, Tx	$35	1.91	$8,200
Boulder, Co	$34	0.98	$8,200
Chicago	$8	1.53	$3,280
Davis, Ca	$7	0.99	$5,200
Gainesville, Fl	$27	1.05	$6,200
Honolulu	$16	1.70	$6,940
Los Angeles	$9	1.03	$5,200
New York	$3	1.00	$4,264
Tucson	$24	1.13	$7,480
Santa Fe	$24	1.14	$8,200
San Francisco	$6	1.00	$5,200
Seattle	$38	0.71	$8,200
Tully, NY	$25	0.93	$7,120

Source: BP Solar, August 2000.
Costs vary because of local tax breaks and subsidies.

In most places, a PV roof will not provide you with the cheapest energy (though it may by 2005). When you buy a dress or shoes for your children, however, do you simply look for the cheapest? No, because you know that quality counts too. When we think "cheapest is best" for the power we buy, we are being cheap with the Earth as a whole, and with future generations. Think of it as a downpayment on a sustainable future. In the Japanese prefecture of Fukui, members of Ecoplan Fukui are hoping to persuade residents to invest in a joint solar rooftop system as a way of spreading the cost.

Install a Ground-Source Heat System

How about a solar floor? The temperature in the soil around your house is relatively constant year-round, and you can use this to heat your home using a ground-source heat (or geo-exchange) system. (Don't confuse this with geothermal energy, which comes from hot rocks deep underground — see p. 34). The technology started in the 1940s, and has become so efficient that for many homes it is the most effective way to save energy and reduce CO_2 emissions.

The system involves installing a series of pipe loops in 12" wide trenches, 10 feet apart, 4 to 6 feet below the surface, carrying a fluid that can conduct heat. It can also use a series of deep bore-holes, or a water system to extract heat from a lake or pond. The fluid absorbs the heat from the ground, and an electrically-powered heat-exchanger provides 90% of your heat through a system of ducts, reducing your costs by 50-70% in the winter. In summer, the system can be reversed to provide you with cooling, reducing your costs by 20-40%. Using a gizmo called a desuperheater it can also preheat your water, reducing your costs by up to 50%.

A ground-source heat system for an average 2000 sq ft house might require a 2.5-ton system (30,000 Btus), and cost $6,250 for a ground system (less for a lake system; more if your house has no ducts). With the reduced heating and cooling bills, payback will take 5 – 12 years. In colder Canada, you might need a 4-ton system costing $10,000.[2] If you have made your home super-efficient, a smaller system will do. A desuperheater will cost $700, and reduce your hot water costs by around $100 a year. In terms of CO_2 emissions, assuming you are down to 7 tons from your home-heating needs, a 50% reduction will prevent 3.5 tons of CO_2 from entering the atmosphere, and leave just 3.5 tons remaining.

Ground-source heat systems are in place in homes, schools and businesses all over North America, working silently, invisibly and without pollution, using the sun's stored energy.

Bill Eager

Residence with grid-connected PV panels in Gardner, MA.

7

Buy Green Power

Turbines at Ponneguin wind farm, Colorado.

Land &Water Fund of the Rockies

Your energy comes into your house through a simple wire. Turn on the switch and, as if by magic, the fridge works, the lights come on.

But what happens at the other end of the wire to produce the power? Some wires lead to a whopping big coal-fired power plant where ancient biomass is poured into a huge hopper and turned into electricity. As carbon burns, it mixes with oxygen, pouring millions of tons of carbon dioxide into the atmosphere.

Depending on where you live, your wires may lead to a nuclear power station, with its legacy of radioactive waste, or a natural gas or oil-fired power plant, which burn carbon and release CO_2. In the Pacific Northwest and other parts of the continent, the wires may lead to a hydroelectric dam which blocks the migration routes of salmon, and likely involves a reservoir that drowned once important wildlife habitat.

The energy industry of North America has powered the 20th century and underpinned the high standard of living that so many of us enjoy. We should be grateful to them. But, as with all things big, the industry has become stuck in its ways.

In North Dakota, wind turbines could provide electricity for much of the USA, while providing the state's farmers with a useful supplemental income. The North Dakota power industry, however, is controlled by four large coal corporations whose plants have been "grandfathered" out of the need to comply with the 1990 Clean Air Act, and which appear to influence the political process. Result: King coal reigns, no wind turbines.

Globally, the generation of electricity is responsible for 30% of the greenhouse gases. If we can crack this one, we will be well on the way to solving the overall problem.

> **"By spending just $30 per year on wind power, an average Colorado family can cut its household carbon dioxide production by about 10%." — Colorado Wind Power**

Where does your energy come from?

Canada	Hydro 56%, coal/oil/gas 29%, nuclear 14%
California	Gas 31%, hydro 22%, coal 20%, nuclear 16%, renewables 11%
Colorado	Coal 95%, hydro 4.7%
Florida	Coal 45%, gas 21%, nuclear 17%, oil 15%
Illinois	Coal 50%, nuclear 48%
Michigan	Coal 70%, nuclear 28%
New Jersey	Coal 49%, nuclear 34%, gas 7%, oil 6%, hydro 2%
New Mexico	Coal 90%, gas 9%, hydro 0.7%
New York	Nuclear 34%, hydro 25%, coal 19%, gas 13%, oil 9%
North Dakota	Coal 90%, hydro 10%
Ohio	Coal 90%, nuclear 9%
Oregon	Hydro 93%, coal 3%, gas 3%
Pennsylvania	Coal 58%, nuclear 36%, oil & gas 4%, hydro 2%
Texas	Coal 50%, gas 37%, nuclear 13%
Vermont	Nuclear 76%, hydro 24%

Source: www.eia.doe.gov/cneaf/electricity/st_profiles/toc.html

Go for It – Buy Green Energy!
In California, you can sign up for certified green power from over 20 suppliers, including municipal utilities in Sacramento and Los Angeles. The "Green-E" logo certifies that 50% or more of the electricity comes from renewable sources, and the rest is at least as clean as system energy. In New England, you have a choice of four suppliers. In New York, you can sign up with the Solar Pioneer program. In Colorado, you can sign up with Colorado Wind Power. In Austin, Texas, you can sign up for energy from landfill gas and wind. In Wisconsin, you can sign up for wind energy from Madison Gas and Electric and Wisconsin Electric. The list goes on, and it's growing every month. You can also sign up for green energy in Pennsylvania, New Jersey, Connecticut, parts of Alberta, Washington, Oregon, Minnesota, Iowa, Michigan, Ohio, Tennessee, Kansas and Texas. The Green Power Network (see box) will help you locate your nearest green energy supplier.

If you live in a state or province that does not allow independent green energy companies to operate, write to your state or provincial legislature and urge them to get on with it. If you live in a state or province that uses a lot of hydroelectric energy, such as British Columbia or Quebec, it is still worth lobbying for renewable energy, because every surplus kilowatt hour of green energy can be exported and used to displace coal-powered energy elsewhere.

If you are able to buy green energy, this will reduce your CO_2 emissions still further. If you live in Colorado, buying four 100 kWh blocks of wind a month at a premium of $2.50 per block will bring you 4,800 kWh a year, and prevent 3.8 tons in CO_2 from coal-fired power from entering the atmosphere. Early support is essential to help the green energy industry get established. Green energy does cost more at this stage of the game, with premiums varying from 4 cents per kWh in Austin to 2.5 cents per kWh for wind power in Colorado, but customers who pay extra to get green energy started are making a tremendous gift to the Earth. Not only are you showing your neighbors that it can be done: you are showing the whole world. They are watching in China and India, where they burn enormous quantities of coal, but where there is plenty of wind energy capacity, too. Think of it as planetary tithing, or a pay-down on your children's inheritance.

All calculations assume an average 1.6lbs of CO_2 per kWh

Resources

Canada's Renewable Energy Project Analysis software: www.retscreen.gc.ca

Canadian Renewable Energy: www.newenergy.org

Center for Energy Efficiency and Renewable Technology: www.ceert.org

Colorado Wind Power: www.cogreenpower.org

EPA Green Power: www.epa.gov/globalwarming/actions/cleanenergy/greenpower

Green Electricity: www.environmentaldefense.org/programs/Energy/green_power

Green-e Logo for Energy: www.green-e.org Tel: 888-63GREEN

GreenMountain.com: www.GreenMountain.com

Green Power discussions: www.green-power.com

Green Power links: www.green-power.com/links

Green Power Network: www.eren.doe.gov/greenpower

How dirty are your state's power plants?: www.ewg.org/dirtypower/dirty.html

Power Scorecard (PA & CA): www.powerscorecard.org

8

Switch to a More Organic, Vegetarian Diet

You've probably heard the joke about cows farting and how they contribute to global warming. Well...it's true.

For the ultimate dinner table conversation, here's how it goes. The cows munch on their grass, or whatever carbohydrate they are being fed, and their food makes its way down to their famous four stomachs. In one of those stomachs, called the rumen, it is broken down by bacteria, releasing methane (CH4) as a by-product. The methane escapes from both ends and makes its way into the atmosphere, where it works as a greenhouse gas that is 23 times more potent than CO_2 over 100 years, and 62 times more potent over 20 years.

We say "cow", but as George Orwell taught us, not all animals are equal. When it comes to farting and belching, cows and beef cattle are far better at it than pigs and sheep, which contribute only 3% to the cows' 97%. As for turkeys and chickens? So far, no one has measured them.

This giant body burp (some 65 to 85 million tons of methane per year) is only part of the problem. Your future milk, cheese, steaks and burgers also produce manure — enormous great piles of it. This releases methane too, contributing another 5% to the global cloud of methane. Taken overall, human activities release 375 million tonnes of methane a year, causing 17% of the overall climate change impact. 25% of this comes from livestock waste and belching, so cows represent 4% of the total problem.

And that's not all. For many steak and burger eaters, your contribution to global warming may have started far away in Mexico, Brazil or Costa Rica, where the rainforest was cut down to provide pasture for cattle, releasing its millennium-old store of CO_2 into the atmosphere. Put it all together, plus the energy needed to grow the grain to feed the cattle to make the beef that sits on your plate, and you can see that, well, meat is a pretty big climate-changer.

"I'm not predicting the end of all meat eating. Decades from now, cattle will still be raised, perhaps in patches of natural rangeland, for people inclined to eat and able to afford a porterhouse, while others will make exceptions in ceremonial meals on special days like Thanksgiving, which link us ritually to our evolutionary and cultural past. But the era of mass-produced animal flesh, and its unsustainable costs to human and environmental health, should be over before the next century is out." Ed Ayres, 1999.[1]

Beef in the Greenhouse

	Beef per person per year	Obesity
USA:	260 lbs	1 person in 5
Canada:	210 lbs	1 person in 8

1 lb beef releases 0.5 lbs methane = 10.5 lbs CO_2 equivalent

260 lbs beef = 130 lbs methane = 1.4 tons of CO_2 equivalent per beef eater

Resources

Brown Box (CSA) farm programs: www.biodynamics.com/usda

Global methane emissions: http://cdiac.esd.ornl.gov/trends/meth/methane.htm

"May All Be Fed: Diet for a New World" by John Robbins. Includes many excellent recipes.

Organic Hub: www.organichub.com

Ruminant Livestock and the Global Environment: www.epa.gov/ruminant/sustain.htm

Slow Food Movement: www.slowfood.com

Vegetarian restaurants around the world: www.vegdining.com

Vegetarian and Vegan Resources: www.vegsource.com

There are three things we can do personally to reduce the planet's burden of methane and CO_2 emissions from the production of meat:

1. Eat Less Meat and More Vegetarian Food.
If you are not sure how to produce meals that are not focussed around meat, treat yourself to a good vegetarian cookbook, and check out vegetarian restaurants. As well as reducing greenhouse gases, there are good health reasons for eating vegetarian food, including a reduced risk of breast cancer, ovarian cancer, prostate cancer, colon cancer and heart disease. Don't worry; well-prepared vegetarian food tastes superb. If you are not ready for a pure vegetarian diet, try reducing your meat by 50% next year, 70% in two years, and 90% in three years time. For every pound less meat you eat, your climate change emissions will fall by 10.5 lbs of CO_2 equivalent.

It's not just the methane emissions that are taking their toll: "Producing 1 lb of feedlot beef requires 7 lbs of feed grain, which takes 7,000 lbs of water to grow. Pass up one hamburger, and you'll save as much water as you save by taking 40 showers with a low-flow nozzle. In the US, 70% of all the wheat, corn and other grain produced goes to feeding herds of livestock."[2]

2. Eat Organically Grown Food.
There are a hundred reasons why we should buy organically grown food, from protecting local species to encouraging more birds, bees and butterflies. For our purposes, the reason involves rebuilding the carbon in the soil. The world's cultivated soils contain twice as much carbon as the atmosphere, in the form of a billion creepy crawlies, bacteria and other organisms. Nature's instinct is to fix carbon in the soil through living creatures. With chemically fed monocrops, farmers neglect to rebuild the soil through crop rotation, compost and manure, causing those creatures to die. Their bodies oxidize or rot, and their carbon is released into the atmosphere, turning into CO_2. What's left washes into creeks and rivers and turns into methane. When you buy food that has been grown organically, you are voting for the soil, and for those wonderful, living carbon-creatures.

3. Eat Locally Grown Food.
The average mouthful of food has traveled 1200 miles in trucks, ships and planes before it lands on your plate. Meanwhile, small family farms struggle to survive — and yet one acre of prime land can produce 60,000 lbs of celery, 50,000 pounds of tomatoes, 40,000 pounds of potatoes, 40,000 pounds of onions, 30,000 pounds of carrots ... or 250 pounds of beef. In Victoria, B.C., Canada, Brian Hughes farms 30 acres organically at Kildara Farm on the Saanich Peninsula. Once a week, he and his family and helpers pick their crops, pack them into brown boxes and deliver them to 275 customers who receive their fresh, locally grown, organic food. By buying locally through brown box farm programs, at farmers markets or through "buy local" initiatives, you support your local economy while keeping some of that carbon out of the atmosphere. Eat your carbon — don't burn it!

Carolyn Herriot

9

Invest in Solar Funds

The shift to energy efficiency, clean energy and the solar revolution is going to require huge amounts of money. The European Commission has said that it wants to attract $30 billion for investment in renewable energy projects by 2003. Where's it all going to come from?

There's no shortage of money in the world - it's sloshing around all over the place. Millions of ordinary people invest in stocks and mutual funds. But do you know where your money is, or what it's doing? You may be strongly anti-smoking, but unless you know otherwise, it's probably being invested with tobacco companies.

That's where "socially responsible investment" (SRI) comes in. Since the 1980s, funds have been set up to serve the needs of people who want to avoid investing their money in things such as tobacco, the military, pornography, cruelty to animals, bad labor practices, child labor, or abuse of the environment. There are over 60 SRI funds in North America, and they're doing just fine. Companies that practice social responsibility also practice financial responsibility. In the USA, one investor in eight is contributing to SRI funds that collectively hold over $2 trillion dollars in investments; 79% of the portfolios focus on companies' environmental records.

With fuel cells, wind turbines and photovoltaic technology being so hot, can you invest in them? The answer is yes. Merril Lynch International has a $250 million New Energy Technology fund which invests in renewable energy, new automotive technologies and energy storage devices. The New Alternatives Fund has almost half of its portfolio in renewable energy. In Canada, the Sentry Select Alternative Energy Fund has a portfolio of 25 renewable energy, fuel cell and energy management companies. In Europe, the Franco-Belgian banking group Dexia has launched an equity fund aimed at reducing global warming, offering investors the chance to earn carbon emission credits in addition to normal equity returns. By investing with these funds, your savings will accelerate the birth of the solar/hydrogen age. See:

Investing with your Values: Making Money and Making a Difference,
by Hal Brill, Jack Brill, and Cliff Feigenbaum (New Society Publishers, 2000).

Who are the Movers and Shakers?

- AstroPower — solar cells and modules
- Avista — hydrogen fuel cells
- Ballard (British Columbia — hydrogen fuel cells)
- BP Solar — Photovoltaics
- DTE Power (one of Plug Power's owners)
- Enron — renewable energy
- Go-Green.com — green energy utility
- GreenMountain.com — green energy utility
- Kyocera, (Japan) — solar
- Plug Power — hydrogen fuel cells
- PowerSource — green energy utility
- Seimens — solar manufacturing
- Spire — solar equipment
- Vestas Wind systems (Denmark) — wind turbines
- ZeTek Power (UK, Belgium) — hydrogen fuel cells

"I think fuel cells will be a howling success. You'll see them spread as fast as PCs and cell phones."
— Amory Lovins, Rocky Mountain Institute

Who's Who in Socially Responsible Investment?

- Campaign ExxonMobil: www.campaignexxonmobil.org
- Calvert Group: www.calvertgroup.com
- Clean Edge: www.cleanedge.com
- Clean Environment Fund (Canada): www.cleanenvironment.com
- Coop America: www.coopamerica.org
- Dexia: www.dexia.com
- Domini : www.domini.com
- EcoMall: www.ecomall.com (click on 'Investments')
- *Good Money Magazine*: www.goodmoney.com
- Green Century: www.greencentury.com
- *Green Money Journal*: www.greenmoney.com
- Merril Lynch New Energy Technology: www.mlim.co.uk/fund-centre/ff-973793.asp
- New Alternatives: www.newalternativesfund.com
- Sane BP: www.sanebp.com
- Sentry Select Alternative Energy Fund (Canada): www.sentryselect.com
- Social Funds: www.socialfunds.com
- Social Investment Forum: www.socialinvest.org (USA); www.uksif.org (UK)
- Social Investment Organization (Canada): www.socialinvestment.ca
- South Shore EcoBank: www.eco-bank.com
- VanCity Savings Credit Union (B.C.): www.vancity.com/vancity/csr

Invest in SRI Funds
Switch some of your investments to socially responsible funds. Visit their websites, and show them to your broker. Look at their performance over the years. Ask your broker, banker, or credit union about funds that have been set up to combat global warming and support renewable energy. Ask about the companies that are at the heart of the new energy economy (see box). If they scoff at the idea, take your money elsewhere.

And Encourage Your Friends ...
If you belong to a church or an investment club, suggest they do the same. Some religious groups are encouraging their congregations to buy green power because they see it as a way of loving God's Creation. In Maine, Interfaith Power and Light is reaching out to Catholics, Methodists, Jews, Congregationalists and others to create a buying pool for green energy. In Massachusetts and California, Episcopal Power and Light is doing the same (see #14). The Interfaith Center for Corporate Responsibility helps church groups make sound investment decisions (475 Riverside Drive, #50, NY 10115).

Use Your Voice
Use your voice to argue for socially responsible divestment. What if you discover that your college or your pension fund is investing your savings in oil companies, or companies that are still part of the problem, and proud of it? In 1999, UCLA's undergraduate student government approved a resolution urging the University of California Regents to divest itself of holdings in businesses that contributed to global warming, targeting Exxon, Ford and General Motors, three members of the Global Climate Coalition (which opposes attempts to reduce the use of fossil fuels). Ford and GM have since left the Coalition, perhaps influenced by the way future car-owners were thinking. Check out where your pension fund is being invested, and discuss it with your colleagues.

Exercise Your Right as a Shareholder
As an investor in a company, you are a shareholder with voting rights at the company's annual general meeting, and you have the right to make resolutions. If you get together with other shareholders, you can sometimes make your influence felt. In 1999, KPMG did a study for Greenpeace that demonstrated if it was to build a 500 MW solar factory, it could reduce the price of solar energy four-fold, thanks to efficiencies of scale.[1] This is such a major price-break that it could trigger a worldwide solar revolution, while making money for BP's shareholders. In April 2000, a group of BP activist shareholders called Sane BP asked the AGM to stop drilling for oil in the Arctic Wildlife Refuge, and to transfer the investment to BP Solar, a wholly owned subsidiary. They won 13.5% of the shareholder vote, considered quite extraordinary for an activist resolution. Three months later BP announced that it would expand its solar activities.

10

Live More Sustainably

> "Listen up, you couch potatoes: each recycled beer can saves enough electricity to run a television for three hours."
> — Denis Hayes, *Earthday Guide to Planet Repair*

What a fabulous world, full of everything a person could ever want to buy. On one level, it's wonderful. People who arrive in Europe or North America from Africa, Asia or Latin America are astonished. A hundred and ninety varieties of breakfast cereal! Five thousand different styles of shoes! For thousands of years our ancestors lived without all this, but today the advertising tells us "This is what life's about."

But where does it all come from? Everything has to come from somewhere — from Earth's fields, forests or oceans; and it has to go somewhere when it's finished. Each North American indirectly consumes 121 pounds of matter every day. Our "stuff" all requires energy to be processed, manufactured and delivered, and if it's not recycled, it produces methane emissions when it sits in a landfill. Manufacturing a car produces 5.22 tons of CO_2.[1] Even a mere daily paper, when you calculate the energy needed to pulp the paper, produces 263 lbs of CO_2 in a year.[2] The more we buy, the more CO_2 is released, at least until we make the transition to the solar/hydrogen world. And even then, will the forests, fields and oceans ever be able to support such incredible consumption? What if everyone in the world wanted to live this way? So many trucks, carrying so much stuff. There has to be a better way.

Reduce, Reuse, Recycle...and Buy Less Stuff

Recycling reduces the amount of energy needed to make new things. Every recycled bottle saves 1lb of CO_2 when it is used to make a new one. Every recycled newspaper saves 0.25lbs. It all adds up. Another way to recycle and get that shopping fix (if you need one) is to shop at thrift and second hand stores and yard sales, and to give old things a new lease of life by refurbishing them. Finding a great deal on an old piece of furniture beats paying the full price at the mall, any day.

Live More Simply

The average American is responsible for 11 tons of CO_2 through the manufacture and delivery of the things he or she consumes.[3] At the end of your life, what will you remember? Will it be the cupboards full of clothes, or the cars you owned? All over North America, there's a quiet revolution taking place called "voluntary simplicity", which has people questioning what they're doing. By taking stock of their lives, re-organizing their priorities and spending less on stuff, they are producing fewer emissions (unless they spend their time flying to exotic places). In its place, they're discovering nature, art, their own local communities, and time for meaningful activity. In the big picture, they are trading matter for spirit. They're helping to change the world.

"Give us This Day our Daily 121 Pounds of Matter"

Material	lbs/day	
Stone & cement	27	Stone needs energy to be cut and shipped. Cement releases CO_2.
Coal	19	Coal releases CO_2 and methane.
Minerals	17	Minerals need energy to be mined, shipped & processed; 7% of global energy.
Oil	16	Oil releases CO_2 and methane.
Wood	11	Wood releases CO_2 when burned; old growth forests store CO_2.
Range grass	10	Grass converts into methane through cattle.
Metals	8	Metals need energy to be mined, shipped, and processed.
Natural gas	1	Gas releases CO_2 and methane.
Total	**121 lbs per day = 22 tons of matter per year per person**	

Source: *Stuff: The Secret Lives of Everyday Things*

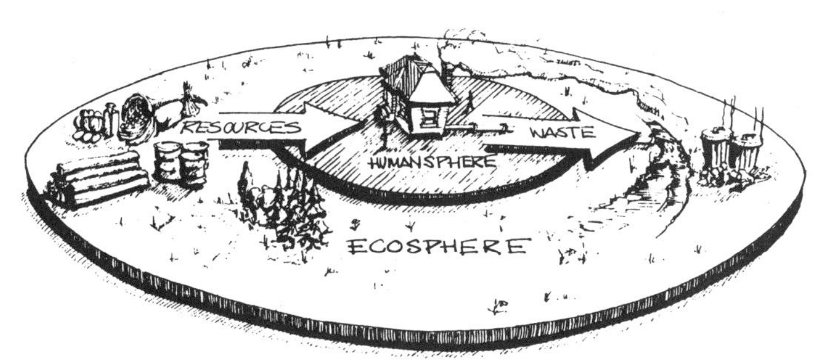

From *Our Ecological Footprint,: Reducing Human Impact on the Earth*, by Mathis Wackernagel & William Rees (New Society Publishers).

Buy Carbon Neutral Products

As the months go by, you will begin to see companies advertising their products as "carbon neutral" or "climate-safe" (see #35). This means they have calculated the carbon cost of their operations, and reduced their emissions to zero either by efficiency and buying green power, or by that plus paying into a carbon-offset fund (to invest in solar energy, efficiency, or tree-planting) to offset their remaining emissions. The first is better, but both are to be commended.

In Britain, you can buy carbon-neutral cars, carbon-neutral holidays, carbon-neutral home-delivered organic food, and take out a carbon-neutral mortgage with the Norwich and Peterborough Building Society to cover the carbon emissions of your house for the first five years, courtesy of Future Forests, which plants trees to offset carbon emissions for individuals, households and companies. In North America, you can buy carbon neutral organic food from Earthbound Farm, carbon neutral yoghurt from Stoneyfield Farm, carbon neutral health products from Shaklee, carbon neutral carpeting from Interface and carbon neutral airline tickets from Triplee.com — and there's more in the pipeline (see #35).

Plant Trees

Let's say you have used a carbon calculator to assess your carbon emissions, shifted to more sustainable travel habits, traded in your clunker or SUV for a more efficient vehicle, invested in more efficient appliances, upgraded your home so that it burns less fuel, installed a solar hot water system, solar PV roof and ground-source heat system, switched to a utility that will sell you green energy, and maximized your recycling. What's next, if you are still producing CO_2 emissions? The answer is — buy carbon offsets. Climate Partners and TripleE will purchase offsets for you in initiatives such as carpooling and school energy retrofits, that help reduce other people's emissions. American Forests will plant trees for you ($1 per tree) on the basis that one tree will absorb one ton of CO_2 over 40 years. They plant 3 trees to make sure that one survives, so 5 tons of CO_2 would need 15 trees and cost $15. Fifty tons would need 150 trees, and cost $150. In Canada, the Tree Canada Foundation runs a very similar program.

Before long, we will be seeing sophisticated Community Carbon Offset Projects which help people calculate their emissions, reduce them by the means described above, and buy carbon offsets to become carbon neutral. When governments start to introduce carbon taxes and rebates (see #73), the whole process will become very familiar. It will be a sign that we are turning the corner to a more sustainable world.

Resources

American Forests: www.americanforests.org 1-800-873-5323

Calculate Your Own Ecological Footprint: www.rprogress.org/interactive

Ecological Footprints: www.ecologicalfootprint.com

Future Forests (UK): www.futureforests.com

Greensense—Resources for Sustainable Living: www.greensense.com

New Road Map Foundation: www.newroadmap.org

North West Earth Institute: www.nwei.org

Simple Living Network: www.simpleliving.net

Tree Canada Foundation: www.treecanada.ca

Climate Partners: www.climatepartners.com

TripleE.com: www.triplee.com/environment

Stuff: The Secret Lives of Everyday Things, by John Ryan & Alan Durning

Your Money Or Your Life, by Joe Dominguez & Vicki Robin

YES! Magazine, A Journal of Positive Futures: www.yesmagazine.org

11

Set up a Local Group

The world is changing. Only a few years ago, it was normal for commentators to say that political power lay with the government, the political parties, the private sector, and perhaps the labor unions. Today, a whole new force is emerging in the shape of NGOs (non-governmental organizations) that are intervening from the local to the global stage — and making a difference — as citizens join together to manifest their values and effect changes.

Over the next few years, we are going to see climate action groups emerging in towns, cities, churches, colleges and rural areas all over the world as people get together to share their concerns and find ways to act. There will be radical groups that practice civil disobedience outside coal-fired power plants, practical groups that organize carbon reduction trainings for businesses, and quiet groups that meet once a week to write letters to their governments and corporations.

Left to itself, renewable energy will gradually gain ground against fossil fuels, perhaps reaching 8% of the global market by 2025. This is nowhere fast enough given the rapid destabilization of the climate, the melting of the Arctic ice, and the threat to the tropical rainforests. We need to increase this figure dramatically to 80% by 2025 if we are to stand a chance of cooling Earth's fever. Make no mistake: it is the NGOs and action groups that make the difference. Where there are effective groups, change happens. Where there are not, the status quo prevails.

Starting a group is easy. All that it takes is calling some friends together, organizing a public meeting and announcing your group. Starting an *effective* group involves:

- learning about climate change and its solutions
- building relationships with businesses, politicians, the media and other non-profits
- choosing actions which can bring you measurable results
- building your membership
- funding your work (eg. the Canadian Climate Change Action Fund www.climatechange.gc.ca)
- sustaining the long-term enthusiasm of your members.

Being part of a effective group can be a truly memorable experience. Many dedicated groups lose members through burn-out, however, so whatever you are doing, remember the Fourth Law of Sustainability: if it's not fun, it's not sustainable. Take time to play, and listen to each other. We have to care for each other, as well as for the Earth.

Actions that can Cool the Climate

- Create a local Carbon Calculator to help people analyze their emissions, and give CO_2 reduction awards to households, schools and businesses.
- Set up a local website with information on energy efficiency and renewable energy.
- Organize a monthly speaker's evening, and set up a Speakers Bureau (see #12).
- Hold a monthly video party to educate your friends and neighbors.
- Create a mobile display and take it around to schools, colleges and shopping malls.
- Organize a lunch-time study circle for businesses, and help them become carbon-neutral (see #34).
- Create an educational quiz show: "Who Wants to Live on a Sustainable Planet ?"
- Set up a green energy buying pool and promotional campaign (see #47).
- Campaign for transit, bikes and light rail, as alternatives to the automobile.
- Lobby your city or town council to join the Cities for Climate Protection Campaign (see #21).
- Campaign for the best state or provincial legislation to promote efficiency and renewable energy (see #61 - #70)

Photo Credit

"Captain Carbon Chaos" team at work in a Victoria school, Canada.

Educate Yourselves

The subject is complicated, and if you want to run a good campaign you have to get on top of it. One way is to set up a Climate Change Study Circle. At the time of writing, no-one has created the curriculum for such a circle, but the Northwest Earth Institute (www.nwei.org) runs study circles on voluntary simplicity and bioregionalism that could serve as a model.

We recommend these books:

- *Carbon War: Global Warming and the End of the Oil Era,* by Jeremy Leggett (2001)
- *Global Warming: The Complete Briefing,* by John Houghton (1997)
- *Greenhouse: The 200-Year Story of Global Warming,* by Gail Christianson (1999)
- *Laboratory Earth,* by Steven Schneider (1998)
- *Natural Capitalism,* by Paul Hawken and Amory & Hunter Lovins (1999)
- *Power Surge,* by Christopher Flavin
- *The Earthday Guide to Planet Repair,* by Denis Hayes (2000)
- *The Change in the Weather : People, Weather, and the Science of Climate,* by William K. Stevens (2000)
- *The Heat is On,* by Ross Gelbspan (1999)

These websites:

- The Heat is Online: www.heatisonline.org
- Climate Solutions: www.climatesolutions.org
- Union of Concerned Scientists: www.ucsusa.org
- Intergovernmental Panel on Climate Change: www.ipcc.ch
- Environmental Protection Agency: www.epa.gov.globalwarming
- Pembina Institute (Canada), Climate Change Solutions: www.climatechangesolutions.com
- The New Energy Revolution: www.davidsuzuki.org/energy

And these videos:

- *Turning Down the Heat : The New Energy Revolution* (Canadian NFB) www.nfb.ca
- *What's Up With the Weather ?* (Nova/Frontline) www.pbs.org/wgbh/warming

Educate the Media

Journalists are busy people who are constantly researching different stories. Most are poorly informed about climate change, and this influences the way they write their stories. Arrange to meet them, get into the habit of sending them stories they might find useful, and suggest they sign up for an environmental email news service (see box).

Free Email Resources

The Daily Grist: www.gristmagazine.com

Energy Efficiency and Renewable Energy Network: www.eren.doe.gov

Environmental News Network: www.enn.com (daily)

Planet Ark: www.planetark.org (daily)

TRENDS in Renewable Energy: www.solaraccess.com (weekly/daily)

Media Lists

- Center for Environmental Citizenship: http://congress.nw.dc.us/cgi-bin/media.pl?dir=cec
- Gebbie Media Directory: www.gebbieinc.com
- Results Canada: www.results resultats.ca/en/media_li.htm

"Each of us has the ability to act powerfully for change; together we can restore that ancient and sustaining harmony."
— David Suzuki

12

Set up a Regional Group

Alongside local climate change groups, regional groups are popping up all over the place. It is spring in the world of climate change solutions! Here are some examples:

Climate Solutions, Pacific Northwest

Climate Solutions (www.climatesolutions.org) grew out of the merger of two groups, the Atmosphere Alliance and the Energy Outreach Center, both based in Olympia, Washington. In April 2000, Climate Solutions launched the Northwest Clean Energy Challenge, a grassroots campaign to mobilize 100,000 citizens in the region to voice their support for clean energy to the governor and the electric utility, and call on the region's utilities and elected officials to install clean, renewable energy. They do research, organize conferences, publish a newsletter, and encourage people to send postcards, emails, and letters to their politicians. Climate Solutions' long-term goal is to see the Pacific Northwest become a world leader in developing practical and profitable solutions to the climate change crisis, focused on the rapid acceleration of clean energy technologies and more livable, less auto-centered communities. They have been invited by Governor Gary Locke to work with Washington State's economic development agency to put together a group of players to hammer out the policy recommendations that will grow the clean energy industry. They promote a Bicycle Commuter Contest, a Neighborhood Connections program that is creating safe cycling and walking routes in Olympia, and have produced a school curriculum guide to sustainable transportation.

Cape and Islands Self Reliance, Cape Cod

The Cape and Islands Self Reliance Corporation (www.reliance.org) was set up as a non-profit organization in 1980 to promote community self-reliance in energy, water and food. They played a pivotal part in protecting the rights of cities, towns and citizens in Massachusetts when the *Electric Utility Restructuring Act* was passed in 1997, paving the way for Community Electric Franchises to be established throughout Massachusetts. These franchises enable customers to aggregate through local governments, non-profits and buying clubs to seek better prices and create energy efficiency programs using funds that were previously diverted to the utility companies. Since 1980, Self Reliance has helped 4,000 homeowners, businesses, churches and agencies to invest in energy efficiency, saving $3 – $4 million every year and preventing 23,000 tons of CO_2 emissions. They spearheaded the development of the Barnstaple County Energy Management Plan, working with a 150-citizen task force that revealed the social and environmental costs of generating "off-Cape" electricity to be $21 million a year.

The Green House Network Speakers Bureau

The National Volunteer Speakers Network consists of over 50 engineers, songwriters, college professors and business people in 20 states who are available at no charge to speak to campus, faith, business and local groups about global warming, climate change policy and the path to a clean energy future. They are united by a belief that global warming presents a very real and present danger to the health of the planet's human populations and natural ecosystems, and that the United States needs to take immediate steps to reduce emissions of greenhouse gases. A few of the speakers are climate experts; the remainder are educated citizens. All are committed to promoting open discussion and debate about global warming policy. To arrange for a speaker, call 503-697-4015. www.greenhousenet.org

> **"The problems of the world cannot possibly be solved by skeptics or cynics whose horizons are limited by the obvious realities. We need men who can dream of things that never were."**
> **— John F. Kennedy**

"What can Minnesotans do to curtail global warming? The problem seems immense, but the solution is straight-forward: Use energy more efficiently and shift to renewable fuels." — David Morris, Institute for Local Self-Reliance

Citizens for Pennsylvania's Future (PennFuture)
PennFuture (www.pennfuture.org) works on global warming, smog, acid rain, and the depletion of natural resources. It campaigns to get dirty power plants to clean up their act and fosters the transition to cleaner, renewable energy. The goal of their Green Power Project is that by 2010, 10% of Pennsylvania's electricity will be generated from renewable energy sources. They helped win $650 million in state funding for the environment, persuaded the government to implement full-cost bonding to make coal companies pay for the environmental damage caused by mining, and created a Renewable Energy Caucus in the state legislature.

Minnesotans for an Energy Efficient Economy (ME3)
Minnesotans for an Energy-Efficient Economy (www.me3.org) is a coalition of 13 organizations that work to improve the quality of life, the environment and the economy of Minnesota by promoting energy efficiency and the sound use of renewable energy. Their work ranges from promoting ecological tax reform (see #67) to encouraging smart growth instead of sprawl, warning Minnesotans about the impacts of climate change, working with farmers to help them reap the benefits of wind energy, and campaigning to ensure that the restructuring of Minnesota's electric industry protects the environment and public health and supports the transition to clean energy. They also engage in research and public education as part of their vision of a sustainable Minnesota.

The David Suzuki Foundation, Vancouver
The David Suzuki Foundation (www.davidsuzuki.org) works towards the vision of a sustainable world, focusing on climate change, forestry and fisheries. With help from many volunteers, they do research, lobby for policy changes, run a Climate Action Team, and put on speaking tours and exhibits.

Setting up a Regional Group
It's all about people. Establishing an effective group involves holding a vision of what your group could achieve, and then connecting with people, one-to-one, building the support you need. To give your group a broad foundation, we suggest that you network with:

- Business leaders
- Municipal leaders
- Climatologists
- Political leaders
- Economists
- Power utilities
- Environmental groups
- Renewable energy proponents
- Farmers
- Schools and colleges
- Forestry activists
- Sustainable transportation activists
- Geographers
- Tax-shift proponents

More Regional Groups that are Building a Sustainable Future

Alliance for Florida's Future: www.icflorida.com/community/groups/nrdcalliance

Citizens for Renewable Energy, Ontario: www.web.net/~cfre

Colorado Renewable Energy Society: www.cres.gen.co.us

Cool Texas: www.cooltexas.net

Iowa Renewable Energy Association: www.irenew.org

Midwest Renewable Energy Association: www.the-mrea.org

New Jersey Future: www.njfuture.org

NorthEast Sustainable Energy Association: www.nesea.org

Northwest Council on Climate Change: www.nwclimate.org

Northwest Energy Efficiency Alliance: www.nwalliance.org

Ontario Clean Air Alliance www.cleanair.web.net

Southern Alliance for Clean Energy: www.tngreen.com/cleanenergy

13

Organize a Schools Initiative

> **"Studies have shown that schools incorporating passive solar features, such as daylighting, use less energy, student grades have improved, and attendance is higher" — Energy Smart Schools**

Most schools, alas, are part of the problem. With a few exciting exceptions, they waste energy, water and valuable resources such as paper, produce a lot of waste, have sterile concrete playgrounds where there could be greenery, and do little to encourage their students to cycle or walk to school. Generations of children think this is normal and don't question it.

But no longer! If you are a parent, student, teacher, administrator, or member of a school board, there are a number of ways in which your school can become a leader in the green energy revolution, exciting your students and saving money. There are renewable energy field trips, solar science fairs, renewable energy clubs — and more:

- At Akron Westfield Community School in Iowa, 20 students took part in a project to install a 600 kW Vestas wind turbine at their school. The students took part in the wind analysis, schematic design, construction documentation, and construction administration, and learned as they went along. Three other Iowa schools have also erected turbines.
- At the Merit Academy in Santa Cruz, CA, a group of students became involved in making a 250 watt electric ice-cream maker powered by a hydrogen fuel cell. What started as a debate about the Gulf War turned into a project with the Schatz Energy Research Center to build the fuel cells, assemble the machine, and tour local schools to demonstrate it. See www. meritworld.com/hydrogen.html.
- Solar schools are blossoming all over America. Olney Elementary, Maryland, has installed a 1 kW thin-film PV grid-connected array on its roof and adopted a solar energy curriculum funded by BP and the Maryland Solar Schools program. The local BP gas station held a "Solar Cents" promotion in November 1999, when for every gallon of gas pumped at the station, they donated a penny to the school's solar project. Waunakee High School, Wisconsin, has installed a 2 kW solar system and integrated it into the science and business curriculum, with the students designing and adding a 1 kW addition every year until they reach 20 kW.

Educational Resources

Climate Change Awareness and Action Kit: www.pembina.org/pubs/ccaa.htm
Climate change teaching materials: www.davidsuzuki.org/energy/classroom.htm
Earthday and Schools: www.earthday.net/howto/teachers-corner.stm
Energy efficiency in schools: www.ase.org/programs/schools.htm
Energy Quest: www.energy.ca.gov/education
EnergySmart Schools: www.eren.doe.gov/energysmartschools
Evergreen Canada (green schoolyards): www.evergreen.ca
EPA's Student Center: www.epa.gov/students
Florida Solar Energy Center on-line curriculum: www.fsec.ucf.edu/ed/teachers
Global Solar Partners (130 solar schools): www.solarpartners.org
Globe Program: www.globe.gov
Green Schools: www.ase.org/greenschools
Green Teacher magazine: www.greenteacher.com
Learning from Light: www.aep.com/Environmental/solar
Merit Academy hydrogen ice-cream project: www.meritworld.com/hydrogen.html
Roofus' Solar Home: www.eren.doe.gov/roofus
Schools Going Solar: www.hydrogenus.com/upvg/schools
SchoolsGoingSolar.org: www.schoolsgoingsolar.org
Smart Moves for Washington Schools: www.climatesolutions.org/smartmoves.html
Sunwind Solar Car Kits and Solar Education: www.sunwindsolar.com
Teachers' Resources for Green Energy: www.green-e.org/teacher
Teaching about Climate Change:, eds. Tim Grant/Gail Littlejohn. New Society, 2001
Watts on Schools: www.wattsonschools.com

Greening School Grounds: Creating Habitats for Learning, Tim Grant and Gail Littlejohn, Editors (New Society Publishers)

Become Energy Smart
America's schools spend more than $6 billion each year on energy, $1.5 billion of which could be saved by better building design, energy-efficient technologies, and improved operating methods. The EnergySmart Schools campaign helps schools reduce their energy consumption and costs and re-invest the savings, while increasing teacher and student awareness about energy and climate change. At Daniel Boone School in Tennessee (1100 students, built in 1973), they installed a ground-source heating and cooling system (see #6), reduced their energy costs by 34%, and are saving $82,000 a year. In Portland, Oregon, the Multnomah County School District used a $20 million State Energy loan to kick-start a 10-year program to retrofit nine buildings that will save $2.28 million a year.

Kick The Car Habit
In British Columbia, the "Way to Go" program (www.waytogo.icbc.bc.ca) works with schools, students, and parents to find alternative ways to travel to school instead of always being ferried by car. Schools use a kit to help them survey their travel habits, determine the best routes to school, educate students about bicycle and pedestrian safety, and implement alternative travel strategies. The first Wednesday in October is International Walk to School Day (www.iwalktoschool.org and www.walktoschool-usa.org).

Dig Up The Concrete
"Greening School Grounds" (www.greengrounds.org) is a Tree Canada Foundation project to assist students, teachers, and parents to bust up the concrete and plant trees and shrubs in schoolyards across British Columbia creating educational landscapes that also absorb CO_2 emissions. This also creates shade and reduces bullying by softening the overall environment. Evergreen Canada is working in 1,000 schools across Canada, turning grey into green. The Los Angeles Unified School District is tearing out thousands of acres of asphalt at hundreds of campuses. After each school develops a landscaping plan, they replace asphalt with grass and trees. They have also launched an ambitious "sustainable schools" program through which each campus will produce its own energy, collect its own water, and feed its own students. There's also a "Cool Schools" Program to plant 8,000 trees in LA schools (www.ladwp.com/coolschools).

Switch On The Daylight
Using fewer artificial lights reduces electricity consumption and the need for cooling. Studies have found that daylighting (using natural light) in schools can improve student test scores by 5% to 21%.[1] In North Carolina, schools built with daylighting have cut their energy consumption by 22% to 64%, with a payback of less than three years, and their students have outperformed those in non-daylit schools in standardized tests by as much as 14%.[2]

Destination Conservation

Destination Conservation is a non-profit energy-saving organization based in Alberta that is active in 973 Canadian and some American schools. They will help you set up a School Team in partnership with the School District and corporate partners, and involve the principal, teachers, students, parents and custodians in making energy-saving and lifestyle changes that will produce annual savings of 20-30% on your school's utility bills. The students perform an energy audit and search for "energy bandits" — places where energy is wasted or leaked. After the first year, they take the energy dollars saved and invest them in further upgrades, such as efficient light bulbs and double-glazed windows, while learning about energy, sustainability and climate change. See www.dc.ab.ca

- Savings by one school board during 5 years with DC: $733,000
- Average first year saving without any capital costs: $2,600
- CO_2 emissions saved by 140 Canadian schools 1993-1997: 13,065 tonnes

14

Organize a Church Initiative

> **"God has charged us to be loving and responsible stewards of creation — but global warming threatens all of creation."**
> **— Rev Milton Jordan, United Methodist Church, Texas**

All around North America, church groups and congregations are organizing around environmental issues, and around climate change in particular. There are prayer and study groups, Earth Sabbaths, and training for church leaders on climate change. Someone has even written a Twelve Step Recovery Program for Energy Addicts. (Step One: "We admit that we are powerless over our compulsive and insistent use of energy, and that our energy use has become unmanageable.") Church congregations are taking very specific actions to save energy, buy green energy, and lobby their politicians.

Buy Green Power

In the Diocese of California, under the leadership of the Rev. Sally Bingham and the Regeneration Project, 41 Episcopal churches have responded to the threat of global climate change by switching to green power from GreenMountain.com, and undertaking energy audits of their buildings. They see their actions as a practical expression of their stewardship of God's creation, and their concern for what we are doing to the Earth. Inspired by their action, churches are following suit in New Jersey, Connecticut, Iowa, Pennsylvania, Maine and Los Angeles, and the California Council of Churches has started California Interfaith Power and Light to spread the initiative. When 33 parishioners of St John's in Clayton, San Francisco, signed up for green energy, the church received a check from GreenMountain for $1,150. In July 2000, the General Convention of the Episcopal Church, held in Denver, CO, with 15,000 participants, was powered entirely by wind energy — the first in North America to do so. (See www.theregenerationproject.org)

The Sisters, the Lamas, and the Wind

In Richardton, North Dakota, the 24 Benedictine Sisters of the Sacred Heart Monastery have invested in two 100-foot wind turbines to save money and reduce their CO_2 emissions. The sisters get no financial help from the Roman Catholic Church and raise lamas on the prairie to help pay their bills. In their first year, the wind turbines generated 37% of their electricity and saved them $12,105. In their second year, they generated 44% and saved $15,836. North and South Dakota have enough wind power potential (255,000 MW) to power two-thirds of the US electricity grid, but there is a lack of interest in new kinds of energy that could threaten the state's coal industry.

> **"Jesus was an active person. Praying about clean water and air is fine. But taking action to make sure that the air and water are clean, that's where we put our faith into action."**
> **— Rev. Sally Bingham, Grace Cathedral, San Francisco**

Publications

Canticle for Creation: Prayer, Information, and Action for the Earth. Quarterly bulletin inserts. Center for the Celebration of Creation, 8812 Germantown Ave., Philadelphia, PA 19118

EarthLight—The Magazine of Spiritual Ecology www.earthlight.org

Global Climate Change: A Guide for Christians, by Rev. Jim Ball, Evangelical Environmental Network: een@esa-online.org 1-800-650-6600

Simpler Living, Compassionate Life: A Christian Perspective $14.95. Earth Ministry

The Greening of Faith: God, the Environment, and the Good Life by John E. Carroll, $19.95 (University Press of New England)

Reverend Sally Bingham, The Regeneration Project, San Francisco.

Rev. Sally Bingham

It is for the preservation of God s creation that we gather together to purchase electric power that has the least possible adverse effect on this fragile earth, our island home.
Maine Interfaith Power and Light

Make Your Church Efficient
Churches are saving energy through efficiency upgrades. The National Council of Churches is helping churches to save money on their utility bills and reduce their CO_2 emissions by making their buildings more energy efficient. With help from the Energy Stewardship Congregations program, church leaders are completing church energy audits, undertaking retrofits, and using the money they save for other purposes. (See www.webofcreation.org/energystewardship/checklist.html)

Educate Your Congregation
The National Council of Churches Eco-Justice Working Group has organized a major Global Warming Campaign in 16 states in partnership with the National Religious Partnership for the Environment, the US Catholic Conference, the Evangelical Environmental Network, and the Coalition on the Environment and Jewish Life. The campaign works to educate congregations about climate change, use less fossil fuel, and make a commitment to care for all God s creation through worship, teaching, congregational lifestyle, and national and global involvement. There are information and strategy packets, booklets, science briefing papers, a 2-day training on climate change, organized visits with elected officials, and a 30-second Public Service Announcement with Maya Angelou that congregations can place with local TV stations. (Contact Ed Dreby, 14 New Jersey Ave, Mt Holly, NJ 08060; www.webofcreation.org).

Pass a Shareholder Resolution
Does your church have a portfolio of investments? If it does, you may be unwittingly supporting companies that are working to sabotage the Kyoto Protocol through the Global Climate Coalition which is financed by fossil fuel interests. Many churches are choosing to place their investments with socially responsible funds and to make nominal investments with companies they do not like, giving them the ability to file activist shareholder resolutions, helped by the Interfaith Center on Corporate Responsibility. (See: www.domini.com/ICCR/htm

Resources

Catholic Conference Environmental Justice Group: www.nccbuscc.org/sdwp/ejp

Coalition on the Environment and Jewish Life: www.coejl.org

Earthday and Faith: www.earthday.net/howto/faith.stm

Earth Renewal: www.earthrenewal.org

Episcopal Power and Light: www.theregenerationproject.org

Energy Star for Congregations: http://yosemite.epa.gov/appd/essbhp.nsf/pages/Congregations

Evangelical Declaration on the Care of Creation: www.creationcare.org/Resources/Declaration/declaration.html

Evangelical Environment Network: www.creation-care.org

Environmental Ministries of Southern California: www.hometown.aol.com/PeterEco

Indigenous Environmental Network: www.ienearth.org

Interfaith Council for Environmental Stewardship: www.stewards.net

Interfaith Global Climate Change Campaigns: www.webofcreation.org/ncc/Regional

National Religious Partnership for the Environment: www.nrpe.org

North American Coalition for Christianity and Ecology: www.nacce.org

Target Earth: www.targetearth.org

The Earth Ministry: www.earthministry.org

Web of Creation: www.webofcreation.org

15

Organize a College Initiative

It is in our colleges and universities that we prepare the leaders, teachers, artists, business people, farmers, and scientists of tomorrow who will either solve the problem of climate change, or live with its devastating consequences. Where else can it be more important for students, faculty, and administrators to show leadership?

Accept The Challenge

At Tufts University in Massachusetts, students, faculty, and staff established the Tufts Climate Initiative in 1998 with a commitment to steer Tufts to a cleaner energy path and meet or beat the US Kyoto goal of a 7% reduction in emissions by 2012. They have started work on an inventory of the university's greenhouse gas emissions, retrofitted one of their buildings, fitted solar panels on another building, set up a lightbulb exchange program, joined the EPA's Green Lights Partnership (see #32), organized a symposium, and researched the changes that will be needed to reduce emissions in transport and other areas. Along the way, they are discovering a mass of barriers, including lack of interest, lack of training, lack of information, failure to account for long-term financial costs and benefits, and student preoccupation with other things – but hey, what's new? If this thing was easy, we'd have done it yesterday.

Buy Green Energy

In April 2000, 6,000 students at the University of Colorado turned out in unprecedented numbers to vote by a 5-to-1 margin to increase their student fees by $1 per semester for four years to buy green power from the Ponnequin Windfarm, operated by Public Service Company of Colorado.

> **"Colleges and universities wield incredible power, and yet in terms of the environment, most have not wielded it well. With only a few noteworthy exceptions, most colleges and universities fail to educate their students in the environmental ramifications of their fields of study"**
> **— Teresa Heinz, Heinz Family Foundation**

The issue was placed on the ballot after 1,300 students signed a petition in favor of purchasing wind energy, supported by a week-long campaign that included information tables, full-page ads in the *Colorado Daily*, presentations to classes, help from the Land and Water Fund of the Rockies, and a mock windfarm made from hundreds of pinwheels that was erected in front of the main campus library. The wind power purchase will save 1,400 tons of CO_2 from entering the atmosphere each year, making the University of Colorado the largest university wind purchaser in the US.

Travel To College Sustainably

When students and staff insist on driving to their studies, many colleges experience a car-parking crunch, while pumping out CO2 emissions. At Camosun College in Victoria, British Columbia, instead of tearing up more green space, they increased the cost of parking, used the income as a subsidy for transit passes, and included a year's U-TREK transit pass in the student fees. At the University of British Columbia, Vancouver, students and staff are cooperating through TREK to find alternative ways to travel.

Switch that Computer Off

Tufts University has 4,300 university-owned computers. Most have no energy-saving features, and many people don't even switch them off at night.

- Percentage of Tufts students who usually leave their computers on: 80%.
- Energy used by a computer: 150 watts/hour (monitor = 60–80 watts/hr, CPU = 60–80 watts/hr).
- Energy used by 4,300 computers over typical working year: 1.3 million kWh.
- CO_2 emissions from 4,300 computers: 975 tons/year.
- CO_2 emissions if one computer in nine is left on all the time: +382 tons/year.
- Myth: Switching a computer off will shorten its lifetime.
- Reality: Switching it off may lengthen its lifetime.

Source: Tufts Climate Initiative

"You have 4,000 universities in the country spending $190 billion on goods and services annually. That's greater than the GDP of all but 20 nations. If schools were practicing renewable energy and buying environmentally sound products, it would have a huge impact." — Anthony Cortese, President, Second Nature

Turn Your Campus Green

In February 1994, the Campus Earth Summit brought together delegates from countries all over the world to craft the Blueprint for a Green Campus, based on the principle that students have the power to demand a more environmentally responsible campus and curriculum, and that faculty and staff can turn out environmentally literate citizens and demand environmentally sound goods and services. The Blueprint provides a detailed 10-step process that can be used to green-up any campus:

1. Integrate environmental knowledge into all relevant disciplines.
2. Improve undergraduate environmental studies course offerings.
3. Provide opportunities for students to study campus and local environmental issues.
4. Conduct a campus environmental audit.
5. Institute environmentally responsible purchasing policies.
6. Reduce campus waste.
7. Maximize energy efficiency. Yale's switch from incandescent to fluorescent lighting will save $3.5 million over ten years.
8. Make environmental sustainability a top priority in campus land-use, transportation, and buildings.
9. Establish a student environmental center.
10. Support students who seek environmentally responsible careers.

Divest From Companies In The Global Climate Coalition

The Global Climate Coalition (GCC) is an oil, coal, and auto company lobby group that has spent millions on advertising, lobbying, and false science trying to convince policy-makers, the media, and the public that there is no scientific rationale for reducing CO_2 emissions and that we have nothing to fear from increased emissions. Cool the Planet is a national student movement whose members are trying to ensure that the GCC melts away before the polar ice caps do by petitioning their universities' regents or trustees to divest from companies that are GCC members. At the University of Washington, students passed motions through both undergraduate and graduate senates, held a rally, sent postcards to their Board of Regents, and used donated cell phones non-stop for an entire day to call Board members emphasizing their concern. The Board finally agreed to start a letter-writing campaign urging the companies it invests in to stop funding the coalition. Staff in the science faculty were particularly supportive, since they had been so outraged to see their work attacked by industry; 53 professors signed a statement saying that global warming is a human-induced problem that needs to be addressed.

Resources

Center for Environmental Cizitenship: www.envirocitizen.org

Brown is Green: www.brown.edu/Departments/Brown_Is_Green

Campus Audit: http://iisd.ca/educate/learn/audit.asp

Campus Environmental Report Card: www.nwf.org/campusecology/reportcard

Cool the Planet: www.cool.policy.net

Greening the Ivory Tower, by Sarah Hammond Creighton (MIT Press Books)

Greening of the Campus IV: www.bsu.edu/provost/ceres/greening

National Wildlife Federation—Campus Ecology: www.nwf.org/campusecology

Penn State's Green Destiny Council: www.bio.psu.edu/Greendestiny

Second Nature: www.secondnature.org

Student Environmental Action Coalition: www.seac.org

Toolkit for Sustainable Development on Campus: http://iisd.ca/educate

TREK (student trip reduction): www.trek.ubc.ca

Tufts University Climate Initiative: www.tufts.edu/tie/tci

University of British Columbia Campus Sustainability Office: www.sustain.ubc.ca

University Leaders for a Sustainable Future: www.ulsf.org

University of Waterloo—WATGreen: www.adm.uwaterloo.ca/infowast/watgreen

US green campus groups: www.envirocitizen.org/enet/environation

16

Organize a Car-free Sunday

Let's face it, we drive an enormous number of cars. The world's transportation sector is responsible for 33% of all CO_2 emissions — and cars and light trucks make up around 80% of that total. But even if every vehicle was a small solar car made from 100% recyclable materials, powered by solar-derived hydrogen or human manure, there would still be too many of them. They are blocking up our towns and cities, making us forget what life could be like without the car, and killing us in unbelievable numbers.

We (the authors) believe that when the sustainability revolution truly takes off, it will be because it offers people a world that is much more attractive than today's. Our cities will have more parks and gardens, more greenways, more city-wide pedestrian walks, more worksharing — and car-free neighborhoods which make people sigh with delight, at the dawning of the solar age.

There is a pleasure in walking around a pedestrian neighborhood. Without fear of being knocked down, we are free to gaze at the sky, to admire the flowers, to stop and talk to passing strangers.

In Europe, cities and towns all over the continent are pushing cars out of their centers. You leave your car at the edge of town and take a park-and-ride bus, or walk. At first, the merchants were afraid they'd lose business; now they love it. With the noise and danger of cars gone, people enjoy their shopping more, and linger longer.

Walking around is such an old tradition. In Valencia, California, residents love the 14 miles of pedestrian *paseos* that criss-cross the town. At Victoria Beach, on Lake Winnipeg, holiday-makers relish the sandy, shaded, car-free lanes where time slows down and friendships grow.

Every neighborhood could be like this if we improved our public transport, adopted car-sharing, created better provisions for cyclists, created more pedestrian routes, and weaned ourselves off our addiction to the car.

Numbers That Die

Number of people killed by cars every week in the USA:

798

Equivalency to the frequency of a major plane crash:

Once every two days

Number of children under 18 who are killed by cars every week while walking:

19

% of pedestrians killed by cars:

15%

% of federal transport safety money spent to protect pedestrians:

1%

Reduction in pedestrian accidents after Seattle introduced a traffic-calming program:

75%

"Imagine you live in a city free of the noise, stench and danger of cars, trucks and buses. Imagine all your needs, from groceries to child care, within a five-minute walk of your home. Imagine that the longest commute within your city takes 35 minutes from door to door, by way of a cheap, safe and efficient public transportation system."

— J.H. Crawford, *Car-Free Cities*

Joel Crawford

Sundown at Parma, on Car-free Sunday.

The Car-free High Street, Winchester, England.

Mary Dauncey

Do as the Romans Do

On Sunday February 6th, 2000, Italy went car-free. In Rome and 150 other towns and cities, travelling by bus, bicycle or on foot, millions of Italians reclaimed the traffic-plagued streets of their historic city centers as cars were banned for the day in a fight against pollution. Car-crazy Italians moved around by bicycle, tricycle, rollerblades and electric vehicles. In cities like Rome and Catanzaro, public transport was free of charge. In Rome, all museums run by the city were free of charge. Hiring electric mopeds for two hours cost the same as an ice-cream. "What is this Sunday like?" asked a smiling man riding a small bicycle in Rome. "It's nice. It seems like a banal thing to say, but that's exactly what it's like." (Reuters) "Disbelieving at first, Italians poured into city centers last Sunday to inhale, taste and savor an unprecedented silence and stillness: life without the internal combustion engine." *(Guardian Weekly)*

Car-Free Sundays

On April 1st, 2001, Victoria, British Columbia, enjoyed a Car-Free Sunday along a large stretch of its waterfront. Neighbors came out to stroll, kids played street hockey, a "green walk" explored the neighborhood, and people in wheelchairs were dropped off by minibus.

Which City will be First?

North America is not like Europe. Its city centers are far less dense, and many smaller stores have given way to chain-stores in the malls. The facts of global climate change still stare us in the face, however. Globally, our cars and other vehicles produce 8 billion tons of CO_2 every year. One Car-Free Sunday will not change this, but it can send an important message: our cars are cooking the planet.

As far as we know, there's not been a car-free Sunday for a whole city in North America yet. In anticipation that this might change, here are some suggestions:

- Find lots of volunteers to help make it happen
- Get support and involvement from your city council and their staff.
- Get support from the local media.
- Get public endorsements by local celebrities.
- Negotiate free or half-price rides on the buses.
- Invite people to organize neighborhood festivals on local streets.
- Organize "Meet the Neighbors" community walks and bicycle rides.
- Issue polite notices to motorists on neighborhood streets.
- Organize bus rides to hiking areas outside the city.
- Advertise the availability of bicycle trailers.

Traffic Reduction Kit

The Traffic Reduction Kit, created by Australian eco-activist David Engwicht, enables residents to solve their traffic problems by tackling the root causes rather than treating the symptoms with traffic-calming. Residents make a "Traffic Reduction Treaty" with another street by which both streets agree to reduce their car use, act as a guest in each other's street, and reclaim their streets both physically and psychologically. In a trial of the kit, residents achieved a 34% reduction in their car use.

Resources

Alliance for a Paving Moratorium: www.tidepool.com/alliance

Car-Free Cities: www.carfree.com

Car-Free Cities by Joel Crawford (International Books), 2000: www.modfirsts.com

CityStreets, New York: www.citystreets.org

Traffic Reduction Kit: www.lesstraffic.com

World Car-Free Day Consortium: www.ecoplan.org/carfreeday

17

Start a Car-Share Organization

Guy Dauncey

The Victoria Car Share Cooperative.

Car sharing started in Switzerland and Germany in the late 1980s, offering people the convenience of using a car without the hassle of owning one. Normally, when you buy a car, you pay for the purchase, insurance, and road tax up-front, but the actual cost of driving is very low, which encourages people to drive for even the shortest trips.

Car sharing changes those assumptions. It gives you access to a car without making you pay the full cost of ownership. Because you pay by the trip, there is an incentive to walk, bike or use transit, and use the car only when you need it. Car sharing makes it feasible for people to rely on alternative modes of transportation, while reducing their overall costs.

CarSharing Portland started in 1998. By May2001, they had over 400 members who shared 25 vehicles, including a Toyota Prius, a Honda Insight, a pickup and a minivan. Members pay a $25 application fee, a $250 returnable deposit, and a $10 monthly membership fee. Each time they use a car, they pay $2 an hour, plus 40 cents a mile. This includes gasoline, insurance, vehicle maintenance and repairs. A trip of 4 hours and 25 miles costs $18. A one-hour trip to the local store might cost $4. Generally speaking, if you drive less than 7,500 miles a year, car sharing will save you money (see box). On average, Portland's members save $154 a month in avoided transportation costs. 26% of the members sold their vehicles on joining, while 53% avoided buying one.

The cars are parked in different neighborhoods, and members keep a key to a lock-box where the vehicle keys are kept. When you want to use a vehicle, you phone the office, book the car for the period you want, and take it out. When you're done, you return the car to the location you picked it up from, and return the keys to the lock-box. It's as simple as that. If the tank is under 1/4 full, you fill it up and send in the receipt to be taken off your monthly bill. Each shared car replaces around ten private cars, so if car-sharing took off we could have parks, ponds and play-spaces in our towns and cities, in place of parking and pollution.

A study by the Swiss Office for Energy Affairs showed that former car owners reduce their transport energy consumption by 50% when they join a car-share organization, reducing pollution, noise and traffic accidents, and reducing their CO_2 emissions by 1.5 tons per year. In North America, where private vehicles are inefficient and car-share groups buy the most efficient vehicles, members probably reduce their emissions by over 3 tons per year.

The Cost of Car-Sharing

The AAA calculates that the average annual cost of a new car (excluding the actual purchase) is about $5,300 per year, or 53¢ per mile if you drive 10,000 miles per year. Even a 10-year old car may cost $2,500 per year.

Cost of Driving a new small sedan

Miles Driven per Year	2,500	5,000	10,000
Car-Sharing*	$1,007	$2,038	$3,700
Car Ownership	$3,173	$3,363	$3,734
Saving	$2166	$1280	- $34

*Car-sharers pay per hour as well as per mile, so the cost varies with the hours used.

Thanks to CarSharing Portland

"I gave up my car to join the co-op, and I really see the money I'm now saving. I get all the benefits of owning a car without that huge debt load." — Mike Darche, Victoria Car Share co-operative

Start a Car Share Organization
There are plenty of resources to help you get started, including a listserv for discussions in the North American Car-Sharing Network (see box). There are active car share groups in Aspen, Boulder, Calgary, Edmonton, Halifax, Montreal, Quebec, Ottawa, Toronto, Kitchener, Waterloo, Boston, Traverse City, San Francisco, Seattle, Vancouver, Victoria and Honolulu.

Starting a car-share organization or cooperative involves learning how other groups work, finding people who share your enthusiasm, obtaining a start-up grant, doing your financial and administrative homework, finding a loan partner, building community relationships, choosing a suitable location, enrolling members, and finally, buying the cars. It may take one to two years from your first meeting to the launch.

The Future of Car Sharing
In Europe, car sharing is growing by leaps and bounds. In 1999, there were 200 car-share organizations serving 130,000 people in 450 cities. The largest group, Mobility CarSharing Switzerland, had 1,400 cars in 300 communities with over 30,000 members.

In spite of this rapid growth, car sharing represents only the tiniest fraction of overall car use. According to those who have studied it,[1] the future of car sharing lies with two progressions:

1. First, the adoption of technologies that make membership and administration easier, such as Internet scheduling, on-board electronics that recognizes drivers by smart cards, computerized information gathering, and the use of GPS to track vehicle locations.
2. Second, the development of wider mobility packages. In Germany, Drive Stadtauto members in Berlin and Potsdam have a "Mobil Card" that gives them 15% off public transportation and enables them to use taxis, pay for the home delivery of food and beverage, and reserve a bicycle or a canoe, all without the need for cash. In Switzerland, Mobility CarSharing's Zuger Pass Plus offers members a discounted combination of car-sharing, public transit, car rental, taxi, bicycle, and various non-transport related services. Starting in 2001, Swiss National Rail will offer a mobility package called "Easy-Ride Switzerland" to 1.5 million pass holders (35% of Switzerland's adult population), giving access and discounts to car-sharing vehicles, rental cars, transit, trains, taxis and cable cars, all with the same smart card. Almost every public transportation company in Switzerland is a partner. Listen up, all you transportation experts — this is the future of transportation!

Car Sharing Resources

CarSharing Portland: www.carsharing-pdx.com

Car-Sharing and Mobility—An Updated Review by Daniel Sperling, Susan Shaheen and Conrad Wagner: www.calstart.org/resources/papers/car_sharing.html

European Car Sharing Network: www.carsharing.org

European Car-Free Cities Network: www.bremen.de/info/agenda21/carfree

Mobility CarSharing Switzerland: www.mobility.ch

North American Car Sharing Network: www.carsharing.net

Vancouver Cooperative Auto Network: www.cooperativeauto.net

Victoria Car Share Cooperative: vvv.com/~carshare

Zipcar, Boston: www.zipcar.com

"At some point in the future, we're going to make hydrogen for cars directly from solar energy." — Fritz Vahrenholt, Deutsche Shell

18

Create a Commotion

This book is very sober. It says, "Look, we've got a huge crisis on our hands. Here are some solutions. Now let's get on with it." It is ninety-nine parts polite and practical, and one part wild and creative — and that part is right here.

Did anything major ever change without conflict, struggle, and mass arrests? Would women have won the vote without the suffragettes? Would South Africa have toppled apartheid without the world-wide boycott campaigns? The wealthy, powerful world of the coal, oil, and auto corporations is hardly going to cap its wells, pack its bags, and stroll off into the history books without a struggle.

All over the world, the oil and gas industry is drilling into fragile environments, sometimes threatening whole peoples with extinction. In Columbia, the U'wa people have said they would rather die than allow Occidental Petroleum to despoil their cloudforest ancestral homeland through oil exploration. Sixty-two of the world's hundred largest economies are corporations, and nine of the ten richest corporations are oil or auto companies, which are largely responsible for climate change.

In the 1930s and '40s, the people of India used civil disobedience and non-violent direct action to tell the British, politely but firmly, to leave. From the 1950s on, the same methods have been used to oppose the nuclear arms race, to end the war in Vietnam, to win equal rights for black people, to stop the expansion of nuclear power plants, to stop logging in ancient forests, and much, much more.

Today, we are struggling to save the world from a global warming apocalypse caused primarily by the burning of fossil fuels, and the failure of political leadership by those who have been paid off by the fossil fuel corporations. Yes, we need public education, progressive policies, and practical alternatives, and without them we'll get nowhere. But we also need protests, parades, and planetary walkathons to awaken the world to the perils of the path we are on.

> **"Never be discouraged from being an activist because people tell you that you'll not succeed. You have already succeeded if you're out there representing truth or justice or compassion or fairness or love."**
>
> **— Doris Haddock (Granny D), who walked across America at the age of 89, to highlight the need for campaign finance reform.**

Resources

- Action Resource Center: www.arcweb.org
- Activists' Handbook: www.Protest.Net/activists_handbook
- *Adbusters* Culture Jammers: www.adbusters.org
- Affinity Groups: www.actupny.org/documents/CDdocuments/Affinity.html
- Alliance for Democracy: www.afd-online.org
- Alliance for Sustainable Jobs and the Environment: www.asje.org
- Coke Spotlight: www.cokespotlight.org
- Corporate Watch: www.corpwatch.org
- Direct Action Network: www.directactionnetwork.org
- *Earth First! Journal*: www.earthfirstjournal.org
- End of Oil Action Coalition: www.tao.ca/~no_oil
- Independent Media Center: www.indymedia.org
- Granny D: www.grannyd.com
- Project Underground: www.moles.org
- Protest.Net: www.protest.net
- Ruckus Society: www.ruckus.org
- Youth for Environmental Sanity: www.yesworld.org

2001 Greenpeace/Lombardi

Greenpeace members dump coal, and faux oil and nuclear waste drums outside Vice President Cheney's residence in Washington.

Make Alliances
In May 1999, steelworkers and environmentalists, realizing they shared a common interest in protecting workers and the Earth from the misconduct of irresponsible corporations, formed the Alliance for Sustainable Jobs and the Environment. It all started with Maxxam Inc.. While Maxxam's Pacific Lumber Company was clearcutting the ancient redwood of northern California, Maxxam's Kaiser Aluminum subsidiary was locking out 3,000 steelworkers. Put two and two together and you have a powerful new combination of over 200 environmental and labor union organizations. The phasing out of fossil fuels and the transition to a solar-hydrogen economy may threaten the jobs of millions of working people, but it will create many more new jobs. By working together, we can ensure that coal miners and oil workers are given the support they need to retrain and find a new place in a sustainable economy. "The alliance has made it possible to dream again" (David Brower, founder of Earth Island Institute).

Form an Affinity Group
If you are involved in any kind of non-violent direct action or civil disobedience, you should think about forming an affinity group (or groups). In the midst of the action, it is easy to feel isolated, scared, or upset. An affinity group is a self-sufficient support system of 5 to 15 people that provides a source of comfort, support, and solidarity to its members. By working and acting together, the group builds a sense of familiarity and trust, like a family. This enables the group to make joint decisions, keep in touch with other affinity groups during a mass action, be there for each other as some members risk arrest, and be there again afterwards when people feel the need to talk about what happened. (For a good description, see www.actupny.org/documents/CDdocuments/Affinity.html)

Hold a Non-Violence Training Session
The Ruckus Society runs training camps where people can learn the skills of non-violent civil disobedience to help environmental and human rights organizations achieve their goals. Activists learn the practice and philosophy of non-violence, how to scale a wall, how to hang a banner, how to work in an affinity group, how to work with the media, and how to understand the complexities of climate change or world trade. Their emphasis is placed firmly on non-violence: "Wherever the location, regardless of the subject, we condemn and do not train activists in any technique that will destroy property or harm any being."

Stage An Action
Now you're ready. Make signs, banners, and large puppets to express your message creatively. Scout out likely targets, such as a crane from which you can drop a banner. Come to consensus on a plan. Then go do it. A civil disobedience action will likely be one of the memorable events of your life. And remember to have some fun while you're at it!

For quick access to all websites listed in this book, see www.earthfuture.com/stormyweather

"Solar, wind are safe and clean. Let's shut down the oil machine! We are here to let you know The time has come for oil to go!"
— Chant sung at the World Petroleum Congress, Calgary, June 2000

19

Educate Your Politicians

"How can we reduce our CO_2 emissions? Let me count the ways."

With apologies to the poet Elizabeth Barrett Browning, a great many ways to reduce CO_2 emissions end up on the floor of a city council, provincial legislature or state senate, or in Ottawa or Washington. Proposals to vote funds for bicycle lanes, or control suburban sprawl, end up in front of city councilors. Proposals to enact clean air legislation or set national renewable energy goals end up in Ottawa or Washington. Or alternatively, if we haven't done our homework, they don't. Many a superb proposal has been killed before it gets to the vote.

They say "Power belongs to those who show up," and it is never more true than when applied to climate change. The coal, oil, gas and motor vehicle corporations have full-time lobbyists working to ensure that the politicians don't act against their interests. Our challenge, as students, parents, teachers, farmers, church leaders, scientists, musicians, and ordinary folk is to talk to our politicians, show them our faces, and make them remember our words.

For many, this is something new. It takes courage to pick up the phone and call your local councilor to ask for a meeting. It takes courage to write a letter to the editor, or to write a longer opinion piece — but this is what makes democracy work. It is when we stay at home feeling left out of the process that others move in to control the show.

There has never been a more exciting time to become involved in local, national, or global issues. The Internet is a stunningly useful resource for learning, sharing, and making a difference. This must be how folks felt in the 16th century when they discovered the freedom of the printing press after the long dark ages when the church controlled every written word. Major corporations have been controlling today's media in much the same way, but now the Internet allows us to create our own independent channels of communication, and to organize together all around the world. The unborn voices of future generations are crying out to prevent their future from being sacrificed on the altar of coal and oil, to appease the greed and desires of the fossil fuel barons.

Run for Office
Politicians are only ordinary people who pluck up the courage to run for office, and succeed. The world urgently needs people who care about the planet to put their names into the hat, and become legislators. Maybe it needs YOU?

> **"Politics, more than any other sphere of life, is where the tough decisions get displaced. It is the only real alternative to warfare invented by human beings. Not only can and does the political terrain offer scope for healing conflict, but it also offers opportunity for creative leaps forward in new policies."**
> **— John Rensenbrink, Maine Green Party**

Three Things You Can Do Today

- Write a letter to your councilor, Member of Parliament or Representative telling her in simple terms that you care about the environment and that you want her to work for solutions to the climate change crisis.
- Arrange to attend a session of your town or city council. When they're done, go up and introduce yourself to one or two, and share your views on climate change. Show them a copy of this book.
- Look up the pages for your political parties on the Internet, and read what their policies are on climate change. You may be shocked — or pleasantly surprised.

Voting for the Planet.

Develop Your Policy Goals

As a group, it is important to work out your priorities. The more specific your policy goals, the greater the likelihood that they will succeed. Better to choose one initiative, whether it be a campaign to get tax credits for investments in renewable energy (see #75) or to stop oil exploration in Costa Rica's rainforest, than to choose too many and over-extend yourselves. There are always too many goals to pursue, so choose the ones that excite you personally or as a group. That way, you will keep your energy high.

Adopt a Politician

The key to success in politics is influence, which means developing relationships. Once you have chosen your campaign goals, try writing to every politician to see who responds. What you want is a political champion, someone who will go to bat for you. When you have found that person, arrange to meet him or her and start building a relationship. If they agree with your goals, ask "How can we help you?" That will shock them; most politicians never hear that question. Once you have a relationship, use it to keep them educated and up to date by sending them relevant information.

For a truly inspiring story about how local groups can work with great success to influence their politicians (on the issue of ending hunger and poverty), read *Reclaiming Our Democracy: Healing the Break Between People and Government*, by Sam Harris, the founder of Results. (http://results.action.org/shirley.html)

Establish an Environmental Scorecard

In Washington DC, the League of Conservation Voters produces an annual scorecard that shows how members of the Senate and Congress have voted on key environmental bills in each legislative session, giving them a score from 0 to 100. Their website shows how your local representatives have been voting and provides an education into the nefarious ways in which the politicians in Washington work. In the Pacific Northwest, Washington Conservation Voters does a similar job, rewarding some politicians for their "Good Green Deeds." These scorecards are very powerful tools — every state and province should have one. Consider creating an Internet-based scorecard for your state, province, or local city council.

Resources

Address Directory for Politicians of the World: www.trytel.com/~aberdeen

An inspiring way to get people involved in Washington State politics: www.callclickvote.com

EarthNet's Legislative Action Center: http://congress.nw.dc.us/cec

FECInfo: www.tray.com

Global Legislators for a Balanced Environment: www.globeint.org

Governments on the World Wide Web (17,000 entries): www.gksoft.com/govt

League of Conservation Voters (Washington DC): www.lcv.org

Local politicians come alive!: www.yourpolitician.com

Project Vote Smart: www.vote-smart.org

See how one person is using the Internet to put people in touch with local, national and global politicians: www.islandnet.com/sunshine/govern.htm

Washington Conservation Voters: www.wcvoters.org

Congress at Your Fingertips: www.capitoladvantage.com/publishing/standard.html

20
Link Up with Other Groups

"There are no passengers on spaceship earth. We are all crew."
— Marshall McLuhan

Twenty years ago, the Danish electric utilities had a vision of Denmark being powered by nuclear energy. Energy was their business and that was their plan. It failed to happen, however, and by 2001, Denmark was obtaining 15% of its electricity from wind and had set a goal for 50% by 2030. Denmark also exports 50% of the world's wind turbines. Over the last 20 years, the Danish wind pioneers and their non-profit groups have changed the entire direction of Denmark's energy policy.

Their success was far from straightforward, however. Almost every vested interest in Denmark fought to oppose them, including the utility companies, the Woodland and Nature Administration, and Denmark's planners. The reasons they were able to overcome this wall of opposition were because:

- they worked co-operatively through the Association of Wind Turbine Owners (Danske Vindkraftvaerker) to lobby for the right to distribute their power into the grid at a fair price;
- they published monthly data on the performance of the turbines, forcing the less efficient technologies to improve or leave the market;
- the Danish Parliament (the Folketing), elected by proportional representation and dominated by a center-left-green coalition, constantly overruled objections from the utilities and its own ministries.

Denmark's success was anything but a normal market progression. If it had been left to the natural tendencies of the so-called market, with the advantages it gives to the existing energy establishment, Danes would have been investing their resources in nuclear power today, not in wind. It is only because the wind energy pioneers worked together on a national level that they were able to achieve today's successes. (For the details, see www.windpower.dk)

The same is true elsewhere. In Freiburg, the "solar capital" of Germany (see #24), the city's interest in reducing CO_2 emissions started with a grassroots campaign to close down the local nuclear power stations. In Sacramento, CA, the campaign to close the Rancho Seco nuclear power plant left a hole in the utility's future power supply and led to the emergence of the Sacramento Municipal Utility District as the premier solar utility in the USA (see #43).

For local and regional groups, the existence of strong national groups is essential. For national groups, the existence of strong local groups from Fort Worth to Edmonton helps them to send a more forceful message to the politicians in Washington DC and in Ottawa, and to industry leaders. We need each other, to work together.

Labor Organizations for Climate Action?

At the time of writing, there is no such group. We need one, however, to work on the following campaigns:

- Persuading employers to buy green energy
- Investing labor union pension funds in SRI funds (see #9)
- Lobbying for a tax shift to raise pollution taxes instead of income taxes (see #67)
- Making energy-efficiency proposals to companies to save $$ and jobs
- Establishing a Sustainable Economy Transition Fund to retrain displaced coal and oil workers
- Arguing the case for increased jobs through energy efficiency and renewable energy
- Arguing the case for a sustainable economy and a sustainable world

Guy Dauncey

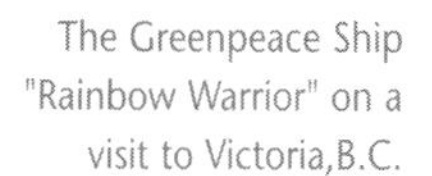
The Greenpeace Ship "Rainbow Warrior" on a visit to Victoria,B.C.

There are well over 300 organizations working for energy efficiency, renewable energy, sustainable transportation, and other solutions to global climate change in North America. Imagine what we could achieve if every local group worked hand-in-hand with a national group. Here are some of the leading national groups (see also #98). For a complete list, see www.earthfuture.com/stormy/weather.

Alliance To Save Energy
A coalition of 60 business, government, environmental, and consumer groups focused on residential, business, commercial, federal, and schools energy efficiency. Active in consumer education, energy policy reform, appliance standards, and international development. (www.ase.org)

American Green Network
Working in Washington DC to promote a sustainable path to America's clean energy future. Keeps policy makers, media, and the public informed about renewable energy and energy efficiency. Active member of Sustainable Energy Coalition; supports the House Renewable Energy Caucus and the Senate Renewables and Efficiency Caucus (www.americangreen.org).

Business Council for Sustainable Energy
An association of leading companies and trade associations. Advocates policies that support sustainable energy technologies, such as fuel cells, solar, cogeneration, wind, and natural gas combined cycle systems. Campaigns for more global clean energy financing. (www.bcse.org)

Environmental Defence
Campaigns on many issues, including a Clean Car Pledge, Green Car Guide, green electricity, and electricity labeling. Their 135,000-member Action Network can fax and email politicians at 24 hours notice. (www.environmentaldefense.org)

Friends of the Earth, US
The US office of the largest international environmental network in the world, with affiliates in 63 countries. Takes on a host of issues including green tax reform, an end to fossil fuel subsidies, and an end to sprawl and highway expansion. (www.foe.org)

Greenpeace
With offices and members all over the world, Greenpeace has a very active climate change campaign. Their efforts include the Greenfreeze (see #5), the TWINGO (see #57), SaneBP (see #9) and the Corporate Top 100 initiatives (see #31), as well as direct actions and other campaigns. (www.greenpeace.org)

Pembina Institute, Canada
Works to increase corporate eco-efficiency, promote sustainable energy paths, lobby for government policies that address climate change, encourage green economic policies, educate the public, and develop practical solutions to climate change. (www.pembina.org and www.climatechangesolutions.org)

Public Interest Research Group (PIRG)
Campaigns to Stop Global Warming, and for a Clean Energy Agenda - clean air, clean energy, clean cars, clean investments (as well as many other issues). Encourages people to write to their politicians. (www.pirg.org/enviro/global_w)

Sierra Club Global Warming Campaign, US
Campaigns to raise the Corporate Average Fuel Efficiency standards for cars, SUVs, and light trucks; "Earthrise" campaign postcards to send to senators and representatives. (www.toowarm.org)

Union of Concerned Scientists
An independent nonprofit alliance of concerned citizens and scientists whose scientists and engineers conduct studies on renewable energy, clean vehicles and the impacts of global warming (among other issues), and provide information on environmental science to government and the media. 40,000 members participate in an Action Network, making their voices heard on national and state legislation. (www.ucsusa.org)

21

Join the Cities for Climate Protection Campaign

For the past ten years, cities around the world have been assuming global leadership in the quest to cool down climate change and usher in the solar age.

There is nothing impossible about eliminating greenhouse gases, as 300 cities — from Portland, Oregon to Tampere, Finland — that participate in the Cities for Climate Protection Campaign already know. The campaign requires leadership and community support; once underway, it is not excessively complicated and all the initiatives that will make a difference have been tried, tested, and found to work.

So what kind of thing are we talking about? Nothing unexpected, but when you put them together, these greenhouse-gas-saving initiatives have a remarkable effect. They will save money and bring many side benefits while making your community safer, more attractive, and more ecologically sustainable. For the most part, they will be politically uncontroversial, which is useful if you are a mayor or councilor and conscious of the people who voted you into office. In practical terms, we are talking about:

- Initiatives to make your town or city's own operations more energy efficient (#22)
- Initiatives to buy clean, renewable energy (#23)
- Initiatives to make your whole community more energy efficient (#24)
- Initiatives to plant trees and support urban agriculture (#25)
- Initiatives to improve transportation and encourage cycling, transit, and walking (#26 and #27)
- Initiatives to reduce solid waste and increase recycling (#28)
- Initiatives to capture methane from your landfills (#58)
- Initiatives to plan sustainable communities and reduce sprawl (#29 and #30).

Nor will you be alone in your efforts. In the USA, if your jurisdiction participates in the Cities for Climate Protection (CCP) Campaign, you will be provided with technical assistance, training, grants, working policies, publications, case studies, and a connection with mayors, councilors, and city staff in cities and towns that are working on similar strategies. The campaign is run by the International Council for Local Environmental Initiatives (ICLEI), 15 Shattuck Square, Suite #215, Berkeley, CA 94704 (510) 540-8843 www.iclei.org/us; iclei_usa@iclei.org.

In Canada, you will receive almost identical benefits through Partners for Climate Protection (PCP), a CCP partner run by the Federation of Canadian Municipalities, 24 Clarence St, Ottawa, K1N 5P3, (613) 241-5221; www.fcm.ca. The Canadian government has created a $2 billion assistance program to help cities green-up their municipal infrastructures.

"Local leaders are realizing that global warming could represent a serious danger to their cities and counties. The good news is that local communities can reduce their contribution to the pollution that causes global warming".
— Sharon Sayles Belton, Mayor of Minneapolis, 1999

- The City of Halifax, Nova Scotia, converted to high-efficiency street lamps and reduced its yearly operating costs by $550,000, preventing 2,000 tonnes per year in CO_2 emissions.
- The City of Ottawa switched to smaller engines in its medium and heavy duty trucks, which saved $464,000 between 1990–1994, plus 2,000 tonnes of CO_2.

"There is something fundamentally wrong with treating the Earth as if it were a business in liquidation."
— Herman Daly

Resources

Cities for Climate Protection (U.S.): www.iclei.org/us

Cities for Climate Protection (worldwide): www.iclei.org/co2

EPA Climate Change Outreach Kit for state and local leaders (slideshow, posters, brochures, exhibits): www.epa.gov/globalwarming/publications

European case studies: www.cities21.com/egpis

Greenhouse Gas Emissions Software developed by ICLEI/CCP and Torrie Smith Associates: www.torriesmith.com

International Council for Local Environmental Initiatives (ICLEI): www.iclei.org

Partners for Climate Protection (Canada): www.fcm.ca

Pembina Institute, Canadian resources: www.climatechangesolutions.com

Toronto Atmosphere Fund: www.city.toronto.on.ca/taf

Write A Local Action Plan

When you join the CCP or the PCP, you will be invited to develop a Local Action Plan:

Step One: Develop a municipal energy profile and establish a base-line inventory for greenhouse gases, so that you know where you stand (software available, see box).

Step Two: Identify measures that will reduce CO_2 emissions.

Step Three: Evaluate the measures and decide which to include in your action plan.

Step Four: Choose your CO_2 reduction target. Most communities joining in the 1990s chose the "Toronto Target," which aims to reduce emissions to 20% below the 1990 level by 2010.

Step Five: Develop a Local Action Plan

That makes it sound quite easy. Lots of towns and cities jump right in, but community support is needed for a plan to have long-term success. The three keys to success are a majority on council, enthusiastic support from the city's staff, and a non-profit organization that can take initiatives and rally support in the community.

In Minneapolis-St. Paul, an executive steering committee was established in 1993 to direct the Twin Cities' joint CO_2 Reduction Project. The committee included the two mayors, councilors, executives from the corporate world (Northern States Power, Minnegasco, Honeywell, and 3M), representatives from the Metropolitan Council and local environmental groups (the Isaak Walton League and the Sierra Club), and a Technical Advisory Committee. They set a goal of 20% below the 1988 level by 2005 (7.5% by 1997), and established six strategies covering municipal action, transportation efficiency, urban reforestation, energy efficiency, energy supply, and precycling/recycling. By 1998, St. Paul had reduced its municipal energy use by 15%, reduced its annual CO_2 emissions by 4,650 tons, and is saving $275,000 on its annual energy bill.

In 1990, Toronto pledged to reduce its CO_2 emissions by 20% by 2005. The city established an Energy Efficiency Office and, by selling a surplus jail-farm for $23 million, established the Toronto Atmospheric Fund to make loans and 50:50 grants for a huge variety of CO_2-reducing initiatives. In 1992, the fund loaned the city $14 million for street lighting conversion, which is saving the city $2 million a year in electricity bills. A recent $2 million loan to the Better Buildings Partnership for energy efficiency retrofits will save local businesses $3 million annually and reduce CO_2 emissions by 40,000 tonnes a year.

Remember the Big Picture

Most cities in the Cities for Climate Protection campaign aim to reduce their emissions by 20% by 2010. According to the IPCC, we need a 60%-80% reduction in emissions if we want to stop the planet's surface from warming by more than 2C (4F). No one expects a city to do this on its own, but if we work together as a world on every front, we can do it.

22

Make Your City's Operations More Efficient

How Much Energy Are You Losing?

Portland, Oregon was the first US city to adopt a CO_2 reduction strategy, setting a goal of 20% below the 1990 level by 2010. Between 1990-1995, the city reduced its energy use by 15%, resulting in $1.2 million annual savings. In Maryland, Montgomery County has been achieving 30-40% energy savings in its building retrofits and 40% energy savings in its new facilities, saving $2.3 million and reducing 7000 tons of CO_2 emissions over four years. Who was it who said that reducing CO_2 emissions would bankrupt the economy?

How Efficient Is Your Fleet?

In 1993, Denver introduced a Green Fleet Program mandating a reduction in the size of the fleet, a 1% annual reduction in fuel expenditures, and a 1.5% annual reduction in CO_2 emissions. The full-sized sedans were downsized to compacts or mid-sized sedans, and fuel efficiency was built into the purchasing bid specifications. By 2005, the program will be generating annual savings of $106,000, and will have reduced CO_2 emissions by 22%, in spite of a 19% increase in overall miles traveled.

How Green Are Your Buildings?

Buildings use a lot of energy, and not just in the lighting. In 1999, San Francisco passed a Resource Efficiency Building Ordinance mandating green design practices for all new and existing buildings owned or leased by the city. This includes toxics reduction, the use of recycled-content materials, improved air quality, energy and water efficiency, construction and demolition waste diversion, and building commissioning. The city's design staff have a mandated training program, and a Resource Efficient Task Force has been set up to guide the process (ddowers@puc.sf.ca.us).

What About All That Office Equipment?

Copiers, computers, laser printers — they all eat up energy and pump out CO_2 emissions. In 1983, Volusia County Council, Florida, adopted a "Reduce, Reuse and Recycle Procurement Policy," which has replaced old copiers and computers with Energy Star models and copiers that print on both sides of the paper. The new machines cost $100 more, but they quickly recoup the cost through energy savings. Because of this, the County's Request for Proposals specifically reserves the right to buy Energy Star equipment regardless of price, over-riding the one-year budget rule in the name of life-cycle costing. In Portland, the city's use of Energy Star equipment may be producing savings of up to $35,000 a year.

> **"Our experience shows that making buildings energy efficient requires more thinking, not more money."**
> **— Ronald J. Balon, Senior Energy Engineer, Montgomery County**

Resources

- American Council for an Energy Efficient Economy: www.aceee.org
- Canadian Association of Energy Service Companies: www.ardron.com/caesco
- Energy Star equipment and buildings: www.energystar.gov
- Energy Star Purchasing Initiative Toolkit: www.epa.gov/nrgystar/purchasing
- Environmental Purchasing Guide (superb): www.pprc.org/pprc/pubs/topics/envpurch.html
- Environmental Purchasing (Goteborg, Sweden): www.cities21.com/egpis/egpc-056.html
- European Good Practice Guides: www.cities21.com/europractice
- King County, Seattle Environmental Purchasing Program: www.metrokc.gov/procure/green
- National Association of Energy Service Companies: www.naesco.org
- Public Employees for Environmental Responsibility: www.peer.org

Three sub-compact fluorescent coil light bulbs.

DOE

How Wasteful Is Your Lighting?
In Newcastle, Australia, the City Hall's Function Center was costing $105,000 a year in electricity through the use of 380 100-watt incandescent bulbs. They're 3% efficient, and waste 97% of their energy as heat. Over a two-year period, the city invested $20,000 in a switch to 18-watt compact fluorescent bulbs, reducing the Center's energy consumption by a staggering 80%, saving 100 tons of CO2 per year, and recovering the cost within 2-3 years. In 1995, they set up the Newcastle Green Energy Project with a revolving loan fund to finance further energy saving programs.

How Warm Is Your Pool?
All that warm water can eat up a lot of energy, but with some thought, you can reduce it by up to 50%. Denver invested in four pool covers for use when the pools were closed, reducing their energy costs from $13,200 to $3,500, with a 3.6 year payback. Solar water heating is another sound investment. In the District of Columbia, Marie Reed Elementary School installed a solar system for their public, all-electric indoor pool, cutting the energy bill by 50% from April to October, with a 4 year payback. Pools can also save money by installing energy efficient pumps, motors, and conventional heating systems. See solar heating: www.seia.org

How Can You Manage and Finance All This?
That's often the tricky part. Regina, Saskatchewan set up an internal $250,000 fund from which the city's departments can borrow at the current interest rate to finance their energy management, building retrofit and fleet conversion initiatives, and repay the loan through their savings. Toronto has its $23 million Toronto Atmospheric Fund. St Paul finances its projects through no-interest loans of up to $1 million a year from Northern States Power. There's also an approach called performance contracting (see box) in which an Energy Service Company (see below) provides your city or county with a service package. In 1992, Redlands, California, signed a 7-year lease with Honeywell to replace old equipment that cut the city's energy use by 50%, saving them over $600,000 in energy and maintenance costs in the first year alone.

Just One Phone Call Away
An Energy Service Company (ESCO) is a business that develops, installs and finances projects to improve energy efficiency and maintenance costs over a 7-10 year period. The services are bundled into the project's cost and repaid through the savings. When an ESCO undertakes a project, the company's compensation, and often the project's financing, are directly linked to the energy saved through performance-based contracting. A 50% reduction in your city's energy bills might be just one phone call away. Typically, an ESCO will:

- develop, design, and finance energy efficiency projects,
- install and maintain the energy-efficient equipment involved,
- measure, monitor, and verify the project's energy saving, and
- assume the risk that the project will save the amount of energy guaranteed.

Energy Performance Contracts

If comprehensive energy performance contracts were used to retrofit private and public buildings throughout Canada, here's what it would look like:

- $50-70 billion in capital investment, paid for through energy savings,
- 1 million person-years of new employment,
- a permanent reduction of $5 - $10 billion in annual energy costs, and
- a 50 million tonnes per year reduction in CO_2 emissions.
- For Canada and the USA, the reduction would be 550 million tonnes per year.

Source: Federation of Canadian Municipalities

23
Buy Green Power

> **"Many people told me a deregulated electric market would be a disaster for the environment. But I believe with every change, there comes an opportunity. Over the long term, local governments can help create a market for investment in renewable energy."**
> **— John Moot, Chula Vista City Councilman**

You may not know it, but a green power revolution is sweeping across North America. It started in California with Santa Monica, which voted in 1998 to switch to the purchase of 100% green power for all of its city operations. Since then, many cities have followed suit, including Chicago, Palmdale, Chula Vista, San Jose, Oakland, Portland and Santa Barbara, while other cities, which own their own utilities, have started to generate their own green power.

Switch to Green Power
A growing number of municipalities in California are voting to issue Requests for Proposals for the delivery of green power, and choosing the one that suits them best. Santa Monica has a contract with Commonwealth Energy worth $2.2 million a year to supply them with geothermal energy, at an additional cost of $110,000 to $215,000 per year. Santa Barbara is buying 100% of its energy from Go-Green.com, which buys solar, wind and other renewable energy on a daily green power spot sales market. Oakland has become the world's largest municipal green power purchaser with the purchase of 9 MW of renewable energy from ABAG Power, a power purchasing program of the Association of Bay Area Governments. There is a growing number of green power utilities to choose from in the US. California, Pennsylvania, Massachusetts, Rhode Island and New Jersey are all open to competition, with Montana, Illinois, Delaware, Arizona and Michigan soon to follow.

Encourage Solar Energy
The City of Chicago is set to become a national leader in the generation of solar energy through a partnership with the Utility PhotoVoltaic Group, the US Department of Energy, Commonwealth Edison and Spire Corporation (a leading supplier of solar manufacturing equipment). Commonwealth Edison and the City have agreed to buy $8 million worth of PV systems for local installation, in exchange for Spire establishing a PV production and assembly business in Chicago which will produce 3 MW of solar panels a year, generating more than 10 million kWh of electricity and displacing 12,500 tons of CO_2 per year. Chicago has also joined with 47 other local government bodies in a load aggregation effort through which 80 MW out of 400 MW will come from renewable energy by 2005, starting at 3% in 2001 and increasing over time.

Since writing this, California has run into a slight problem, causing the market for renewable energy to stall. Normal service will hopefully be resumed soon.

What Is Green Energy?

And can you trust the label? The non-profit Center for Resource Solutions has established the Green-e Renewable Electricity Certification Program to identify "green" electricity products. The Green-e logo means that:

- At least 50% of the electricity supply comes from the sun, water, wind, biomass or geothermal energy. Hydro facilities less than 30MW or dams certified as low impact.
- The non-renewable part of the product has air emissions equal to or lower than the traditional mix of electricity would have had if you did not switch.
- The company abides by the Green-e Code of Conduct..
- The product does not contain any nuclear power other than that contained in system power purchased for the eligible product's portfolio.
- One year after restructuring, the product must contain at least 5% new renewable electricity, increasing to 10% in the next year, and continuing until 25% of the content is from new renewable resources. See www.green-e.org.

Green-E logo for energy

"It's exciting to think that we can power city facilities using energy from the sun and wind. It is the responsibility of every city to look at how we can help create an energy policy that can be sustainable and environment-friendly." — Harold "Rusty" Fairly, Santa Barbara Council member

Learn from Europe's Green Cities
Växjö, in southern Sweden, is one of 900 cities that belong to Europe's Climate Alliance, whose member cities have committed to protect the Earth's atmosphere and co-operate with indigenous partners to protect the world's rainforests. In 1999, Växjö agreed to reduce its emissions from fossil fuels by 50% by 2010, and to stop using fossil fuels. The city's combined heat and power plant is switching from oil to forest biowastes (recycling the ash to the forest), and local buses and cars are being adapted for ethanol, plus a host of other initiatives. The small town of Hufingen, Germany (pop'n 7,000) plans to meet 25% of its energy needs from locally constructed cogeneration (district heating) plants, biogas, wind, micro-hydro and energy efficiency, reducing CO_2 emissions by 37% by 2000, and 68% by 2010. Vienna, Zurich and Saarbrucken are using progressive power rates to benefit those who use off-peak power and charging more to large users, with considerable success.

Combine Your Buying Power
"Community Choice" legislation enables municipalities to combine their buying power to obtain better prices and purchase renewable electricity for their residents and businesses, giving them the same privilege that large consumers such as auto manufacturers and oil refineries enjoy. Massachusetts and Ohio have community choice programs, but New Jersey and Pennsylvania have prohibited municipal aggregation, under pressure from the large utilities. The concept was developed in Massachusetts by Cape and Islands Self-Reliance (see #12), which negotiated a compact under which 21 towns and two counties signed a combined contract with Select Energy which included energy efficiency and renewable energy. In Ohio, the residents of Parma, a suburb of Cleveland, voted overwhelmingly for community choice in a local referendum.

Generate Your Own Green Energy
There are 2,000 municipally-owned utilities in the US, and every one offers the opportunity to generate renewable electricity. Seattle City Council has voted that 100% of Seattle's new energy needs will be met from renewable sources, and Seattle City Light is now seeking 100MW of renewable energy. The small town of Cedaridge, Colorado, obtains 1 MW of free, clean power from a micro-hydro turbine inserted into its water supply pipeline; municipal utilities in Iowa and Minnesota are installing wind turbines. In Eugene, Oregon, the Emerald Public Utility District burns methane from its landfill to produce electricity; Austin Energy offers a Green Choice option with energy from wind and landfill gas. The Sacramento Municipal Utility District (SMUD) and the Los Angeles Department of Water and Power (LAPWD) both offer major solar electricity programs (see #43). SMUD also offers a "Greenergy" option with power from geothermal energy and landfill gas, while the LAPWD offers a "Green Power for a Green LA" Program with energy from the sun, wind and hydro. The best source of up-to-date news is the Green Power Network (see box).

Resources

American Local Power Project: www.local.org

European Climate Alliance: www.klimabuendnis.org/kbhome/kb_memb.htm

Go Green Power (California): www.gogreenpower.org

Green Power Network: www.eren.doe.gov/greenpower

Green Power Project (Pennsylvania): www.pennfuture.org

Ratepayers for Affordable Green Electricity: www.citizen.org/CMEP/RAGE

Renewable Energy Policy Project: www.repp.org

Solstice, Sustainable Energy On-Line: www.solstice.crest.org

Växjö, Fossil-Fuel Free: www.kommun.vaxjo.se/agenda21/environmentalinfo/engfff.html

24
Make Your Whole City Efficient

"**RECO has reduced the amount of energy the average home uses in this city by more than 15%, and we've done it without costing the city treasury a dime." — Lesley Stansfield, RECO Administrator, San Francisco.**

What if your whole city had the will to make its operations more efficient and climate-friendly? What if you knew there were citizens, schools, churches, colleges, local societies, banks, engineers, farms and businesses all deeply concerned about the future impacts of global climate change, and keen to do something about it? What then might be possible? In time, can a whole city become environmentally sustainable with almost zero CO_2 emissions?

That's the question the citizens of Freiburg (180,000) in southern Germany asked in the 1980s, after a coalition of farmers and students prevented the building of a nuclear power plant at Wyhl. Then the vision of the Freiburg Solar Region began to take shape. Today, the region is full of low energy houses, solar collectors, photovoltaics, wind, hydro and biomass energy sources. The International Solar Energy Society has made Freiburg its home. The world's first solar house with long-term hydrogen storage is there. A local company, "SolarFabrik," makes 25% of Germany's photovoltaic cells, and a Communal Photovoltaic Facility has been set up, enabling every citizen to become a shareholder. Europe's largest solar research facility, with 350 staff, is also there. Freiburg has 10,000 people working in the environmental sector, and ecotourism is taking off. Freiburg's goal is a 25% reduction in the city's 2 million tons of CO_2 emissions by 2010, which they aim to achieve by energy conservation (51%), cogeneration (28%), renewable energy (14%) and traffic measures (7%). The city has invested $5.6 million in energy efficiency and renewable energy, saving $18 million, and a car-free development has been built at Vaubon (see #27).

A Greenpeace report, Unlocking the Power of Our Cities, shows that 66% of Britain's electricity production could be generated by solar photovoltaics if it were used on all new buildings, and substituted for existing building surfaces such as facades, glass roofing, parapets, glazed stairwells and roofs. Each square meter of photovoltaics on a building in the UK will, over its lifetime, displace a tonne of CO_2.

In North America, the cities of Austin, Berkeley, Chicago, Okotoks, Portland and Toronto come closest to Freiburg in their breadth of vision, but a surge of interest is arising all over the continent as people wake up to the nature of the crisis and the devastating consequences that await us as a planet if we fail. Cities that are well advanced in energy, CO_2 reduction and other urban sustainability programs are being invited to participate in a worldwide "Solar City" partnership program, to demonstrate greenhouse gas reduction.

Resources

Berkeley's RECO: www.ci.berkeley.ca.us/energy/RECO.html

Building Codes Assistance Project: www.bcap-energy.org

Energy Codes: www.eren.doe.gov/EE/buildings_energy_codes.html

Fort Collins REACH Program: www.fcgov.com/utilities/reach.asp

Portland's BEST program: www.ci.portland.or.us/energy/bestmain.html

Portland's Local Action Plan on Global Warming: www.ci.portland.or.us/energy/co2.htm

Portland's Residential Conservation program: www.ci.portland.or.us/energy/residential.htm

Portland's Sustainable City Principles: www.ci.portland.or.us/energy/principles.htm

San Francisco's RECO: www.eren.doe.gov/consumer info/refbriefs/engycode.html

Solar City partnership: www.solarcity.org

Sustainable Okotoks (Alberta): www.town.okotoks.ab.ca/text/sust2.htm

Unlocking the Power of Our Cities: www.greenpeace.org/~climate/renewables/reports/unlock.html

Portland City Hall, whose energy efficiency has improved by about 25%

Learn from your Peers

The City of Oslo, Norway, controls its local electricity rate. In 1982, it raised the rate by 1 ore (1/6th cent) per kWh and used the $10 million a year revenue to initiate cost-effective energy conservation projects that would otherwise go unfunded. The fund has grown to $100 million and has provided loans to 6,000 projects, with a 15-year repayment period.

In Portland, Oregon, the city's Energy Office runs a program called Business for an Environmentally Sustainable Tomorrow (BEST). The office's staff help local businesses apply for energy retrofit funding, and develop transportation alternatives for their employees. From 1990-1995, with assistance from ICLEI, they helped 300 businesses save $3.5 million, avoid 1.3 million miles of travel, and eliminate 15,200 tons of CO_2.

In Fort Collins, CO, with help from ICLEI, the city's REACH program (Residential Energy Assistance Through Community Help) provides free weatherization services for low-income city residents through voluntary contributions from city utility customers' monthly payments.

New York has approved guidelines allowing homeowners to install rooftop grid-connected PV systems and net-metering, and Boulder, CO, has adopted a solar-access ordinance, setting area-wide rules that protect a building's access to sunlight. Honolulu and Kauai have adopted building energy efficiency codes based on Hawaii's Model Energy Code, designed to reduce overall energy use.

San Francisco has raised the efficiency of its homes by 15% since 1981, for no cost. The Residential Energy Conservation Ordinance (RECO), modeled on Berkeley's RECO, prescribes a range of energy-saving measures such as added insulation, weather-stripping and low-flush toilets. The ordinance is triggered whenever a house is sold, a metering conversion takes place, home improvements are made above a certain cost (including hotel rooms), or a building inspection is needed, and an Order of Abatement can prevent the transfer of the property unless the owner complies. There was opposition from the real estate community at first, but with publicity, training workshops and the involvement of the private sector, RECO has become an accepted part of doing business in San Francisco.

Where there's a Will, there's a Way

Osage, Iowa has raised efficiency standards in its building code, and refuses to supply utilities to buildings that don't meet the standard. Since 1984, every building in town has complied. Santa Barbara County rewards developers who exceed the minimum energy-saving requirements of the building code by allowing them to jump ahead in the queue for approvals.

A Tale of Three Cities

Fort Collins, Colorado (population 108,000)

City-wide CO_2 emissions per person per year, 1990	15.5 tons
Business-as-usual emissions per person per year, 2010	24.5 tons
Greenhouse Gas Reduction Target: 30% below 2010	17.2 tons
Kyoto Protocol: 6% below 1990 by 2010	14.6 tons
Needed to prevent 2 C global temperature rise: 75% below 1990	4.1 tons

Source: Fort Collins Greenhouse Gas Reduction Plan

Freiburg, Germany (population 180,000)

City-wide CO_2 emissions per person per year, 1987	10.8 tons
Business-as-usual emissions per person per year, 2010	11.4 tons
Climate Plan, cost-positive potential for reduction: 25%	8.1 tons
Climate Plan, technical potential for reduction: 60%	4.3 tons
Needed to prevent 2C global temperature rise: 75% below 1990	2.7 tons

Source: Freiburg Comprehensive Climate Plan

Portland, Oregon (population 550,000)

City-wide CO_2 emissions per person per year, 1990	18.3 tons
1997 forecast for 2010 following CO_2 initiatives: 10% below 1990	16.4 tons
CO2 Reduction Strategy goal: 20% below 1990	14.6 tons
Needed to prevent 2C global temperature rise: 75% below 1990	4.5 tons

Source: Portland CO2 Reduction Strategy

25
Fill Your City With Greenery

Sacramento Municipal Utility District

Trees planted as part of the Sacramento Shade Program help reduce air-conditioning costs.

Trees give us such pleasure. They unlock something spiritual, and connect us to something deep in our ancestral memory, when the whole Earth was covered with trees.

> *I think that I shall never see*
> *A poem lovely as a tree.*
> *Poems are made by fools like me*
> *But only God can make a tree*
> — Joyce Kilmer, 1888-1918

Trees also provide shade and cool the air. They shelter us from the heat of the sun, a service they perform silently and freely. As they grow, they absorb carbon dioxide, storing the carbon and releasing the oxygen, a process they have fulfilled for as long as they have been on Earth.

There are some who argue that trees are untidy and just make litter; they must have had tree-deprived childhoods. For most, trees are a wonder. They are also a necessary part of any strategy to reduce carbon dioxide emissions because they store carbon, cool the air and reduce the need to generate electricity for air-conditioning.

Question: What are the cheapest ways to reduce the summer peak electricity load?

(a) Turn off the air conditioning
(b) Generate more energy
(c) Improve energy efficiency
(d) Plant trees

Answers: (a) and (d). Generating new energy costs 10 cents to reduce a 1kWh load. Improving efficiency costs 2.5 cents. Planting trees costs 1 cent. 100 million additional mature trees in US cities would save $4 billion per year in energy costs (Arbor Day Foundation).

How much CO_2 does a tree absorb? That depends on the tree. American Forests estimates that the average tree will absorb 46 lbs of CO_2 per year, and an average acre of forest will absorb 4.5 tons of CO_2 per year. Tree Canada estimates that an average Canadian tree will absorb 20 lbs of CO_2 in an urban setting and 23 lbs in a rural setting. The numbers vary because trees vary, and forests vary.

In many places, we are losing ground. In the Puget Sound region, from 1972-1996, overall heavy tree cover fell by 37%. In Chesapeake Bay, it fell by 34%. In Washington DC, it fell by 64%. In Chattanooga, it fell by 50%. In Fairfax County, VA, it fell by 41%. New York lost 20% of its urban forest (175,000 trees) in the 1990s.

Gaining Ground: Sacramento, City of Trees

Annual electricity use in Sacramento for air conditioning:	1,300 GWh
Annual cost to Sacramento residents:	$105 million
Annual electricity savings from shading and cooling by Sacramento's urban forest:	157 GWh
Annual dollar saving to Sacramento residents:	$19.8 million
CO_2 removed annually by Sacramento's urban forest:	238,000 tons
Number of trees planted between 1990-2000 by Sacramento Shade, a partnership between the Sacramento Municipal Utility District and the Sacramento Tree Foundation:	500,000

Source: www.smud.org

"Properly placed shade trees can lower residential air conditioning bills by as much as 40%."
— Sacramento Shade Program

Start a City Tree-Planting Program

If you want to embark on a program of tree-planting in your town or city, American Forests can help you get started through its Global ReLeaf campaign, and Tree-Canada can help similarly. First, invite interested members of the public to join you on a Task Force with local councilors, city staff, and representatives from conservation groups, nurseries, and your local utility. American Forests will show you how to conduct an Urban Ecosystem Analysis. They'll show you how to use their CITYgreen Software to assess the best approach and build your strategy, including budgeting, setting a tree-canopy goal, developing policies, and carrying out neighborhood planting bees. They will also help you figure out the benefits, from reduced storm drain runoff to city-wide air conditioning and carbon sequestration, data that may win over your city's finance department.

In Dade County, Florida, residents of Richmond Heights worked with Cool Communities, a local non-profit group, to get more trees planted in their unshaded neighborhood. Using aerial images generated by CITYgreen to make their case, they were able to involve 600 volunteers and get 200 trees planted in a five-block area in just three weeks. In Tucson, "Trees for Tucson" aims to plant 500,000 trees by working with homeowners, schools and neighborhood groups. According to the city's calculations, each tree will save 27 kWh annually through cooling and evapo-transpiration, and 61 kWh through direct shading.

Cities need strong tree protection bylaws. Building permits can be made conditional on tree protection. Development proposals can be denied if they do not show enough commitment to protecting trees. Gainesville, Florida, has a strict tree protection ordinance, while nearby Ocala has a loose one, without teeth. Gainesville has twice the tree cover, and its residents spend $126 less each year on electricity.[1]

Imagine neighbors working together on every street in your city to plant trees. Imagine a city map which showed which neighborhoods had more trees and which had less. Imagine an urban tree tax-shift with higher taxes for neighborhoods with less tree canopy and lower taxes for those with more. We have to think big if we are to bring down the Earth's temperature.

Grow More Food

The more food we grow locally, the less energy is needed to ship food in, the more CO_2 can be absorbed by the soil, and the better our nutrition. The City of Seattle provides community garden space for over 45,000 urban gardeners, in more than 1,600 garden plots around the city, encouraged and assisted by Seattle Tilth. See www.seattletilth.org

Resources

American Forests: www.americanforests.org

Canopy (Palo Alto): www.canopy.org

City Farmer's Urban Agriculture Notes: www.cityfarmer.org

Friends of the Urban Forest (San Francisco): www.fuf.net

How to calculate carbon in urban trees: infoghg@eia.doe.gov 1-800-803-5182

International Society of Arboriculture: www.isa-arbor.com

National Alliance for Community Trees: www.actrees.org

National Arbor Day Foundation: www.arborday.org

National Tree Trust: www.nationaltreetrust.org

Shading Our Cities: A Resource Book for Urban & Community Forests, by Gary Moll (Island Press)

Tree Canada Foundation: www.treecanada.ca

Treefolks (Austin): www.treefolks.org

Treelink - Community Forestry Resource: www.treelink.org

Urban Trees sequestration spreadsheet: www.eia.doe.gov/oiaf/1605/techassist.html

26
Reduce, Reuse, Recycle

Ellie Roelofsen, from Victoria, B.C., with the total garbage she threw out from October 1999 to January 2001.

Guy Dauncey

There has never been a culture that consumed as much as we do, whether it's the 25,000,000,000 styrofoam cups Americans use every year, or the 900 million trees cut down every year to provide us with paper. By March 2001, American Forests had planted 17 million trees. Our right hand is destroying what our left hand is so desperately trying to restore.

We create an unbelievable amount of waste, not just from the things we throw away as individuals, but also in their mining, shipping and manufacturing. The amount of waste generated to make a laptop computer is 4,000 times its weight; the materials needed to create a single sheet of paper weigh 98 times more than the paper. When you tally it up, counting everything we do, it comes to two tons of material consumed every day for every man, woman and child in America,[1] and probably the same in Canada. In 1970, Americans recycled 7% of their garbage and sent 113 million tons to landfills. In 1998, they recycled 28%, but threw away 150 million tons. We are not making progress.

Why is this mentioned in a book on climate change? Because energy is used at every stage and most of it comes from burning fossil fuels. The more we consume, the more energy is used, and the higher are the CO_2 emissions. It is imperative that we reduce the amount of stuff we consume and re-use things as often as possible before we discard them. In Seattle, the King County Solid Waste Division has started a "Re-Use" project in partnership with Goodwill. Residents pay the regular disposal rate, and leave re-usable goods, including a surprising amount of furniture, in a 20-foot container.

Consuming less is the top imperative. Recycling far more than we are currently doing is the next essential. We have to evolve from a throw-away society into a zero-waste, closed-loop society. Cities and towns can help by educating people on the critical importance of reducing, recycling and composting. They can aim for 60% recycling by 2008, as Seattle is doing. They can go further, and aim for zero waste to the landfill by 2020, as a third of the councils in New Zealand are doing.

Recycling reduces CO_2 emissions

- Steel made from recycled steel uses only 39% of the energy needed to make virgin steel. 97.9% of automobile steel and 58.2% of the steel in tin cans is recycled.
- Paper made from recycled paper uses only 26-45% of the energy needed to make virgin paper. 40% of paper and paperboard is recycled. For each ton of paper recycled, 17 trees are left standing to gather carbon.
- Aluminum made from recycled aluminum uses only 5% of the energy needed to make virgin aluminum. 52% of aluminum packaging and 64% of aluminum cans is recycled.
- Plastic made from recycled plastic uses only 43-25% of the energy needed to make virgin plastic and releases far fewer toxic chemicals. Only 5% of plastics is recycled.

Source: Natural Resources Defense Council

"Recycling means jobs. Landfilling resources wastes jobs. It is time to plan for zero waste — for our communities and for future generations."
— San Francisco Mayor Willie L. Brown, Jr.

How Far Can We Go?

The current recycling rate in American municipalities is around 28%. If that increased to 65%, it would reduce greenhouse gas emissions by 25 million tons a year. Several New Jersey communities recycle 65%, and Hornby Island in British Columbia recycles 66%, so it is clearly possible. Here are some options that can speed up the re-use, and recycling of materials:

- Nebraska's Materials Exchange Program provides an elegant way for businesses and schools to exchange unwanted materials such as computers, chemicals and wooden pallets. www.knb.org
- Mount Vernon, Iowa's "Pay as you throw" program cuts the amount of trash going to the landfill by 40% and almost doubles the recycling rate by giving people a cash incentive to reduce and recycle. www.epa.gov/payt
- Guelph, Ontario uses a city-wide wet-dry system, recycling 55-60% of the dry waste and 60-65% of the wet waste. www.compost.org/guelph.html
- Dublin, CA's Construction and Demolition Recycling Ordinance requires a waste management plan as part of the permitting process for projects over $100,000, demonstrating that 50% or more of the materials will be diverted from the landfill. www.dublinca.org
- San Jose's Source Reduction and Recycling Procurement Policy is expanding the market for goods made from recycled materials. www.sjrecycles.org
- Oakland/Berkeley's Recycled Market Development Zone channels loans and economic development support to businesses using recycled materials. www.ciwmb.ca.gov/RMDZ
- A 1995 study of the Tri-Cities by the Institute for Local Self-Reliance showed that a ton of waste returned $40 in municipal revenue when landfilled, $120 when recycled, and $1,110 when remanufactured. www.ilsr.org

"The earth provides for every man's need, but not for every man's greed."
— Mahatma Gandhi

Resources

Wasting and Recycling in the US, 2000, by Brenda Platt and Neil Seldman (Institute for Local Self-Reliance). $25 from GrassRoots Recycling Network, PO Box 49283, Athens, GA 30604-9823.

Composting Council of Canada: www.compost.org

Global Recycling Network: www.grn.com

Grassroots Recycling Network: www.grrn.org

EPA, Climate Change and Waste: www.epa.gov/globalwarming/actions/waste

EPA's Jobs Through Recycling: www.epa.gov/jtr

National Recycling Coalition: www.nrc-recycle.org

National Waste Prevention Coalition: http://dnr.metrokc.gov/swd/nwpc

Re-use Development Organization: www.redo.org

Recycling and GHG Reduction Spreadsheet: www.eia.doe.gov/oiaf/1605/techassist.html

Recycle City — Turn Dumptown into Recycletown (educational game): www.epa.gov/recyclecity

Reduction and Re-Use / The Cygnus Group: www.cygnus-group.com

ReThink Paper: www.rethinkpaper.org

Re-use Development Organization: www.redo.org

Use It Again, Seattle!: www.cityofseattle.net/util/useitagain

Waste Prevention World: www.ciwmb.ca.gov/wpw

Waste Reduction CO_2 spreadsheet: www.environmental-expert.com/freeware/warm.htm

Waste to Wealth — Waste Reduction Record Setters: www.ilsr.org/recycling/wrrs.html

Zero Waste America: www.zerowasteamerica.org

Zero Waste Canada: www.targetzerocanada.org

Zero Waste New Zealand: www.zerowaste.co.nz

27

Build a Sustainable Transportation System

So now we come to the big one: the massive transportation system we have built up around us. Globally, transportation pours out 21% of the world's CO_2 emissions from human activities. In Ottawa, it is responsible for 31% of the emissions. In Portland, 37%; in Fort Collins 51%; Austin 30%; Berkeley 45%; Miami-Dade 38%; Saint Paul – Minneapolis 32%; San Francisco 25%.

And the problem is growing, as people move to the suburbs, increase the length of their commutes into work, and then hit the road at the weekend in search of those glorious parks and mountains which are shown every night on TV in the sports utility vehicle adverts. Gasoline releases 20 lbs of CO_2 for every gallon burnt (2.5kg of CO_2 for every liter of gasoline), so for a car that does 20 mpg, that is a pound of CO_2 for every mile traveled – and the roads are full, full, full, right across the country. And yet for most trips, cars are still so convenient, with their lush sound systems and room for all the stuff people like to ferry around.

How ever are we to turn this colossus around? The carbon we are burning is melting the Arctic ice that has been stable for millions of years — and that is just the smallest harbinger of things to come.

There is, on the horizon, the prospect of a different, more sustainable transport system. In this future world, fast, efficient trains and transit systems will link our communities and carry us on the majority of trips. Cycling will become safe and easy. Neighborhoods will be redesigned as pedestrian communities with their own village centers. Cars and trucks will be powered by hydrogen derived from algae or the sun, and be charged the full price of the roads and parking space they use. Personal electronic 'smart' cards will give us entry to a world rich in transportation options. As well as the travel flexibility we seek, such a system will clean the air, reduce the terrible litany of deaths and injuries on the road, bring city centers back to life, and restore the culture of the street which brings so much pleasure to civilization.

Such a system is already taking shape in parts of Switzerland and Denmark; it can happen here too, if there is the will. It requires vision, persistence — and a willingness to invest in bike-routes, transit and other travel alternatives, instead of roads.

Toronto's St. Roch School — Walking Wednesday.

Greenest City's Active & Safe Routes to School Program

Farewell, Free Parking

In California, some businesses pay their employees a monthly travel allowance and then charge the full market price for parking (known as 'cashing out'). Typically, 10 – 30% of employees switch to walking, cycling, living nearby, ridesharing, telecommuting or using transit, pocketing the allowance. The company then sells or leases the now empty parking spaces.

In 1991, Washington State introduced a Commuter Trip Reduction law (revising it in 1997), requiring employers with more than 100 employees to reduce their peak-period trips by 25% by 1999, and 35% by 2005. www.ga.wa.gov/ctr

> **"Current pricing fails to return to individual motorists the savings that result when they drive less."**
> **— Todd Litman,**
> **Victoria Transport Policy Institute**

Piece Together the Jigsaw Puzzle
The transition to a sustainable transport system is like a massive jigsaw puzzle, with many pieces. But like a jigsaw, there must be a picture — a Transportation Masterplan. Here are some of the pieces which are beginning to make a difference around the world:

Portland, OR, has a Trip Reduction Incentives Program which has enticed 92% of the surveyed users (city employees) to reduce their drive-alone trips to work, and increased transit work trips from 60% to 83%. Like many European cities, they have electronic timetables at their bus shelters. Portland has numerous other transport components in its CO_2 Reduction Strategy.

Freiburg, Germany, has laid 250 miles of bicycle path, introduced a single transit pass covering 90 different train, bus and tram routes, ripped up roads and slashed the price of transit by 30% (www.iclei.org/egpis/egpc-023.html). They have created pedestrian shopping areas — which the merchants like because shoppers stay longer when they don't have to feed the meter — and built a car-free settlement at Vaubon, where residents pay extra if they want to use a car.

Victoria and Toronto have "Safe Routes to School" projects, which enable parents to feel ok about letting their children walk to school, instead of driving them(www.waytogo.icbc.bc.ca and www.goforgreen.ca).

In Maplewood, New Jersey, residents helped to develop an innovative jitney bus service to get commuters to the Manhattan train, instead of building an expensive parking structure. It proved so popular that other communities along the line are developing services. (Thanks to ICLEI.)

When Hasselt, Belgium (68,000 residents, 200,000 commuters) was short of funds and losing people, the town abandoned plans to build a third ring road, closed one of the two existing ring roads and planted trees in its place, laid more pedestrian walkways and cycle routes, increased the frequency and quality of the bus service – and announced that public transport would be free. Over the course of the next year, the use of public transport increased by 800% and business flourished, resulting in fewer accidents and road casualties and more social activity. With the increase in business, the mayor slashed local taxes, and the town is gaining new residents 25 times faster than it was losing them.

Edinburgh, Scotland, is planning to introduce electronic road-pricing on all roads leading into the city, after several years of investing in park & ride, public transit and rail. The city will borrow the finance for the investments, and repay it through road tolls.

Resources

Association for Commuter Transportation: http://tmi.cob.fsu.edu/act

Automobile Trip Reduction Programs for Municipal Employees: www.iclei.org

Better Environmentally Sound Transportation (Vancouver): www.best.bc.ca

Calculating local auto-subsidies: www.iclei.org/co2/auto/cars.htm

Car-Free Cities: www.carfree.com

Center of Excellence for Sustainable Development: www.sustainable.doe.gov

Community Guide to Traffic Calming: http://pti.nw.dc.us/task_forces/transportation/docs/trafcalm

Citizens Guide to Traffic Calming: www.lgc.org/clc

Greenest city: www.greenestcity.org

'Packing Pavement': www.tampabayonline.net/bguard/home.htm

Portland's Traffic Management: www.trans.ci.portland.or.us/Traffic_Management

Rides Program (commuter trip reduction, San Francisco): www.rides.org

Surface Transportation Policy Project: www.transact.org

Telework Guidelines: www.ci.portland.or.us/energy/teleworkguide.html

Transportation Alternatives (New York): www.transalt.org

Transportation for Livable Communities: www.tlcnetwork.org

Victoria Transport Policy Institute: www.vtpi.org

28

Encourage Cycling and Walking

"Every time I see an adult on a bicycle, I no longer despair for the future of the human race."
— H.G. Wells

The most cost-effective way to reduce your town or city's CO_2 emissions from vehicles is to encourage people to walk or cycle, instead of driving.

The trend to more driving and longer trips is not an inevitable aspect of progress. It is a learned response to zoning, financing and transportation decisions made over the years that have emphasized building roads for cars, and paid very little attention to the alternatives.

The financial realities behind our love affair with the car are quite an eye-opener. The taxes that motorists pay at the pump are used to fund major highway expenditures. Two-thirds of local road costs are funded through local taxes, which everyone pays, non-drivers included. When you add the cost of police time for traffic management and law enforcement, emergency services for road accidents, subsidized parking, and the cost of sicknesses caused by air pollution, it really adds up. Studies have shown that the average subsidy per car is around $1,700 per year – that's 15 cents a mile, or an extra $3 to $4 on a gallon of gas.[1] If people had to pay the true cost of driving, they would think twice before jumping into a car. At the very least, they would buy more fuel-efficient vehicles.

A city parking space costs around $10,000; $20,000 in a parkade. If 5,000 commuters switch to cycling as a result of a decision to invest in bicycle lanes, this frees up $50 to $100 million worth of parking spaces for more profitable uses. Assuming a round trip of 10 miles, a bicycle-commuter prevents 1 ton of CO_2 from entering the atmosphere every year.[2]

Pedestrians, meanwhile, bring culture to a city while reducing CO_2 emissions. When was the last time you stopped for a chat when passing a friend in your car? A truly walkable city puts care into the design of its pedestrian routes, widens the sidewalks, builds sidewalk bulges at intersections, creates open-air pedestrian shopping centers, and encourages pedestrian advocacy groups.

Cycling into the Future

In Koga, Japan, the city has bought 20 bikes for official use by city staff.

In Sutton, London, city staff receive $1 per mile cycling allowance.

In Palo Alto, CA, city staff receive 7 cents per mile cycling allowance.

In Los Angeles, and in Saanich, British Columbia, all new commercial developments must include bicycle parking, showers and clothing lockers.

In Denmark, 25% of all trips are done by bike.

In 1999, 3.2 million Americans biked to work at least once a week, a 100% increase since 1985.[3]

A Pedicab in New York

George Bliss

Lindsay Hill

Cycling the CamelTrail in Cornwall, UK.

Establish a City-Wide Cycling Strategy

A good way to develop a cycling strategy is to establish a Cycling Task Force and appoint a Cycling Coordinator, with a protected budget. Each $1 invested in cycling saves $2 in car-related costs, so the logical size for the budget should be "twice the goal". If the goal is 10% of trips in the city to made by bicycle, the budget for cycling should be 20% of the transportation budget. It is essential that the city's engineers get out of their cars and into the saddle to give them a feel for the obstacles and difficulties that cyclists face. If there is not already a cycling advocacy group in your town or city, it would be a sound investment to help cyclists set one up.

Some of the elements of a bicycle-friendly city include bicycle lanes on all major roads; priority bicycle routes through quiet residential streets; bike racks on all the buses; bicycle-controlled traffic signals; cycling maps; bicycle deliveries; cycling education programs; free electric charging stations for electric bikes; and an active cycling advocacy group. A cycling city is a quieter, safer, cleaner, more affordable, more climate-friendly, more people-friendly city.

In 1999, *Bicycling Magazine* named Montreal the best city in North America for cyclists, followed by Portland (Oregon), Tucson, Seattle and Toronto. Montreal has 230 miles of bike paths and lanes, many of which are marked by concrete barriers and permanent pylons, with routes across busy city intersections painted bright blue. At the end of May each year, 10,000 children take part in "Le Tour des Enfants," a 12-mile ride for kids under the age of 12, ending in a massive barbecue. The following weekend, the city hosts "Le Tour de l'Ile," the largest organized bicycle ride in the world, when 45,000 cyclists cruise a 40-mile loop through the city and residents celebrate along the route with parties and yard decorations, the proceeds going to cycling advocacy groups and bike programs. Outside Montreal, there is La Route Verte, a 1,800-mile network of bicycle paths and lanes through the countryside that is due for completion in 2004.

Resources

America Walks: www.americawalks.org

Association of Pedestrian and Bicycle Professionals: www.apbp.org

Bike Cart Age: www.bikecartage.com

Cycling Links: www.bts.gov/smart/links/bike.html

Improving Bicycling Conditions in Your Community: www.bikeplan.com

La Route Verte, Quebec: www.velo.qc.ca

League of American Bicyclists: www.bikeleague.org

Mainstreet Pedicabs: www.pedicab.com

National Center for Bicycling and Walking: www.bikefed.org

Organizing a Bicycle Commuter Contest: www.climatesolutions.org/commuter-contest.html

Pedestrian and Bicycle Planning: a Guide to Best Practices: www.vtpi.org

Walkable Communities: www.walkable.org

Why Bicycle Trailers?: www.bikeroute.com/WhyTrailer.htm

"Modern society will find no solution to the ecological problem unless it takes a serious look at its lifestyle."
— Pope John Paul II

29

Stop Sprawl: Plan for Smart Growth

If we continue to allow our cities to spread into the surrounding farmland and green space, we will never stop the flow of greenhouse gases from the increased number of vehicle trips and the loss of carbon-storing soil and forests.

In the US, green space is being lost to low density sprawl at 365 acres an hour. The lower the housing density per acre, the more roads have to be built, the harder it is to make transit work, the more it costs to run our communities, and the further people have to travel to get anywhere. We cannot control greenhouse gas emissions unless we also control sprawl.

There is an alternative: old-fashioned, compact, pedestrian-oriented communities, the way towns used to be in the old days, where you live surrounded by green space, walk into town to do your shopping, meet your friends, and pause to chat in a neighborhood café. The kind of community where you can cycle around with ease, and find work in the local economy. Vermont is full of this kind of community, and Vermonters love them. The solar community of Civano, outside Tucson, is being built this way, and sales are booming (see www.civano.com).

Well-planned, compact growth also costs less for roads, utilities and schools (see box). Who knows – it may even be that people who live in old-fashioned towns produce fewer greenhouse gas emissions because they spend less time buying "stuff" to fill up the emptiness of life in a barren, amenity-deprived suburb.

There are seven major solutions to sprawl. There is not space to amplify them here, but you will find everything you need in the Resources box.

1. Establish Comprehensive Planning and Urban Containment Boundaries.
Map your region for its ecological assets. Develop an overall plan of where you want to be in 50 years, set urban containment boundaries to stop growth eating into the rural areas, and plan long distance greenways. Portland (OR) started doing this in 1973 and has benefited enormously, helped by the "1000 Friends of Oregon" group. Delaware, Maryland and New Jersey are also moving in this direction.

2. Write Model Development Codes.
Development Codes can be used to stipulate the need for prior ecological assessment, compact pedestrian communities with neighborhood centers, affordable housing, greenways, community-based economic development, energy and water efficiency, and a greenhouse gas reduction strategy. Set your development cost charges high, and then reduce them for each aspect of sustainability that a developer meets. Woodsong, North Carolina, and other new urbanism developments are moving in this direction.

"It's a simple but powerful fact that the faster the rate of sprawl, the faster the rate of disinvestment in the inner city."
— Zack Semke, Coalition for a Livable Future, Portland

Smart Growth is Smart Economics

- Compact, well-planned growth consumes about 45% less land, and costs 25% less for roads, 20% less for utilities, and 5% less for schools.[1]
- Fresno, CA, doubled in size 1980 – 1998. Increase in city revenues per year: $56 million. Increase in cost of services per year: $123 million.[2]
- Cost of transportation for cities that prioritize freeways and sprawl: 12% - 16% of gross regional product. Cost of transportation for cities that prioritize transit and more compact development: 4% - 8% of gross regional product.[3]

3. Prevent the Loss of Farmland and Green Space.
Protect farmland through designated protective zoning. Use "Purchase of Development Rights" programs to compensate farmers for not developing their land, and use tax dollars and bonds to buy threatened green space.

4. Encourage the Formation of "1000 Friends" Groups.
1,000 Friends of Oregon is a nonprofit citizens group that was founded in 1975 to protect Oregon's quality of life through the conservation of farm and forest lands, the protection of natural and historic resources, and the promotion of livable communities. Similar groups exist in Florida, Hawaii, Minnesota, New Mexico, Washington, Wisconsin and Pennsylvania, where they provide leadership, publish Sprawl Report Cards, educate the public, and go to bat in defense of good planning. See www.friends.org

5. Redevelop urban areas.
Tax land instead of houses, to encourage more intensive use of valuable urban land (site-value taxation), as Pittsburgh does. Establish downtown tax incentive "renaissance" zones, as Grand Forks, North Dakota, is doing. Ask your banks to use "Location Efficient Mortgages," as Seattle is doing. Use codes and homebuyers' incentives to encourage the redevelopment of urban areas.

6. Stop proposals that encourage sprawl.
Say no to the big-box retail stores, whether it's K-Mart, Wal-Mart or Home Depot, as 86 communities across the US have done. So too has Norway, where a nation-wide ban on out-of-town big-box stores is in place, supported by 443 out of 450 municipalities. Ireland has also banned out-of-town big box stores. Sprawl-Busters can offer resources, help, and instances of successful legal and zoning decisions. See www.sprawl-busters.com

7. Share the revenues.
Make a pitch for regional consolidation, as Indianapolis and Montgomery County, Maryland, have done. Develop methods to assess the true costs and benefits of development proposals, as New Jersey and Delaware are doing. Eliminate the hidden subsidies that encourage sprawl. Share the revenues from new property taxes with neighboring communities, as Minneapolis-St. Paul does.

Smart Growth Resources

(see also Solutions #66 & #77)

Stop Sprawl Resources for Activists: www.sierraclub.org/sprawl/resources

Austin, TX, Smart Growth Matrix: www.ci.austin.tx.us/smartgrowth

Bay Area Greenbelt Alliance: www.greenbelt.org

Better, not Bigger: How to Take Control of Urban Growth and Improve Your Community, by Eben Fodor: www.newsociety.com

Car-Free Settlements in Europe: wwwistp.murdoch.edu.au/research/car free.html

Center for Livable Communities: www.lgc.org/clc

Congress for the New Urbanism: www.cnu.org

Location Efficient Mortgages: www.locationefficiency.com

Minneapolis smart growth resources: www.me3.org/projects/sprawl

New Urban News: www.newurbannews.com

Smart Growth America: www.smart-growthamerica.com

Smart Growth Network: www.smartgrowth.org

Sprawlwatch Clearing House: www.sprawlwatch.org

Sustainable Subdivision Report Card: www.earthfuture.com/consultancy/real estate.asp

30

Build a Sustainable City

"Roll up! Roll up! Join your city's CO2 reduction strategy"

It doesn't sound very sexy, does it? Not as exciting as a fun-run to raise funds for children with leukemia. The world is slowly cooking in its own heat, but it is a lot easier to get people motivated about a sick child than a sick planet. People know that it's happening, but they feel overwhelmed by the size of the problem.

It is easier to motivate people with a compelling vision of a sustainable city. So pause, and picture your town or city in the year 2025. Urban creeks that have been "daylighted" from their underground journey, with winding footpaths by their sides. Schoolyards full of trees and gardens. Rooftop gardens. Pedestrian and cycle paths that criss-cross the city. Residential streets that have been traffic-calmed, creating mini-parks. Pleasant and efficient transit and light rail, using smart-cards which double up to rent a bicycle or a small neighborhood vehicle. Community banks that lend to local businesses, and encourage job creation. A downtown filled with artists, musicians, festivals, galleries, and gorgeous architecture. Fountains, water sculptures, neighborhood wetlands. Water-parks for children. Solar roofs. Solar windows. Solar walls. Electricity from renewable sources. Sewage that is treated in aquatic bird sanctuaries, or in greenhouses filled with plants. Recycling systems that catch 95% of the city's wastes. Materials exchange networks. Houses and offices that are constructed from "green" building materials. A city managed as a single, connected ecological organism. And trees everywhere, cooling the city with their leaves, catching the breeze. Suburbs that have been transformed by narrowing the streets, adding granny suites and neighborhood centers, inserting pedestrian trails and mini-parks, and turning down the streetlights. Outside the city, protected by an urban containment boundary and accessible by a network of greenways, the farms, forests and river banks where you can see where your food comes from, and rediscover the dreams of childhood. A sustainable city.

Resources for a Sustainable City

Center for Urban Ecology (Australia): www.urbanecology.org.au
Center of Excellence for Sustainable Development: www.sustainable.doe.gov
Center for Sustainable Cities: www.uky.edu/~rlevine
Cities as SuperOrganisms: www.oneworld.org/guides/thecity
"Cities of Exhuberance" *YES! Magazine*, Summer 1999
Civano, Tucson: www.civano.com
EcoCity Builders: www.preservenet.com/EcoBuild.html
EcoCity Cleveland: www.ecocleveland.org
European Sustainable Cities Project: www.sustainable-cities.org
Five E's Sustainable Cities: www.eeeee.net/sd03087.htm
Green Communities Assistance Kit: www.epa.gov/greenkit
Hamilton-Wentworth (Canada) Vision 2020: www.vision2020.hamilton-went.on.ca
London 21 Sustainable Network: www.london21.org
Oakland Ecopolis: www.wtp.org/ecopolis.html
San Francisco's Sustainable City Plan: www.sustainable-city.org
Santa Monica's Sustainable City: www.ci.santa-monica.ca.us/environment/policy
Sustainable Chattanooga: www.chattanooga.net/sustain
Sustainable Seattle: www.scn.org/sustainable
Sustainable Urban Neighbourhood (UK): www.urbed.co.uk/sun
UN's Sustainable Cities Program: www.unchs.org/scp
Urban Ecology: www.urbanecology.org
Whyalla EcoCity (Australia): www.whyalla.sa.gov.au/enviro

Rivercity Company

Chattanooga's new riverside walk

See the City as a Whole

Believe it or not, every attribute described above will help lower CO_2 emissions. This is how closely the vision of a sustainable city and the need to reduce fossil fuels are linked.

In Chattanooga, "the dirtiest city in America," citizens began organizing in the early 1980s, starting with a series of free downtown concerts and a program called "Vision 2000." Organized by the Lyndhurst Foundation (note the role of the non-profit society), city residents were invited to imagine a future for their city. They named 40 goals and proposed 233 solutions. Here's some of what has happened since then: A 22-mile riverside walk along the Tennessee River that cuts through the heart of downtown, with playgrounds, performance spaces, fishing piers, and leaf-shaded walkways. A river from which you can eat the fish. The $45 million Tennessee Aquarium, which focuses on freshwater ecosystems and serves as a magnet for the downtown area. Around it, a revitalized shopping district of converted warehouses and 3,500 units of inner-city housing. A free, 17-person electric shuttle bus that picks up commuters from satellite parking areas on the edge of town. A business (Advanced Vehicle Systems) that exports 22-seater electric buses to other cities around the world. City manufacturers who have been obliged by the Clean Air Act to invest $40 million in pollution control equipment. In the pipeline are a grass-roofed trade center, and an eco-industrial park designed to attract zero-emissions industries. There is a long way to go, but there is an excitement in the air that comes from the synergy of all these things.

Freiburg, Växjö, Santa Monica, San Francisco, Seattle, Portland, Toronto: these cities and others are all embracing the vision of a sustainable city, and discovering the power that the vision has to energize their citizens. Now picture the process starting in every town and city. Picture a ten-fold amplification of the process, as every neighborhood, school and public institution develops its own goals and projects. Picture every business signing up with an eco-business initiative such as Cool Companies, Climate Wise (#31), Climate Neutral (#35) or The Natural Step (#40). Picture residents on every street working together to reduce the use of pesticides, install solar roofs and plan local traffic calming. Picture citizens in cities and towns all over the world who have decided to take the future into their hands, and build a sustainable world.

> **"It's not one particular project. It's a series of things that are happening simultaneously."**
> **— Karen Hundt, urban designer, Chattanooga**

Tools for a Sustainable City

- QUEST is a user-friendly computer-based modeling tool that explores alternative scenarios for the future of any given region, including social and ecological policy choices — a Sustainable Sim-City: www.envisiontools.com
- Global Vision's Sustainable City GIS computer simulation: www.global-vision.org/city

EcoCity Magazine: www.ecocitymagazine.com

Green Map System: www.greenmap.com

31

Choose to Become Carbon Neutral

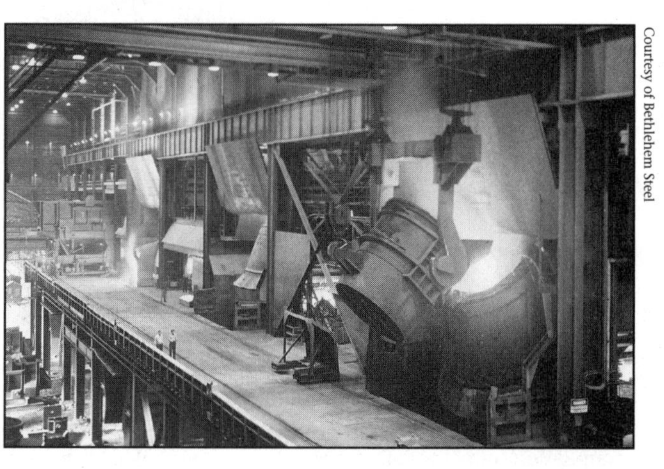
Bethlehem Steel, Burns Harbor facility.

Courtesy of Bethlehem Steel

All over the world factory furnaces are blazing, steam turbines are generating electricity, boilers are heating water, motors are whirring, lights are shining, and computers and office equipment are humming, as companies large and small work around the clock to produce the goods and services we desire. That's a huge amount of power, and most of it comes from burning fossil fuels. Industry releases more than 33% of the world's energy-related CO_2 emissions. How are we going to reduce them?

At the Burns Harbor facility on the shores of Lake Michigan, Bethlehem Steel produces five million tons of steel sheet and plate products per year to make cars, machinery and appliances. Their daily utility bill for electricity, gas and water is $300,000: that's one huge amount of energy. To make most of that electricity, they pump water from the lake into steam turbines and heat the water with waste gases and natural gas to drive a turbine. By modifying one of their steam turbines and making various other steam system improvements, they saved over 40 million kWh of electricity and 85,000 million Btus of gas, with a payback of just over a year. For each subsequent year, they saved $3.4 million.

Different versions of this story can be heard all over North America. When Pratt and Whitney, the jet engine makers, looked at ways to save energy, they realized that the screen-savers on their 5,000 computer monitors were costing them $200,000 a year when the monitors were not in use. After spending $1,600 on educational decals and $2,000 to install them and read the electricity meters, they are saving $234,000 annually in reduced energy and cooling needs and preventing 2,800 tons of CO_2 a year from entering the atmosphere.

Can every company save money and greenhouse gas emissions in this way? The evidence seems to say "Yes," and for a return on investment that will make your financial manager's toes curl. The book *Cool Companies* (see box) is full of stories that demonstrate how a typical company can achieve a 33-50% return on investment on an energy upgrade. The evidence is so compelling that we may soon see shareholders suing their companies if they do *not* invest in energy efficiency and greenhouse gas reduction.

Books That Are Helping the Atmosphere

- *Cool Business Guide.* Pembina Institute, $25. www.pembina.org 1-800-884-3515
- *Cool Companies: How the Best Businesses Boost Profits and Productivity by Cutting Greenhouse Gas Emissions,* Joseph Romm (Island Press 1-800-828-1302 www.islandpress.org).
- *Natural Capitalism: Creating the Next Industrial Revolution,* Paul Hawken, Amory & Hunter Lovins (Little, Brown: www.naturalcapital.org).
- *The Heat is On,* Ross Gelbspan (Addison Wesley) (www.heatisonline.org)

"We view business and the environment as parts of the same total system: the Earth's ecosystem. Therefore, we think through the environmental consequences of every aspect of our business."
— Eckart Pfeiffer,
President & CEO of Compaq

A 'carbon neutral' company is one that has reduced and/or offset its carbon emissions to zero.

Step 1: Decide that reducing your greenhouse gas emissions is a priority, and an opportunity to establish a healthier bottom line. Appoint someone to research the help you can get from groups such as Climate Wise, Climate Savers, Environmental Defense, The Natural Step, and the Rocky Mountain Institute. Establish a policy, and appoint a champion. Carry out a greenhouse gas (GHG) and energy inventory for your company and its products, looking at their full life-cycle cost.

Set clear goals. IBM's goal is to reduce its emissions by 4% each year. Johnson & Johnson's is to reduce its emissions to 7% below the 1990 level by 2010. Dupont's is to reduce by 65% by 2010. Educate your staff, and get them involved in the idea of making a difference on the planet. Set up a rewards program to encourage energy-saving suggestions. Create a GHG Reduction Program, with goals, budget, leaders, and action plans.

Step 2: Maximize your energy efficiency and retrofit your buildings (see #32).

Step 3: Reduce your solid waste and increase your recycling (see #32).

Step 4: Organize your transportation and commuting more sustainably (see #33).

Step 5: Re-engineer your product-line to reduce the other GHGs (see #34).

Step 6: Generate or buy green energy (see #34).

Step 7: Ask your suppliers to reduce their emissions, and buy carbon offsets to reduce or absorb your remaining emissions (see #35).

Climate Wise

Climate Wise is a voluntary EPA partnership, with 500+ participants representing 12% of US industrial energy use. Participants submit a Climate Wise Action Plan identifying cost-effective energy efficiency and pollution prevention measures, and report their results annually. Climate Wise helps its partners identify actions to save energy and reduce costs, and gives them action plan software including case studies, a list of energy efficiency technologies, and tools to quantify results.

Climate Savers

Climate Savers is a partnership between the World Wildlife Fund and Joseph Romm's Center for Energy and Climate Solutions that helps companies adopt a comprehensive portfolio of climate and energy management strategies. Partners include IBM and Johnson & Johnson.

Assume Leadership

If you are a manager, read Joseph Romm's book *Cool Companies*. Ask your fellow managers to read it, and set up a strategy meeting. If you are a junior manager, shop-floor worker, or regular employee, form a lunch-time study group to discuss the books suggested (see box), then ask for an opportunity to present your findings to management. Put up a poster or send an email announcing the group, and try to get someone from each department to join.

Resources

Canada's Voluntary Challenge and Registry: www.vcr-mvr.ca
Climate calculator for businesses (UK): www.bestfootforward.com/carbon-calc.htm
Center of Excellence for Sustainable Devt: www.sustainable.doe.gov
Climate Savers: rebecca.eaton@wwfus.org
Climate Wise: www.epa.gov/climatewise
Cool Business Guide: www.pembina.org
GreenBiz.com: www.greenBiz.com
Greenhouse Gas Technology Information Exchange: www.greentie.org
Greening of Industry: www.greeningofindustry.org
Climate Solutions—Industry (Canada): www.climatechangesolutions.com
LEAP—Long Range Energy Alternatives Planning software: www.seib.org/leap
Partnership for Climate Action: www.environmentaldefense.org/PCA
Pew Center on Global Climate Change: www.pewclimate.org
Rocky Mountain Institute: www.rmi.org
Sustainable Business.com: www.sustainablebusiness.com
The Natural Step: www.naturalstep.org
Voluntary Reporting Technical Assistance: www.eia.doe.gov/oiaf/1605/techassist.html
World Business Council for Sustainable Development: www.wbcsd.ch
100 Top Corporations and Climate Change: www.greenpeace.org

32

Become Energy Efficient and Reduce Waste

Jim Yost

The BigHorn Home Improvement Center retail complex in Silverton uses natural daylighting to improve energy use.

In 1981, Ron Nelson was energy manager of Dow Chemical's Louisiana Division, with 2,400 employees at 20 locations. To reduce pollution, he set up a contest to find energy-saving projects that would provide a minimum 50% return on investment. The first year's 27 projects averaged 173%; the next year's 32 projects averaged 340%. By 1993, 575 audited projects had averaged a 204% return on investment and the overall energy savings were paying Dow's shareholders $110 million a year, reducing greenhouse gas emissions and making a profit at the same time.[1]

Five Stages to an Energy-efficient Building:

1. Upgrade the lights through the EPA's Green Lights program.
2. Upgrade the building for maximum efficiency and occupant comfort.
3. Reduce the heating and cooling needs by making improvements to the windows and roofs.
4. Reduce the size of the building's fan system, and introduce variable speed drives.
5. Upgrade the heating and cooling plant equipment to a lower capacity, properly sized, energy-efficient system.

Source: EPA

When the California manufacturer Verifone decided to renovate and daylight one of its buildings, the improvements saved 60% of the energy and would have paid for themselves in 7.5 years. The workers liked the natural daylight so much, however, that productivity rose by 5% and absenteeism fell by 45%, leading to a payback of under one year.[2]

In Malaysia, Western Digital built a new factory with 10% increased floor space, but their focus on energy efficient design reduced its energy consumption by 44%, with a one-year payback.[3] The rule of thumb is that a business that undertakes a buildings upgrade can expect to achieve savings of $1 per square foot of floor-space per year.

There is probably not a business in North America that is not pouring greenhouse gases needlessly into the air, at a financial cost to the company. When Boeing introduced more efficient lighting to its buildings, it achieved a 90% reduction in energy use with a two-year payback — a 53% return on investment. The higher quality lighting also cuts down glare and helps the workers reduce defects.[4] Measures that IBM have initiated over the past ten years have saved the company $525 million and prevented more than 6 million tons of CO_2 from entering the atmosphere.

The question for businesses is not "Shall we do it?" but "Why, for Earth's sake, are we *not* doing it?" The need to educate managers, staff, accountants, directors, CEOs and shareholders is critical – and there is help available to make it easy.

"Over the long term, it is more profitable to do the right thing for the environment than to pollute it."
— Aaron Feuerstein, CEO, Malden Mills

Change Your Lighting
Lighting consumes 20-25% of all electricity, believe it or not, but upgrades that improve lighting quality while reducing energy use by 50-70% are becoming common.[5] If every facility in the US invested in a full Green Lights retrofit, offering a 30%+ rate of return, it would save 65 million kilowatts of electricity and prevent 87 million tons of CO_2 from entering the atmosphere.[6] ProjectKalc is free software that offers a full analysis of potential lighting upgrades, including a user-modifiable data-bases of costs, labor time and performance for 8000 hardware applications. (www.epa.gov/buildings/esbhome/tools/pkalc.html)

Upgrade Your Building
In the 1990s, San Diego's Environmental Services Department renovated its 3-storey, 73,000 square foot office building, reducing its energy consumption by 70% from 21-22 kWh to 7-8 kWh per square foot. The annual savings was $80,000, or $1.10 per square foot. To calculate your annual savings if you renovate, use the formula: P square feet [size of your facility] x $1.10.

Upgrade Your Motors
Motors, steam, compressed air and process heat use almost 70% of all industrial electricity. To find out how you can upgrade, see the Office of Industrial Technologies Best Practices website: www.oit.doe.gov/bestpractices.

Join WasteWise
WasteWise is an EPA program with 1,000 partners in 53 industry sectors, from large corporations to hospitals and schools. When you sign on, you make a commitment to reduce wastes, set 3-year goals for waste prevention, recycling and the purchase of recycled products, and track your progress. WasteWise partners have prevented the emission of 73 million tons of CO_2 and other greenhouse gases. (www.epa.gov/ wastewise or 1-800-EPA-WISE)

Increase Your Recycling Rate
By reducing waste and by more efficient recycling, businesses can reduce their greenhouse gas emissions in three ways: reducing landfill methane emissions; increasing carbon sequestration in the forest; and reducing emissions from the extraction and processing of raw materials. (See also #10 & #26). When McDonald's eliminated 27,000 tons of packaging material, it saved $12 million over 7 years. The company now uses recycled paper in half of its food packaging. When Interface introduced leasable, returnable carpet squares that are 100% recyclable and compostable, it reduced its energy use by 95%.

Resources

- Alliance to Save Energy: www.ase.org/checkup/business
- *CADDET Energy Efficiency* newsletter: www.caddet-ee.org
- Canadian Energy Efficiency Centre: www.energyefficiency.org
- Canada's Office of Energy Efficiency: www.oee.nrcan.gc.ca
- Energy Efficiency and Renewable Energy Network: www.eren.doe.gov
- Energy Efficiency Resources: www.leonardoacademy.org/efficiencyresources
- Energy Efficiency Toolkit: www.caddet-ee.org/ee_tools.htm
- Energy Ideas Clearinghouse: www.energy.wsu.edu/eic
- Energy Saving Now: www.energysavingnow.com
- Energy Star & Green Lights: www.energystar.gov
- Energy User News: www.energyusernews.com
- EPA—Actions by Industry: www.epa.gov/globalwarming/actions/industry
- E-Source: www.esource.com
- Facility Energy Decision System software: www.pnl.gov/feds
- Industries of the Future: www.oit.doe.gov/industries.shtml
- Mining, Minerals and Sustainable Development: www.iied.org/mmsd
- NICE3—Funding for innovation: www.oit.doe.gov/nice3
- Recycled Paper Coalition: www.papercoalition.org
- Results Center—Best Practices: www.solstice.crest.org/efficiency/irt/trc.htm
- Sustainable Cement Industry: www.wbcsdcement.org
- Waste at Work—Prevention Strategies for the Bottom Line:www.informinc.org/wawgate.htm
- Zero Waste Alliance: www.zerowaste.org

33 Travel Sustainably

"Green business: It's more than a good idea. It's a powerful, practical strategy for shifting society onto a more socially just and environmentally sound course."
— Alisa Gravitz, Co-op America

Very few businesses could exist without transport. For the rest, trucks, trains, planes, and cars are indispensable. Transportation produces over 30% of all greenhouse gas emissions in the USA, 26% in Canada. Clearly, something has to be done.

The problem is easy to ignore; just swallow the higher fuel bills and carry on with business as usual. But wait! As with energy efficiency, savings can be made by reducing emissions:

- Bison Transport, in Winnipeg, cut its fuel use by 5% in 1997 and another 7% in 1998 by driver-training, and installing an on-board driving and fuel management system that gives real-time performance data.
- SGT2000, a Quebec trucking firm, achieved a 5.8% reduction in fuel consumption in its 350 tractors and 1050 trailers by training drivers, buying new equipment, and installing auxiliary heaters and onboard computers.
- Stoneyfield Farm, a New Hampshire yogurt maker, modified its milk delivery schedules to minimize fuel use and is promoting car-pooling in order to reduce its GHGs.
- The three largest grocery chains in southern California are buying 150 alternative-fuel big rigs to replace their heavily polluting diesel trucks (after a lawsuit by environmental groups and the state attorney's office).
- FedEx Express, working in partnership with the Alliance for Environmental Innovation, has asked manufacturers to design and develop a delivery truck which will reduce pollutants by 90% and be 50% more fuel efficient, to replace the present FedEx Express truck.

The International Council for Local Environmental Initiatives (ICLEI) has a useful booklet, *Business Strategies for Transportation Efficiency at Work*, that summarizes the steps a business can take to reduce transportation emissions.

Manage Your Company Fleet More Efficiently

- Increase vehicle efficiency by regular maintenance, driver training, and onboard fuel management systems.
- Specify a minimum fuel efficiency standard in all purchase agreements and introduce a company policy to minimize idling beyond 3 minutes.
- Use software to plan the most efficient route between frequently visited destinations.
- Downsize your fleet, phase in smaller, more efficient and alternative-fueled vehicles, and charge for the use of company vehicles, creating an incentive to use them more efficiently.
- The city of Edmonton, Alberta, increased its average fuel economy by 20% by coaching employees on fuel-efficient driving.

Bucking the Old Travel Habits

At Pioneer Pacific Property Management's Station Tower, located in Surrey, British Columbia, more than 50% of 700 employees from 30 different organizations work together in a program called TravelChoices:

Showers and secure bike lockers are provided for cyclists.

TravelChoices members have free access to local fitness facilities, including exercise equipment, showers, and lockers.

A ride-matching service links potential carpool partners within the complex.

Reserved, preferential parking is available for carpools and vanpools.

TravelChoices members get guaranteed ride home insurance.

The TravelBucks incentive program gives members one TravelBuck for each day they use alternative transportation to and from work which can be cashed in for coffee, transit FareSaver Tickets, ski passes or rental car certificates.

Source: On-Line Transportation Demand Management Encyclopedia www.vtpi.org/tdm

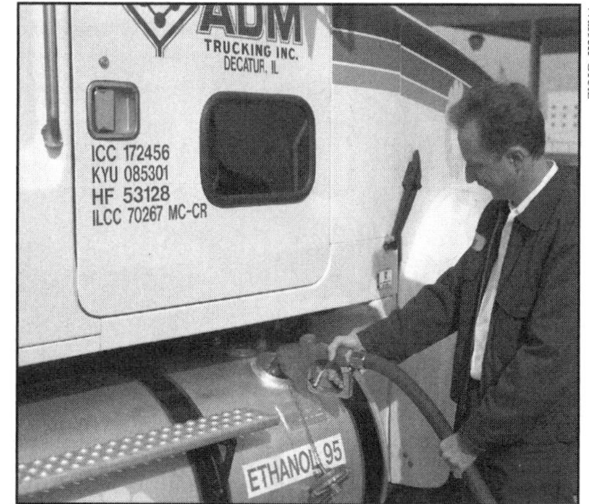

Warren Gretz

This truck's engine has been modified to run on 95% ethanol, a biofuel made from corn.

Move Your Inventory More Efficiently

Switch to rail, barge, or ship instead of truck where possible, and use inter-modal containers. Consolidate loads to avoid empty return trips, and install computers in cabs to transmit up-to-the-minute information. In a rational world, companies in City A and City B that ship goods to each other would share the use of their trucks. The National Transportation Exchange has set up an Internet-based exchange that auctions empty spaces on trucks; UPS is pursuing a similar system. Make your trucks more aerodynamic, and use radial tires. Reward efficient drivers and company divisions. Reduce weight: Anheuser-Busch has reduced the weight of its cans by 13% and bottles by 24% since 1990, allowing more efficient shipping. Consider rationalizing supply chains. Paperexchange.com is working to reduce its transport costs by matching paper buyers with mills that are closer.

Encourage Sustainable Commuting

Appoint an employee transportation coordinator and set up a rewards system that encourages travel by ridesharing, transit, bicycle or on foot, with a guaranteed free taxi-ride home in case of emergency. Provide a lunch-time shuttle to local watering places. Start a vanpool program. Provide zero-interest loans for commuters to buy bicycles. Set up a "bike-buddies" program for new riders, and install bicycle lock-ups and showers where cyclists can change. Make your environment friendly to pedestrians, and make sure the nearest bus shelters are comfortable.

Change Your Parking Policies

Free parking means free polluting: 90% of all car commuters in the US get free parking as a benefit. Start charging for parking and use the income to subsidize the purchase of transit passes. Negotiate with local government to reduce the mandatory parking requirement if you reduce your employee parking, and then set up a "cash out" option for employees whereby they are charged for parking, and offered cash in lieu if they do not park. As the parking spaces empty, lease them out to recoup your costs. Provide free parking for carpools and alternative-fuel vehicles.

Encourage Telecommuting and Flexible Working

Allow employees to work from home. Set up a satellite office closer to employees' homes, owned by the company or shared with others. Do more video-conferencing; spend fewer days traveling to conferences and meetings. Match employees with branch offices closer to their homes, and allow flexible hours, enabling employees to use public transit schedules. Introduce compressed work weeks, giving employees the freedom to work a 10-hour, 4-day week or a 9-hour, 9-day week. AT&T has 36,000 employees who work from home an average of six days per month, together eliminating 80,000 tons of CO_2 annually.[1]

Resources

21st Century Truck Technology Roadmap: www.osti.gov/hvt/21stcenturytruck.pdf

Alliance for Environmental Innovation: www.environmentaldefense.org/Alliance/cleanertrucks.html

British Columbia's "Go Green" program: www.gogreen.com

Business Strategies for Transportation Efficiency at Work: www.iclei.org.us

Fleetsmart Program (Canada): http://fleetsmart.nrcan.gc.ca

Green Fleets (ICLEI): www.greenfleets.org

Greenhouse Gas Reduction Opportunities for Freight Transportation: www.davidsuzuki.org/files/freight.pdf

Sustainable Mobility Project: www.wbcsdmobility.org

Traxis Inc., removable smart card for fleet optimization: www.interlog.com/~traxisto

United Parcel Services: www.ups.com/about/inits.html

34

Reduce Your Other Emissions,

and Buy Green Power

So – you've become more energy efficient (Step 2); reduced your solid waste and increased your recycling (Step 3); and reduced your CO_2 emissions from transportation (Step 4). You are three steps away from being carbon neutral, which means you will have reduced or offset all of your emissions.

To complete the journey, you need to re-engineer your product line to eliminate other greenhouse gases (Step 5); generate or buy green power to displace the fossil fuels you are still using (Step 6); and buy carbon offsets to mop up the remaining CO_2 emissions (Step 7). When you've done all this, you can call yourself carbon neutral. The world may not believe you, because there is a lot of controversy about the merits of carbon offsets, but you will have done your best using today's technologies.

Xerox's Score-Card

Xerox started its Waste-Free Factory program in 1993, with the following goals:

1. Decrease municipal, hazardous and chemical waste by 90%.
2. Decrease air emissions by 90%.
3. Decrease water discharges by 50%.
4. Increase use of post-consumer recycled materials to 25% of purchases.
5. Increase energy efficiency to within 10% of each facility's theoretical optimum.
6. Conduct annual assessments.

A plant becomes "Waste-Free" when it scores 450 points out of a possible 500. The average score in 1993 was 250; by 1996 it was 459. The goal for 2000 was 500.

Source: Cool Companies, www.xerox.com

Eliminate The Other Greenhouse Gases

There are several lesser-known but powerful greenhouse gases that need urgent elimination (see p. 12). Nitrous oxide (N_2O) is produced by catalytic converters on vehicles, and during the manufacture of adipic acid, used in the manufacture of nylon and to add tang to food. N_2O also results from production of nitric acid, used to manufacture synthetic fertilizer and explosives.

HFCs, HCFCs, PFC and SF6 are emitted during a number of industrial processes, such as aluminum smelting, plasma etching in the semiconductor industry, as a dielectric in electrical transmission, and as a protective cover gas in the casting of molten magnesium. The ozone-depleting CFCs which were outlawed by the Montreal Accord are also powerful greenhouse gases, but some are still in use, after being smuggled in from Mexico.

In March 2001, in a world first, members of the Semiconductor Industry Association signed a voluntary agreement with the US Environmental Protection Agency in which they agreed to reduce their use of perfluorocompounds (PFCs) to 10% below the 1995 level by 2010. The commitment will require the member industries (including Intel, Advanced Micro Devices, and NEC Electronics) to reformulate their chemicals and change their manufacturing processes. PFCs are used to clean semi-conductor-making equipment, and etch silicon wafers to create circuitry patterns. Each molecule traps 10,000 times as much heat as CO_2, and they can live in the atmosphere for 2,000 to 50,000 years.

"Over the long haul of life on this planet, it is the ecologists, and not the bookkeepers of business, who are the ultimate accountants."
— Stewart Udall,
former US Secretary of the Interior.

Produce Green Power
A number of companies are generating their own renewable energy. Big Horn Materials in San Francisco has installed a solar rooftop system; Real Goods has installed a 132kW solar array at its Solar Living Center in Hopland, CA. Real Goods customers have prevented more than 530,000 tons of CO_2 from entering the atmosphere by purchasing energy efficient appliances. The Leadville Water Treatment Plant in Colorado has installed a large "Solarwall" (www.solarwall.com), saving $4,300 in energy costs per year, with a 7.1-year payback. The First National Bank of Omaha has installed a hydrogen fuel cell as a source of reliable electricity, reducing its CO_2 emissions from power by 33%.[1]

Buy Green Power
For companies in states and provinces where renewable energy is readily available, a simple phone call can eliminate a lot of CO_2 emissions. IBM, Rocky Mountain Steel Mills, Coors Brewing and Fat Tire Brewing are among 300 companies buying wind energy from Colorado Wind Power. In California, Birkenstock, Real Goods, Toyota, Fetzer Vineyards and the Union of Concerned Scientists' office are powered exclusively by renewable energy. Kinko's is supplying 75 of its California stores with green energy from GreenMountain.com, while Time Warner is buying it from Commonwealth Energy, along with 42 Panda Express and 24 Pick Up Stix restaurants. Even the LA Dodgers are buying green energy, to keep the outfield pure.

Putting it all together
The World Resources Institute is a Washington, DC-based center for policy research and technical assistance on global environmental and development issues, operating with partners in 50 countries. In its Climate Protection Initiative, WRI advances flexible, market-based policy options that foster environmentally sound development. (See www.wri.org/wri/cpi) In 1999, the staff decided to reduce their CO_2 emissions to zero by 2005. Here's how they are doing it:

- Completing a CO_2 inventory (1990: 1500 tons; 1999: 1,633 tons).
- Installing energy-efficient lights and equipment.
- Ensuring that the office equipment is turned off at night. (Saves 2.5 tons of CO_2, $150/year)
- Reducing paper use by 10%. (Saves 4 tons of CO_2, $1,300/year)
- Replacing some air travel with videoconferencing (saves 88 lbs of CO_2 for every 100 miles not traveled).
- Working closely with the property manager, power company, shipping services and the US Postal Service to reduce indirect emissions.
- Exploring urban forestry options for carbon offsets.

Big Mac, Smaller Footprint

Underneath a McDonald's restaurant in the suburbs of Detroit, 32 boreholes plunge 200 feet into the ground, for a ground-source heat system (see #6). Filled with water, they absorb the ambient heat from the surrounding rock, which is extracted by three heat-pumps and used to heat the dining room and children's play area, and cool the kitchen.

- The kitchen, dining room, and play area use efficient lighting.
- The lights have occupancy sensors and photoelectric dimmers.
- The windows are triple-glazed.
- The motors that run the cooking fans are energy efficient.
- The building shell has improved insulation.
- The overall building has 40-50% less CO_2 emissions than a typical restaurant.
- (Maybe it's time for veggieburgers, instead of the beef....See #36)

Source: Cool Companies

Authors' note: Some of these Californian green energy purchases may sadly have bitten the dust, because of the 2000/2001 energy confusions.

35
Buy Carbon Offsets

Let's say your business has worked its way through Steps 1 to 6, and reduced its emissions to a minimum. As the months pass, you will find new ways to reduce them further. The more your staff work as a team, the more ways you will find to reduce your emissions.

You may also have talked to your suppliers, as the World Resources Institute is doing (see #34); to your customers, as Mazda is doing in Britain; or to your distributors, as Shaklee is doing (see opposite), to encourage them to reduce their emissions.

If you can't reduce your CO_2 emissions to zero, you can do it by buying carbon offsets. Carbon offsets are greenhouse gas emissions that have been reduced or sequestrated (stored) somewhere else to balance the emissions that a company, school, church, government, family, or sports team produces. You set up a partnership with a carbon offset organization, work Steps 1 to 6, and buy offsets for the remaining emissions. You can then call your products "carbon neutral," "climate-neutral," or "climate cool." It's a whole new train a-coming down the track.

An offset can be acquired in any project that is measurable and certifiable. It might involve retrofitting a Venezuelan steel mill to make it more efficient, paying to install a solar PV array on a school, or planting trees to absorb CO_2 emissions. The public likes tree-planting, so that often seems like a win-win solution. But what if the trees burn down in a global-warming-induced forest fire, or are killed by a global-warming-induced insect infestation? What if the warmer temperatures mean that your trees begin *emitting* stored carbon, instead of absorbing it?

Similar questions can be asked of most carbon offset projects. The steel mill might have been obliged to upgrade within a couple of years, anyway. The country where you bought your rainforest offsets may delay legislation to protect its forests because it prefers to receive income from the offsets. This is not to say that the process should stop, but these are complex questions to which there are no easy answers. Buying carbon offsets shows social and environmental responsibility, but it can also be an easy way for companies to buy a green mantle without reducing their actual emissions.

> **"Any CO_2 offset program such as Climate Care is open to the charge that it simply offers consumers an unsustainable way to pay their way out of making changes to their behavior and purchasing habits. It is clear that to counter this the Trust must develop an offset program that makes a sustainable change in the CO_2 content of products consumers buy. To do this, the Trust will aim to create as much of its offset portfolio as possible in "front of pipe" renewable energy and energy efficiency projects, with a special emphasis on projects that help develop new technologies and make them economic."**
> **— Carbon Storage Trust, UK**

Carbon Offset Organizations

American Forests: www.americanforests.org

Carbon Storage Trust/Climate Care (UK): www.co2.org

Climate Neutral: www.climateneutral.com

Climate Partners: www.climatepartners.com

EcoSecurities Ltd: www.ecosecurities.com

Edinburgh Centre for Carbon Management: www.eccm.uk.com

Face Foundation (Holland): www.facefoundation.nl

Future Forests (UK): www.futureforests.com

Stoneyfield Environmental Cookbook: www.stonyfield.com/earth/climate_change.htm

Tree Canada Foundation: www.treecanada.ca

Trexler and Associates: www.climateservices.com

Buy Carbon Offsets
In Britain, Mazda formed a partnership with Future Forests for the launch of the Demio, and committed to plant five trees for each car sold to cover the estimated emissions from its first year of driving (1 tonne of carbon). After the first year, Mazda owners are encouraged to continue with the Future Forests program, and become "carbon neutral" drivers. Fiat made a similar arrangement with Future Forests for its 1999 Seicento Citymatic hatchback.

In the USA, Shaklee, Interface, and Boston-based Saunders Hotel Group were the first certified "climate neutral" companies with the Underwood, Washington, Climate Neutral Network. Shaklee made its entire company climate neutral, and is working to encourage its 500,000 independent distributors to do the same. Over the years, it has made its buildings, windows, lights and appliances more energy efficient, and used recycled building materials, but it was still producing 25,000 tons of greenhouse gases a year. By buying carbon offsets, which included subsidizing the rates of a New England green energy utility to encourage new customers, it is reducing its effective emissions to zero.

The Climate Neutral Network and the British Carbon Storage Trust take great care to ensure that their offsets will make a lasting difference, and minimize the risk of additionality ("it would have happened anyway"). To become certified as carbon neutral, a company must conduct a detailed analysis of its climate emissions, create a plan to reduce emissions, and buy offsets to cover the remaining emissions. Both organizations are advised by well-reputed environmental consultants, such as the Rocky Mountain Institute, the Natural Resources Defense Council (USA), and Forum for the Future (UK). Because of uncertainties around tree-planting, the Carbon Storage Trust is suggesting that tree-planting be limited to 10-15% of future carbon offset portfolios.

In New England, the Stoneyfield Farm Yogurt Works reduced its emissions through lighting retrofits, waste heat recovery and other means, and then bought carbon offsets by planting trees in Oregon to absorb 2,000 tons of CO_2, working through Trexler & Associates. They then wrote an "Environmental Cookbook" with a detailed recipe for reducing CO_2 emissions through carbon offsets, enabling others to benefit from their experience.

What else is in the carbon-neutral (CN) pipeline? A CN fuel-card, a CN hotel chain, CN campuses, CN flooring from Interface, CN gas stations, CN hybrid and alternative-fueled cars, CN flights, CN conferences, CN holidays, and CN travel clubs (www.greenglobe21.com and www.triplee.com). This train is here to stay.

Carbon Offset Projects

Project	Price per Ton of Carbon
Rural solar electrification in Sri Lanka (displaces kerosene)	$18.00
Tropical rainforest protection in Costa Rica (sequestration)	$2.00
Forest protection in Chiapas, Mexico (sequestration)	$12.00
Fuel switching in Czech Republic (coal to natural gas cogen)	$16.00
Urban forestry, US (sequestration)	$36.00
Methane recovery from abandoned coal-mine	$2.00
Energy efficiency/solar energy in schools (displaces fossil fuels)	n/a
Tree-planting via American Forests[1]	$11.00
Tree-planting in the UK, via Future Forests[2]	$99.00
Portfolio of wind, biomass, solar and forestry, Climate Care, UK	$50.00

Sources: Stoneyfield Cookbook, et al

36 For Farms

"The adoption of recommended agricultural practices can reverse these trends, and resequester carbon lost from the world's soils." — Rattan Lal, Professor of Soil Science, Ohio State University

Modern farming practices are a major contributor to global warming. A farmer's best friend is the soil, but when the soil is not protected by conservation or no-till farming methods, the carbon in it breaks down and escapes as CO_2, adding to the world's burden of greenhouse gases.

And that's not all. The use of nitrogen fertilizers releases nitrous oxides. Cattle release methane in the process of digestion. Cattle and pig manure releases methane and nitrous oxide when collected in an anaerobic, liquid system. Gasoline used to run farm equipment and dry crops releases carbon dioxide. Trucks that deliver produce to markets release carbon dioxide. Taken together — and globally — farming is responsible for 8% of the world's GHG emissions. The good news is that there are solutions that can make farming more ecologically sustainable (see also #69 & #78).

Practice Soil Conservation

In 1998, the Rodale Institute completed an intensive 15-year research project in which they alternated corn with soy beans and other legumes on an 8-acre plot, enriching the soil on some fields while plowing under immature plants on others. The carbon level in the soil in the experimental plots soared, nitrogen losses were cut by half, and yields were just as good compared with crops that used commercial fertilizer on adjacent conventionally grown fields. In Australia, the Keyline farming system encourages farmers to design their farming landscape following the contour lines, planting tree breaks along contour ridges, retaining rainwater in lakes and ponds, and using simple scratch-plowing instead of turning the soil over, to build deep biologically fertile soil. In Canada, a consortium of power utilities is paying prairie farmers $2 million to adopt practices such as zero-tillage to build the soil and store carbon, which they hope to use to offset their own CO_2 emissions (see #52).

Shift to Organic Methods

Organic farming builds carbon in the soil while working with nature's biodiversity; it also allows insects to control each other and fosters a rich ecological system. Across North America, the demand for organic produce is outstripping supply. In the US, certified organic cropland grew from 403,000 acres to 850,000 acres between 1992 and 1997. In Vermont, nearly a quarter of the state's farmland is managed organically, with farmers being motivated by the desire to get away from chemicals and supply the growing market.

Resources

Agriculture and Climate change: www.globalchange.org/featall/2000winter2.htm

Agriculture— Industry of the Future: www.oit.doe.gov/agriculture

AgSTAR Methane to Biogas: www.epa.gov/agstar

Canadian Organic Growers: www.cog.ca

Community Supported Agriculture: www.sare.org/csa

Conservation Technology Information Center: www.ctic.purdue.edu/CTIC

GRACE Factory Farm Project: www.factoryfarm.org

Growing Carbon: A New Crop that Helps Agricultural Producers and the Climate Too: www.swcs.org/f_pubs_education.htm>

Harvesting Clean Energy for Rural Development: www.climatesolutions.org

International Federation of Organic Agriculture Movements: www.ifoam.org

Keyline Designs: www.keyline.com.au

Organic Farming Research Council: www.ofrf.org

Organic Trade Association: www.ota.com

Rodale Institute: www.rodaleinstitute.org

Rowett Research Institute, Aberdeen: www.rri.sari.ac.uk

Ruminant Livestock Efficiency Program: www.epa.gov/ruminant

Solar Energy in Agriculture: www.seia.org/SolarEnergy/agriculture.htm

The Health of Our Air —Toward Sustainable Agriculture in Canada: http://res2.agr.ca/research-recherche/science/Healthy_Air/1toc.html

Windustry—Harvesting the Wind: www.windustry.org

Warren Gretz

Cows and wind farms co-exist well at this Buffalo Ridge wind farm in SW Minnesota.

Use less nitrogen fertilizer.
Nitrous oxide (N_2O) is a greenhouse gas 310 times more powerful than CO_2 that is released by the use of nitrogen-based fertilizers. N_2O emissions can be reduced by using soil tests to apply nitrogen only when needed, and using organic fertilizers instead.

Capture biogas, or shift to dry manure systems.
When animal manure rots in a liquid, oxygen-free environment, it produces methane. Methane can be captured as biogas to heat farm buildings, or be avoided by the use of aerobic systems such as composting, or by using Canadian hoop houses, as they do in Iowa, which yield dry manure ready to spread on fields.

Reduce your animals' methane.
Fermentation in the stomachs of cattle, buffalo, sheep and pigs produces over 43% of agriculture's methane.[1] Improving the nutrient balance and quality of livestock feed can reduce emissions. Scientists at the Rowett Research Institute in Aberdeen, Scotland, have discovered that adding Brevibacillus brevis to animal feed converts 17% of the methane into CO_2.[2]

Install energy efficient equipment.
Change the lighting in chicken houses to compact fluorescent lamps; switch to large, slow fans instead of small fast ones; add insulation and weather-stripping; use air-to-air heat exchangers to recover fresh air and 90% of the heat.

Solar crop drying.
Solar dryers can dry fruits, vegetables, grains, and herbs. In Kansas City, Bill Ward invented a zero-energy way to dry grain in a silo by boring a hole in the top and adding a small windmill with hollow blades that sucks a draft of air up through the silo, drying and cooling the grain.[3]

Invest in wind turbines.
Farmers in windy areas such as North Dakota and the Canadian prairies can earn $1,500 per wind turbine per year in lease payments, without interrupting their farming. An Iowa farmer who leases a quarter acre of cropland to a utility for use as a wind turbine site can earn $2,000 a year in royalties from the green energy produced, compared to $100 a year from corn.[4] When farmers form their own wind co-operatives, as they do in Denmark, they keep more of the income.

Convert waste biomass to ethanol.
Damp and below-grade grain, nutshells, peach-pits, cotton-gin trash, switchgrass and other agricultural wastes can be used to make ethanol as a fuel for vehicles. To analyze the income and employment benefits from a biomass plant in a rural community, see www.etsu.com/biosem.

Sell your development rights.
By selling your development rights, you can ensure that your farm will remain a farm forever, and help prevent sprawl. This does not apply in Canada, where agricultural land is protected by zoning (see #29 & #66).

Farm fewer animals.
Most of farming's greenhouse gas emissions come from cattle and pigs. Growing grains, vegetables and fruits for human consumption uses less than 5% of the raw materials needed for production of meat, and minimizes emissions all round.

The Carbon Lies in the Soil

Tons of carbon in an acre of healthy soil: 44

Number of distinct RNA genomes found in 1 gram of healthy soil: 4,000

% of original topsoil on US farms lost since settlement: 33%

% of Great Plains soil productivity lost in the first 28 years of cultivation: 71%

Ratio by which this loss of topsoil exceeded the formation of new topsoil: 17:1

% loss of farm productivity due to loss of topsoil: 8-20%

% of carbon in the atmosphere from the loss of soil carbon: 7%

% of US carbon emissions that could be sequestered back into the soil: 8 – 17%

Yearly income that farmers might receive for carbon sequestration per acre: $9-$20

Source: Natural Capitalism, Rattan Lal

37 For Forest Companies

"Let the land dictate the practices, not the mill's need for logs. Be patient, be prudent, be good stewards — of the whole forest...the watersheds, the birds, the plants, the animals."
— Kane Hardwood, Manager of the 126,000-acre FSC certified Collins Pennsylvania Forest

As we enter the ancient forest and feel its cool, dark shade, we enter a primeval ecosystem that has been building its complex web for thousands of years. The relationship between trees and climate change is equally complex, because forests sequester carbon. They do it as part of the natural cycling of carbon that has been occurring on Earth for millions of years. All living things on this planet — whether in the ocean, the soil, or the forest — are carbon-based organisms that take in carbon dioxide to build their tissue. While alive, they store it as carbon; when they die, they release it as carbon dioxide.

If we plant more trees, the thinking goes, they will absorb some of the CO_2 that our fossil-fueled lifestyle is releasing. It takes 40 to 150 years for new trees to absorb a useful amount of carbon; meanwhile, if we don't also reduce our carbon emissions, the climate will continue to warm and we risk losing the carbon through forest fires, insect damage, or increased heat, making the overall problem worse (see also #54).

A tree that is planted in a field or a city will sequester carbon until it is cut down, dies, or burns in a fire (which will be more frequent in a warmer world). In a commercially harvested forest, the story is very different. In the Pacific Northwest, where clearcutting is still the major method of logging, it will be 150 years or more before a new coastal Douglas fir stores as much carbon as an old growth tree that was cut down. Only 60% of the forest's carbon lives in the timber; 40% lives in the soil and branches, which release their carbon when the forest is cleared. If we want to store more carbon, we have either to stop cutting trees, or change the way we cut them, shifting to ecoforestry. The suggestion that cutting a forest allows its carbon to be stored as timber, enabling the replacement forest to absorb more carbon, works on the erroneous assumption that there is no carbon-loss from the soil when the forest is cut. As we said, it's a complex world out there in the forest.

A climate-stable world needs forest practices that maximize the amount of carbon stored in the forest and minimizes the loss of carbon that occurs when trees are felled.

Use Less Timber, Store More CO_2

- Engineered wood products (e.g., "Parallam") allow timber companies to make their timber go twice as far.
- % of world timber harvest consumed in the U.S.: 27%
- Number of homes built every year in the U.S.: 1.2 million
- % of U.S. lumber used in home-building: 72%
- Average board feet of lumber used in a new home: 16,000 (35 trees)
- % reduction in lumber achievable by framing more efficiently: 20%
- % reduction in lumber achievable by using engineered wall framing: 50%
- Wood saved if all new houses were framed efficiently: 3.8 billion board feet (28,000 acres)
- Wood saved if all new houses used engineered wall framing: 9.6 billion board feet (70,000 acres)
- Tons of CO_2 stored over 40 years if all new houses used engineered wall framing: 12.6 million[1]

Sources: NRDC, Society of American Foresters, Natural Capitalism, American Forests

Nina Rastogi

Merve Wilkinson in his forest at Wildwood, British Columbia, which he has managed sustainably since 1938.

Shift to Certified Forestry

The solution lies in forestry practices that have been certified by the Forest Stewardship Council, also known as ecoforestry, a tried-and-tested method of commercial forestry that seeks to maintain the ecological integrity of the forest while harvesting its timber. Ecoforestry never allows a clearcut, and rarely a patch-cut, relying instead on individual tree selection. Over 60 years, an ecologically managed forest will yield more timber of higher quality than one that has been clearcut. The roots and soil are left undisturbed, along with their carbon; the seedlings take root from the genetic stock of the best old trees; the diversity of insect and bird life controls the pests; and because the trees grow more steadily in a semi-canopy, their fiber is tighter. Each year, the annual growth of the forest is carefully removed; over 60 years, the overall yield is higher than if the forest had been clearcut and left to grow for 60 years before being cut again. Because of its methods, it is almost certain that certified forestry stores more carbon and minimizes carbon-loss during harvesting, although sufficient research has not yet been done to prove this.

Around the world, 55 million acres of forest have been certified, including 25 million acres in Sweden, 8 million acres in Poland, 7 million acres in the USA (1 million acres in Maine), and 88,000 acres in Canada (April 2001). Timber from certified forests is in strong demand, especially in Europe, thanks to consumer awareness and campaigns by the Rainforest Action Network, Greenpeace, and the Natural Resources Defense Council. In 1999, Home Depot (which sells 10% of the world's lumber) announced that it would stop selling timber products from old-growth forests and give preference to certified wood. In March 2000, two of the largest U.S. home builders (Kaufman & Broad Home Corporation and the Centrex Corporation) made a similar commitment.

Reduce Your Other Emissions

In Oregon, California, and Pennsylvania, the Collins Companies have certified forests totaling 295,000 acres. They have also joined The Natural Step (see #40), trained all of their employees in the precepts of sustainability, and are in the process of moving their mills and other operations to a condition of zero emissions, zero waste, maximum efficiency, and maximum renewable energy.

In Britain, the DIY retailer B&Q requires its suppliers to record the full environmental impact of their products, and forest products suppliers are required to be certified by the Forest Stewardship Council. B&Q's operations are all becoming carbon neutral through afforestation or energy efficiency initiatives. (See #35)

Resources

Certified Forest Products Council: www.certifiedwood.org

EcoForestry Institute: www.ecoforestry.ca

Forest World: www.forestworld.com

Forest Stewardship Council: www.fscoax.org

Forest Stewardship Council (Canada): www.fsccanada.org

Forest Stewardship Council (US): www.fscus.org

Forest Stewardship Council (UK): www.fsc-uk.demon.co.uk

Merve Wilkinson's Wildwood, BC: www.ecoforestry.ca/Wildwood

Natural Resources Defense Council: www.nrdc.org/land/forests

Rainforest Action Network: www.ran.org

Smart Wood: www.smartwood.org

Sustainable Forest Industry: www.wbcsd.ch/sectoral/forestry

The Collins Companies: www.collinswood.com

38
For Architects, Builders and Developers

A house may seem like such an ordinary thing. It's your home; you live there. You raise tomatoes in the back yard. You relax there with your friends. In the evening, when you come home from work, it keeps you warm.

An office, bank, or factory is not such a huge affair, either. You work there, contribute to the economy, and, hopefully, find fulfillment. So what's the big deal?

Here's the big deal:

- 66% of all electricity produced in the U.S. is used to heat, light, and cool buildings.[1]
- 35% of all greenhouse gas emissions in the U.S. are from energy used in buildings.[2]
- 25% of all wood harvested in the U.S. is used in buildings.[3]
- Only 2% of buildings are efficient.[4]

On the other hand...

- Efficient new buildings can save 70%–90% of their traditional energy use.[5]
- Existing buildings can be upgraded to use 30%–50% less energy.

Our buildings are cooking the world. This may sound like a harsh thing to say, but it is undeniably true. That is why it is so important for architects, developers, designers, builders, and energy engineers to learn about these issues. Mr. Problem? Meet Ms. Opportunity.

In 1978, Holland's fourth largest bank decided it needed a new building. The bank's owners wanted it to be "organic," using lots of sunlight, water, plants, art, natural materials, and energy-conserving techniques. It was completed in 1987. At 540,000 square feet, it is one of the most remarkable buildings in the world. Water trickles down channels in the stair railings and daylight pours in. The architect, Anton Alberts, designed its ten interconnecting towers to feel like a comfortable third skin (after skin and clothes), and its occupants love it. It has no conventional air-conditioning (relying instead on passive cooling with backup absorption chillers), and uses 90% less energy than its predecessor, 80% less than other new buildings in Amsterdam. The energy-saving features cost an additional $700,000, but they save the bank $2.9 million a year in energy bills and paid for themselves in three months. The bank – the International Netherlands Group – is now Holland's second largest bank.[6]

Village Homes, Cool Living

In Davis, CA, is a subdivision with a difference. Village Homes was completed in 1981, and is the coolest, most desirable subdivision in town, where homes sell for $20,000 above the market value in a third of the normal time. Included in its 70 acres are 240 homes, 12 acres of green space, 12 acres of common agricultural land, and 4,000 sq ft of commercial space. Some other characteristics:

- No front roads; car access by back lanes; resident-designed green space in front.
- No storm drains; natural grass swales saved $800/house, which was invested in green space.
- Passive-solar houses and tree-shaded streets save 50%–66% on energy bills.
- The houses are clustered in groups of eight around shared green space.
- 80% resident participation in community activities and harvest festivals.
- Many residents work locally or cycle to work.
- Guestimate of reduced CO_2 emissions per household: 75%.

Source: Designing Sustainable Communities: Learning from Village Homes, Judy and Michael Corbett. Island Press, 2000.

"By skimping on design, the owner gets costlier equipment, higher energy costs, and a less competitive and comfortable building; the tenants get lower productivity and higher rent and operating costs."
— *Natural Capitalism*

Design Green, Efficient Buildings
We owe it to our children and grandchildren, and to the forests and parks we love, to reduce the amount of energy we use. The know-how is literally tumbling out of the Internet (see box), but it does mean re-educating builders and tradespeople, who often have a set way of doing things. By building an efficiency incentive into the design contract, the designer can keep a portion of the energy savings over the first few years as a bonus fee.[7] ENERGY-10 is a PC-based design tool that helps architects and designers identify the most cost-effective, energy-saving measures in less than an hour, leading to savings of 40%–70% with little or no increase in construction cost (see box).

Resources

Advanced Building Technologies: www.advancedbuildings.org
Architects, Designers, Planners for Social Responsibility: www.adpsr.org
Building America (50% less energy): www.eren.doe.gov/buildings/building_america
Canada's R2000 Buildings: www.r2000.org
Center for Maximum Potential Building Systems (Austin TX): www.cmpbs.org
Center for Resourceful Building Technology: www.crbt.org
E-Design, Florida Sustainable Communities: www.fcn.state.fl.us/fdi
Energy Efficient Building Association: www.eeba.org
Energy Star Buildings: www.energystar.gov
ENERGY-10 software: www.nrel.gov/buildings/energy10
Environmental Building News: www.buildinggreen.com
Green Building Information Council, Canada: www.greenbuilding.ca
Greenclips: www.greendesign.net/greenclips
Green Design Network: www.greendesign.net
High Performance Buildings: www.eren.doe.gov/buildings/highperformance
Oikos Green Building Source: www.oikos.com
Partnership for Advanced Technology in Housing: www.pathnet.org
Solar Housing case studies: www.eren.doe.gov/buildings/case_study/casestudy.html
Sustainable Buildings Industry Council: www.sbicouncil.org
US Green Building Council: www.usgbc.org
Green Development: Integrating Ecology & Real Estate, by Alex Wilson/RMI (Wiley)

Design Travel-Smart Buildings
A travel-smart office building rewards its employees for not commuting by single-occupancy vehicle. When Intrawest was building a commercial block in Surrey, B.C., it saved $500,000 on the required parking stalls because Surrey council agreed that its "TravelChoices" program would effectively reduce single vehicle commuting (see #33).

Retrofit Old Towns and Cities
There is a great future in renovating the desolate, parking lot-filled areas of downtown cities, bringing them back to life with dense, eco-designed urban spaces and water-filled parks. Wherever there's a sad, ugly strip, there's an opportunity for renovation instead of allowing sprawl to consume more forest and farmland.

Build Sustainable Communities
A typical modern subdivision forces its residents to drive in order to get anywhere. If we go on building them, there'll be no "there" to get to. By designing attractive, pedestrian, mixed-use, ecologically planned subdivisions, we can lower CO_2 emissions while providing a satisfying traditional neighborhood for people to live in (see #29). It's only a matter of time before someone builds the first carbon-neutral housing project.

39
For Financial Institutions

> "When you look back on your life, will you be able to say, 'Yes, I joined the pioneers of natural capitalism, and we did everything we could to make human industry a friend, not a foe, of the planet?'"
> — Tessa Tennant, Director, SRI Policy & Strategy, Henderson Investors, UK.

The world is flush with money. It rushes around cyberspace at trillions of dollars a day, constantly chasing a better return. A small edge here, a big gain there. Everyone's doing it, from your pension fund manager to the lunch-club investors who meet once a week over a glass of wine. Hedges today, derivatives tomorrow; and all the while the world's temperature is slowly growing, fueled by the money that these investments provide. An oil-well here, a coal-mine there; this is the way global warming is being financed, all the way to the Arctic meltdown, the extinction of the polar bears, the vanishing of the ocean conveyor belt that keeps northern Europe warm.

Just outside the enchanted circle of bankers and investors, however, growing numbers of people are becoming anxious and hungry for something different. When the British Co-op Bank broke ranks with the mainstream banks in 1992 and shifted its policies to become an ethical bank, 84% of its customers thought it was a good move. The bank's deposit base grew by 13% as people brought their money from other banks, and profits climbed from a £6 million loss in 1991 to a £10 million profit in 1992.

Can finance help to slow down global warming and assist in the transition to a sustainable, solar-hydrogen economy? The global re-insurance industry certainly hopes so; they are the ones who are taking the biggest hits each time there is a new climate-related disaster. Investors in the New Alternatives Fund believe so; they are putting their money behind the wind turbines, fuel cells, and solar arrays that will be at the center of the new economy. The European Bank for Reconstruction and Development, the Dexia Group, and the Public Finance International Bank believe so; they have put $58 million into the Dexia-FondElec Fund for Energy Efficiency and Emission Reduction, specifically to address greenhouse gas reductions. And GMAC Mortgage believes so; it has invested over $770 million in energy efficient mortgages.

The opportunities are enormous: lending for energy-efficiency and solar upgrades, investing in sustainable community businesses, screening loans and investments for environmental performance, establishing carbon-offset investment funds, adopting environmental operating policies and principles. The wave is coming; the field of socially responsible investment has grown from $40 billion in the early 1980s to over $2 trillion in 2000 in the US. That's one dollar in eight of all money held in stocks. The public is looking for ways to make a difference, and banks, credit unions, and other financial institutions can provide it.

Where is all the Cool Money?

Number of homes in the U.S.: 100 million[1]

% of homes in America that can be called "efficient": 2%[2]

Greenhouse gases resulting from U.S. homes: 286 million tons (1998)[3]

Typical annual return on investment from a home efficiency upgrade: 30%

GHG reduction from a U.S.-wide efficiency upgrade: 85 million tons per year

% of total U.S. greenhouse gases that this represents: 5%

Typical increase in market value for every $1 saved in energy costs: $20[4]

Typical annual savings in a house that exceeds the Model Energy Code: $170 to $425[5]

Typical increase in market value for such a home: $4,250 to $10,625

Typical cost of a home retrofit: $4,000

Potential size of the untapped U.S. energy efficiency loan market: $392 billion

Current availability of loans to meet such a market: low

Cold weather operation at the Green Mountain Power wind plant in New England

Offer Energy-Efficient And Solar Mortgages

A $400 billion market for energy-efficient mortgages is sitting untapped because few banks have investigated how it works. A homeowner's monthly utility bill will be lower after an upgrade, so her mortgage can be larger, enabling her to finance the upgrade. The Residential Energy Services Network is developing a national market for certified home energy rating systems to assist the process (see box). Banks and credit unions that move now will be ready to service the market explosion when progressive energy codes arrive (see #65). To help families finance their efficiency improvements, see #5, *Set Up a Home Energy Account.*

As the price of solar PV roofs and solar shingles falls, there will be a growing demand for solar mortgages, repaid through energy savings and the income from net metering.

Offer Location-Efficient Mortgages

When a family relies on public transit, walking, or cycling rather than driving, and can shop, work, and socialize in its local community, reducing CO_2 emissions, it spends less on transportation. In Chicago, a family living in a densely populated, transit-rich, urban area will spend $350–$650 less on transport per month than a family living in a typical suburban household. By incorporating the savings into a location-efficient mortgage, families can gain a significant stretch in their borrowing capacity. Currently available in Chicago, Seattle, and LA. Coming soon to Portland and San Francisco.

Open A Community Banking Fund

In 1989, the Chittenden Bank (then the Vermont National Bank) opened the Socially Responsible Banking Fund, allowing its customers to specify that their savings be used within Vermont to support affordable housing, small businesses, sustainable and organic farms, environmental businesses, community building, and education. By 1999, deposits had reached $229 million. From 1997 to 1999, the number of investors grew from 11,000 to 29,000, who were lending to almost 2,000 local businesses and initiatives. There is an indirect link to CO_2 reductions through support for the local, as opposed to the global, economy.

Say Yes To Climate-Friendly Investments And No To Unfriendly Ones

The first socially responsible investment fund to create a "Cool Planet" investment portfolio will be inundated with funds. The companies exist and the public wants to invest (see #9). Conversely, if coal and oil companies can't attract new finance, they can't expand.

And yes, global ethics do matter in banking. In 1996, the Export-Import Bank of America refused to provide export guarantees for the highly controversial Three Gorges hydroelectric dam in China. Increasingly, bank customers, pension fund holders, and insurance fund members want an assurance that their money is not being used for purposes like clearcut logging or fossil fuel development.

Resources *(see also #9)*

ASE Energy Efficient Financing: www.ase.org/consumer/finance.htm

Energy Efficient Mortgage Home-owners Guide: www.pueblo.gsa.gov/cic_text/housing/energy_mort/energy-mortgage.htm

Energy Efficient Financing Info Center: www.nationalguild.com/residential/hers.html

Energy Efficient Mortgages: www.national-hero.com

Location Efficient Mortgages: www.nhi.org/online/issues/103/lem.html Also: www.locationefficiency.com

National Energy Raters Association: www.energyraters.org

Residential Energy Services Network: www.natresnet.org

Solar Bank: www.solarbank.com

Sustainable Banking—the Greening of Finance: www.greenleaf-publishing.com

40

Embrace Natural Capitalism

Warren Gretz

Biodiesel mini motor home run by diesel fuel produced by the "Green Grease Machine."

As humans on this rather small planet, continuing our adventure into the unknown, we have unwittingly pumped too much carbon dioxide into the atmosphere. From the shared planetary bridge, the orders go out: Switch all fuels! Upgrade all efficiencies to the maximum! Max up protection of living carbon systems to absorb excess carbon dioxide! Report back to the bridge within the hour!

This is not the only problem we face as a result of our carelessness with the planet's ecosystem. Other reports are coming in, thanks to the Internet Pony Express. Fresh water supplies running low! Fisheries collapsing! Toxics accumulation approaching dangerous overload! Extinctions increasing! In the "human societies" sector of the Planetary Indicators Console, the green lights say: Democracy healthy! Human lifespan increasing! Living standards improving! But in the sector labeled "planetary life-supports," almost all the lights are flashing red. The human system is thriving, but natural systems are collapsing. On a rational planet, the orders would immediately go out: Rethink all operating systems! Upgrade natural ecosystems protection in all systems! Adjust all values accordingly!

The planet's leadership, however, is not free of politics. The old guard, who grew up in an era when nature was bountiful and waste was not a problem, are not very concerned. To the younger generation, the flashing red lights warn that the prospects for a sustainable future are disappearing before their eyes. The old guard struggles to maintain control, encouraging its friends in the boiler-room to explain that it's too soon to read anything into the signals, and that the planet cannot afford to change course.

Elsewhere on the planet, however, people who see the same signals are wasting no time. They are moving ahead unilaterally, taking urgent steps to adjust their values and rethink their operating systems. Businesses, homes and institutions are sending out the orders: Reduce all carbon emissions and waste outflows to zero! Maximize local ecosystem protection! Re-engineer all systems to maximize ecosystem harmony! Adjust all assumptions, signals and price systems to include natural capital! The captains on the bridge may not be listening, but the process of changing course has begun.

The Natural Step's Four System Conditions

In order for a society to be sustainable, nature's functions and diversity must not be:

1. subject to increasing concentrations of substances extracted from the Earth's crust;
2. subject to increasing concentrations of substances produced by society; or
3. impoverished by overharvesting or other forms of ecosystem manipulation. And
4. resources must be used fairly and efficiently in order to meet basic human needs worldwide.

Source: *The Natural Step for Business* (New Society Publishers)

"It was not until we saw the picture of the earth from the moon, that we realized how small and how helpless this planet is; something that we must hold in our arms and care for."
— Margaret Mead

Join The Natural Step

One of the organizations that has been leading the call to adjust values and re-organize our operating systems is The Natural Step. It's a systems approach, designed by Dr Karl-Henrik Robert of Sweden (with help from hundreds of scientists) to assist companies and organizations to re-design their fundamental approach to business and the environment by re-aligning their operations with nature's laws.

These laws are expressed in four System Conditions (see box), and provide a reliable framework that businesses and organizations can use to begin the process of adjustment. Recognizing that a large business is like a huge oil-tanker that cannot turn around instantly, The Natural Step encourages a step-by-step process so that a company or institution can gradually phase out its toxic chemicals, reduce its CO_2 emissions, reduce its ecological footprint, increase its resource efficiency, change its harvesting practices, and bring its waste stream down to zero.

In Sweden, when the hotel chain Scandic adopted The Natural Step as its pathway to natural capitalism, the company's Environmental Dialogue process (involving 5,000 employees in eight countries) identified and adopted over 2,000 separate steps, from changing the way soap was provided to phasing in "eco-rooms" where 97% of the materials used are recyclable or biodegradable. IKEA, another Swedish company that has embraced The Natural Step, is steadily increasing the percentage of organically grown cotton and certified forest products in its furniture.

The first company in North America to adopt The Natural Step was Interface, the carpet company. Interface's chairman, Ray Anderson, realized that making a carpet that required 2 pounds of fossil fuel for every pound of petroleum feedstock, and that served a useful purpose for 10 years and then sat in a landfill for 20,000 years was not a very smart – or Earth-friendly – way to do business. In addition to developing a carbon-neutral carpet tile – Solenium – that can be remanufactured back into itself, Interface changed the whole way it did business. Instead of selling carpets, it started offering a carpet *service*, leasing squares of carpet and replacing a square whenever it wore out, recycling it into a new one. By these innovations, Interface reduced the flow of materials and embodied energy in its Solenium squares by 97%. Other participants in North America include Nike and the Canadian town of Whistler (see #70).

Join Business for Social Responsibility

Founded in 1992 as a global resource for companies seeking to sustain their commercial success in ways that demonstrate respect for ethical values, people, communities and the environment, BSR has over 1,400 member companies, including American Express, AT&T, Honeywell, Home Depot, Starbucks, Time Warner, Universal Studios, Wal-Mart and Walt Disney, and smaller companies such as Bright Horizons, Fel-Pro, Fetzer Vineyards, Patagonia, Tom's of Maine, and the Stonyfield Farm Yogurt Works. See www.bsr.org and www.cbsr.bc.ca.

Read *Natural Capitalism* and *The Natural Step*

These two books provide an inspiring introduction to the area, and are full of detailed, thought-provoking stories that provide evidence of the turn-around that is underway.

Resources

Greenbusiness.com: www.greenbiz.com

Natural Capitalism: www.naturalcapital.org

The Natural Step (USA): www.naturalstep.org

The Natural Step (worldwide): www.detnaturligasteget.se

Scandic Hotels "eco-room": www.scandic-hotels.com/br/30/30rummet.html

Natural Capitalism: Creating the Next Industrial Revolution, by Paul Hawken, Amory Lovins and L. Hunter Lovins (Little, Brown, 1999)

The Natural Step for Business: Wealth, Ecology and the Evolutionary Corporation, by Brian Nattrass and Mary Altomare (New Society, 1999)

41
Set New Goals

> **"The sunshine that strikes American roads each year contains more energy than all the fossil fuels used by the entire world."**
> **— Denis Hayes, Bullitt Foundation**

The next 15 solutions apply to corporations and utilities that generate, buy, sell, or distribute coal, gas, oil, and electricity. There are many changes happening in this field, with new companies emerging and older ones adopting new technologies. We'll start with electric utilities and work our way around to the oil companies.

Electricity was discovered in 1797 by the Italian physicist Volta, and first transmitted over a distance in 1891, when a cable was erected in Germany across the River Neckar between Lauffen and Frankfurt. Today, we all depend on it. Take it away, and the world's industry and commerce would grind to an immediate halt.

We have been ingenious in the ways we've found to generate electricity. We dam rivers; burn coal, gas and oil; erect windmills; tap the Earth's geothermal heat; split atoms of uranium; install tidal turbines; gather it from sunlight, biomass, and garbage; and generate it from hydrogen-oxygen fuel cells. We can even use a bicycle to drive a small generator.

There are over 3,000 electric utilities in North America: 2000 publicly owned and 1000 owned by private investors. Historically, utilities have chosen reliable, non-interruptible power sources that produce low-cost electricity at large central plants, mainly from coal (USA) and hydro (Canada). With the turning of the new century, however, everything is changing. Restructuring is opening up a host of new problems and possibilities. Utilities are being asked to generate renewable energy and reduce CO_2 emissions. The prices of oil and gas are rising in anticipation of a global shortfall. The price of renewables is falling, and will continue to fall. Appliances and energy technologies are becoming more efficient. Customers who are concerned about the environment want to know where their power comes from and are asking for choices. Weather extremes are causing unexpected power outages; glaciers are melting, threatening future hydro shortages. Businesses that use 24-hour computers are looking for non-interruptible power. People are producing their own energy from fuel cells, micro-turbines, solar collectors, and wind turbines. New power storage devices, such as carbon fiber flywheels, are appearing. Spot-market pricing is changing the way retail utilities bid for power. Gas and coal companies are exploring ways to turn fossil fuels into hydrogen, splitting off the CO_2 and locking it away underground. About the only thing that has not changed is the dream of nuclear fusion, which remains a dream in spite of the enthusiasm of its backers.

For many utility workers, it must be exciting to be present at the start of the biggest revolution their world has seen since electricity was first distributed.

Electricity that Rattles the Climate

US electricity generation 1998: 3,618 billion kwh

Tons of C02 released by US electricity generation: 1,895 million

% of US greenhouse gas emissions that this represents: 35%

US electricity from hydro 1998: 319 billion kwh (9%)

US electricity, non-hydro renewables 1998: 75.3 billion kwh (2%)

Canadian electricity generation 1998: 551 billion kwh

Tons of C02 released by Canadian electricity generation: 100 million

% of total Canadian greenhouse gas emissions that this represents: 14.5%

Canadian electricity from hydro 1998: 329 billion kwh (60%)

Canadian electricity, non-hydro renewables 1998: 6.1 billion kwh (1%)

Sources: Energy Information Agency, Canada's Greenhouse Gas Inventory

Electricity towers at sunset

Warren Gretz

Join Climate Challenge
Climate Challenge is a partnership between the U.S. Department of Energy and the electric utility industry that encourages cost-effective ways to reduce, avoid, or sequester greenhouse gases:

- Efficiency improvements in generation, transmission, distribution, and end-use.
- Increased use of energy efficient electro-technologies
- Fuel-switching to lower carbon fuels, such as natural gas or nuclear
- Fuel-switching to renewable energy
- Transportation actions, including switching to natural gas and electric vehicles
- Forestry carbon sequestration
- Recovery of methane gas from landfills and coal-seams
- The use of fly-ash as a Portland cement substitute.

By March 1997, Climate Challenge had participation agreements with over 650 utilities and utility trade associations representing 71% of 1990's electric generation and utility carbon emissions. Commitments by utilities under the partnership will reduce CO_2 emissions by 172 million tons by 2000 (8% of utility GHGs). Their *Climate Challenge Options Workbook* is so packed full of ways to reduce CO_2 emissions, they should get a medal.

Detroit Edison is an investor-owned utility with two million customers in southeastern Michigan and 10,000 MW of capacity, which comes from eight fossil fuel power plants, a nuclear power plant, and a hydroelectric facility. It is the seventh largest electric utility in the U.S. As a leading member of Climate Challenge, Detroit Edison generates its own solar power and offers it to customers through its *Solar Current* program; installs ground-source heat pumps; encourages customers to switch to more efficient appliances; runs energy efficiency classes in local schools; sponsors a *Solar Schools* program; provides home energy efficiency and retrofit workshops; offers energy efficiency audits and information for businesses; is planting ten million trees; founded Plug Power, a leading fuel cell maker; extracts energy from landfill gas plants; and is experimenting with superconductive cables (see #49).

Offer Energy Services, Not Just Kilowatt Hours
The idea that customers just want cheap power is a myth bred during the years when nobody thought about electricity or where it came from. In today's world, people are looking for utilities that will serve them as individuals and as a community, show how they can become more energy-efficient in their homes and businesses, give price-breaks on the most efficient equipment, provide "6 nines" power (99.9999% reliable), help them to restore salmon runs and eliminate smog, and contribute towards building more sustainable communities. By satisfying these wishes, utilities will generate solid customer loyalty the way companies like The Body Shop and Patagonia have done.

Resources

Canadian Electricity Association: www.canelect.ca

Climate Challenge: www.eren.doe.gov/climatechallenge

Detroit Edison's environmental initiatives: www.dteenergy.com/environment

Lighten the Load—Preventing Power Shortages with Clean, Efficient Energy: www.ems.org/lighten_the_load

Sustainability in the Electrical Utility Industry: www.wbcsd.ch/sectoral/electricity

42
Develop a Distributed Grid

"We are on the verge of a significant transformation of the electric utility industry that will, 50 years from now, perhaps seem as significant as Edison's invention."
— Terry Esvelt,
Bonneville Power Administration.

It is a hot summer afternoon in the year 2005, and you have just installed a 4 kW plug-in solar system on your roof, costing you $5,000. Back in 2000, it would have cost $32,000, but the price has dropped four-fold thanks to economies of scale from large-scale mass production, and you have received a $3,000 rebate from your utility's Public Benefits Fund.

Your solar system is connected to the grid through net metering, and you have bought a domestic flywheel that spins in a vacuum at 20,000 revolutions a minute to store your surplus. Your utility buys and sells power by the millisecond on the spot market and pays the highest price when demand is highest — such as this afternoon.

Buying solar energy from customers like you helps the utility to meet its peak demand, avoid outages, and postpone expensive investments in new generating capacity. Your flywheel contains a computerized microcontroller, and when the price hits a pre-agreed threshold it discharges its energy into the grid, while you are lazing in your hammock with a glass of iced tea.

It's called the distributed grid, and it is promising to revolutionize the way in which utilities collect and distribute power. Large central power plants will still exist, but they will be assisted by millions of smaller solar, wind, fuel cell, diesel and gas-fired microgenerators that sell their power into the grid at the optimum time via computerized telecommunications. The utilities will optimize their power flow by storing surplus energy in flywheels, or in fuel-cell systems such as Regenesys which uses electrolytes to store 5-500 megawatt hours of electricity, enough to supply a small town for hours at a time.

- In the Pacific Northwest, the Bonneville Power Administration is planning such an energy web to replace its mainframe model of power production.
- In New York, the State Energy Research and Development Authority is planning to aggregate customers' back-up generators to meet spikes in energy demand.
- In Spokane, Avista Corporation is test marketing a fuel cell that will enable business and residential consumers to plug in and join the distributed grid.
- From 2002, Plug Power will be selling their GE HomeGen 7000 residential fuel cell that runs on natural gas or propane, and can meet 100% of a household's electricity needs from a box the size of a domestic fridge.

Resources

Avista: www.avistacorp.com

Beacon Power super flywheels: www.beaconpower.com

Bloomberg electricity trading: www.bloomberg.com/energy

DoE's distributed power site: www.eren.doe.gov/distributedpower

FuelCell Energy: www.fce.com

Interstate Renewable Energy Council: www.irecusa.org/connect.htm

Micropower – The Next Electrical Era by Seth Dunn, Worldwatch Institute

NatSource (energy brokers): www.natsource.com

Net Metering: www.eren.doe.gov/greenpower/netmetering

Plug Power: www.plugpower.com

Regenesys energy storage: www.regenesys.com

Silicon Energy: www.siliconenergy.com

Sure Power Corporation: www.hi-availability.com

'The Energy Web' — a speech by Terry Esvelt: www.bpa.gov/corporate/kc/sp/sp012600x.shtml

ZeGen: www.zevco.com/zegen.html

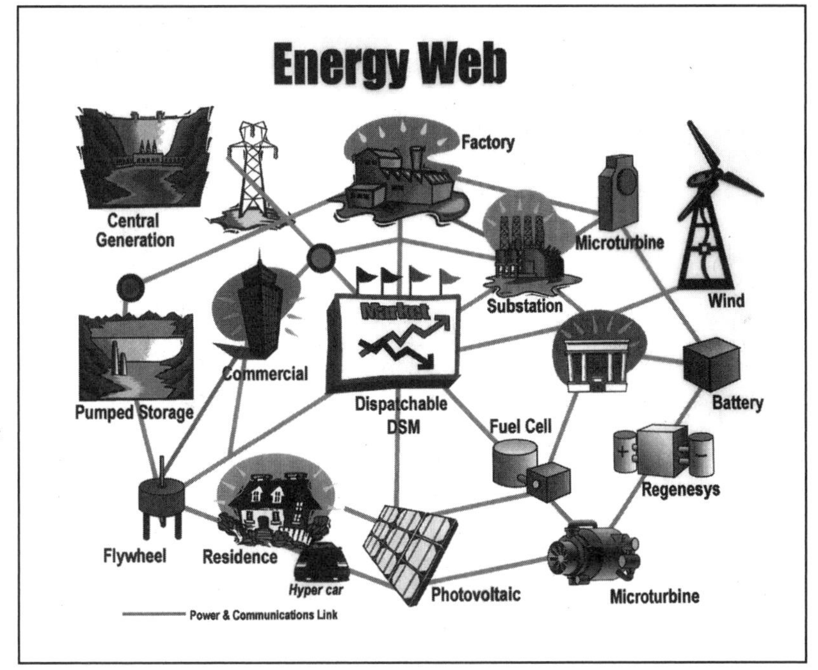

Introduce Net Metering
Net metering enables a customer to send surplus energy from a solar, wind or microturbine generator back into the grid and get paid for it as the meter spins backwards. Different states have introduced different policies; the best allow all energy technologies and all customer classes to qualify, place no restriction on the volume of energy that can be sold, and allow customers to balance their credits with their debits over a whole year (see #63).

Remove the Barriers
A US Department of Energy Report[1] examined 65 distributed power projects, and found that 89% of them experienced significant barriers at the hands of the utilities, which often caused delays. A solar PV grid project in British Columbia said "it was like pulling teeth" trying to get the local utility to support them. Be part of the future, not of the past.

Introduce Smart Meters
Imagine an indoor electricity meter that gives you a colored digital read-out of your real-time power consumption, not just for the house as a whole, but for each room, or even each function. It tells you what the spot price of energy is, carries out pre-arranged switch-offs to save power when it is expensive, receives messages from your utility, allows your utility to bill you without making an expensive personal visit, and gives you a monthly read-out of your CO_2 emissions. Puget Sound Energy has tested Internet-based Silicon Energy thermostats in 100 Kent, WA, homes enabling the homeowners to monitor and adjust their heating systems while they were away. Early trials suggest a possible 10% saving on energy bills.

Encourage Businesses to buy their own Microgenerators
The average computer is designed to withstand a power outage lasting no more than 0.008 seconds. For companies, a power outage can carry an enormous financial cost, making the purchase of an on-site power generator a good investment. First National Bank of Omaha, the seventh largest credit card processor in the US, has installed a fuel cell power system from Connecticut-based Sure Power in its Omaha processing center that provides "6 nines" power (99.9999% reliable). In Bensenville, Illinois, McDonalds has a state-of-the-art, gas-fired, 75 kW Parallon microturbine for use during peak hours when electricity is expensive, that will save them $35,000 a year. If other customers followed suit, it would save the local utility from having to invest in new generating capacity. In Wakefield, Rhode Island, the South County hospital installed a 200 kW PC25 fuel cell that produces a third of the hospital's electricity during peak hours, saving them $60,000-$90,000 a year. The Los Angeles Department of Water and Power is installing a 250 kW fuel cell energy plant in its headquarters building. There's nothing like leading by example.

43

Build Solar Capacity

Craig Miller Productions and DOE

Workers install PV panels on Natatorium in Atlanta

With every month that passes, solar photovoltaic (PV) energy is becoming more efficient, more affordable, and easier to install. In Australia, Pacific Solar offers a plug-in solar roof that can be installed in a day. In the US, Siemens Solar offers a pre-engineered "Earth Safe!" solar kit for the residential market designed for net metering that can be installed by a qualified electrical contractor at $3,395 for a 300 watt system ($11.30 per watt) for utility customers.

In Holland, Greenpeace commissioned a study by KPMG that showed that a 500 MW solar production facility (5 times larger than anything that exists today) would reduce the price of PV modules to $1.25 per watt. When solar hits $1.00 per watt, home-owners in sunny areas will start lining up at the bank for solar mortgages, banks will start providing them, and utilities will use net metering to harvest the excess energy during peak summer hours. By 2005, grid-connected solar energy should make an explosive entry into the market.

Encouraged by solar's progress, the solar PV industry in the USA has set itself the goals of growing by 20-30% a year for the next 20 years and providing up to 15% of new US peak electricity generating capacity by 2020 (3,200 MW; today's solar capacity is around 200 MW). Industry experts estimate that the cost of solar will fall to $1.00 per watt by 2010. The Sacramento Municipal Utility District (SMUD) estimates that solar will fall to $3 per watt as early as 2003. The prices of oil and gas, meanwhile, will be going through the roof because they are not renewable resources, and once production has passed its peak, prices soar. By encouraging early entry into the market, utilities such as SMUD are accelerating the transition to solar financial self-sufficiency.

If federal and state policies were to remove the hidden environmental subsidies that coal- and gas-fired electricity receive and transfer an equivalent subsidy to solar and other renewable energies, the price might fall sooner. As soon as this happens, the "Henry Ford" factor will kick in, and householders throughout the world will be picking up the phone to order a plug-in, net-metered solar roof, financed by a solar mortgage. If we are serious about slowing down climate change, this is what we should be doing.

The Cost of Solar in Sacramento

Price of a SMUD 2 kW solar system in 2000: $4,740 ($2.37 per watt)

Price without the SMUD buydown: $10,140 ($5.07 per watt)

Electricity generated per kilowatt per year on a south-sloping roof: 1800 kWh

Electricity generated per kilowatt per year on an east- or west-sloping roof: 1530 kWh

Typical annual needs of a Sacramento household: 7,200 kWh

Reduced utility bill from a 2 kW system @ 10 cents per kWh: $360

Monthly Green Fee for solar pioneers: $4

Payback period for a 2 kW system (not counting the Green Fee): 13 years

Avoided CO2 emissions from a 2kW system per year: 1.8 tons

Source: SMUD

"All the world's energy needs could be achieved by solar many thousands of times over."
— Roger Booth, Head of Renewable Energy Supply and Marketing, Shell.

Jim Hamm

SMUD solar array in Sacramento

Learn from the Solar Pioneers

The Sacramento Municipal Utility District (SMUD) is the leading utility in North America when it comes to obtaining power from solar and selling it to customers. In 1989, the Rancho Seco nuclear facility was shut down following a successful anti-nuclear campaign. This left SMUD with a hole in its power supply, part of which it met by constructing a 4.2 MW PV plant and launching a PV Pioneers Program. By the end of 1999, there were PV systems on 550 homes, churches, and commercial buildings that were generating 2 MW of capacity. Most of SMUD's rooftop PV units are marketed as Sunslates, which are attached to the roof as regular shingles. SMUD's goal is to install an additional 10 MW by 2002, causing the cost of solar to fall below $3 per watt (9 cents/kWh) by 2003, at which price it is competitive for peak electricity. Customers will pay $2.80 per watt, and with a 20 cent subsidy, the solar residential market will reach take-off point.

The Los Angeles Department of Water and Power is using its Public Benefits Fund (see #64) for a $38 million, four-year program that will pay up to $5 per watt for customers to purchase and install their own PV systems. The goal is to get the price of solar down to $2.50 per watt and launch LA as a solar-powered city, attracting PV manufacturers to locate there. In New York, the Long Island Power Authority has set up a $160 million fund to install 10,000 solar arrays over the next ten years. To see what Commonwealth Edison is up to in Chicago, see #23.

Join the Utility PhotoVoltaic Group (UPVG)

The UPVG is a non-profit organization consisting of 93 US electric service providers that are working together to make solar electricity a mainstream choice. By January 2000, through a partnership with the US Department of Energy, their cost-sharing TEAM-UP project had helped to install 7.5 MW of PV power in 2,300 installations in 31 states, including schools, businesses, and homes.

Educate Your Customers

When three Texas utilities held a two-day workshop on new power generation, including detailed discussions of costs, availability, and viability, customers picked solar, windpower, and energy efficiency by a wide margin over purchasing wholesale power or building natural gas- or coal-fired plants. The secret to success lies with education to make customers aware of the environmental costs of different kinds of energy.

Resources

Earths Safe! solar kit: www.aapspower.com/aapspower/

Hubbert Peak, future oil shortage: www.hubbertpeak.com

Long Island Power Authority's Solar Pioneers: www.lipower.org/pioneer.html

Los Angeles Department of Water and Power: www.ladwp.com

Pacific Solar's Plug&Power solar roof: www.pacificsolar.com.au

Sacramento Municipal Utility District: www.smud.org

Siemens Solar: www.siemenssolar.com

Solar energy primer: www.nrel.gov/lab/pao/solar_energy.html

Solar Energy Resource Maps: www.rt66.com/rbahm/maps.htm

Sunslates: www.astropower.com/developers.htm

Utility PhotoVoltaic Group: www.upvg.org/upvg

44

Offer Wind and Other Renewable Energy

It would not be true to say "everyone's doing it," but in May 2001, 76 utilities in the US and several in Canada were offering green energy to customers willing to pay a premium to receive it, and the number is growing steadily.[1]

For years we have been schooled to think of energy as being all the same, when all that mattered was the price. As people become aware of how their energy is generated and the impact it has on the environment, however, they are willing to pay more for green energy. Buying the cheapest energy, the new thinking says, is like buying the cheapest food for your child: it can cause untold damage in the future for the sake of saving a few dollars today.

In Texas, Austin Energy offers energy from wind and landfill gas (methane) for a premium of 0.4 cents per kWh. In Michigan, Traverse City Light & Power offers wind energy for a premium of 1.6 cents per kWh. Wisconsin Electric offers energy from wind, water and landfill gas for a premium of 2 cents per kWh. The Tennessee Valley Authority offers solar, wind and landfill gas through its Green Power Switch program for a premium of 2.7 cents per kWh.

Eighty or so utilities out of 3,000 may not seem like much, but it is a beginning. As utility restructuring opens up consumer choice, utilities are realizing that unless they generate customer loyalty, they risk losing their market to a new generation of power companies offering solar and wind energy, residential and commercial fuel cells, smart metering, energy efficiency tools and carbon neutral power. A 1999 study by the National Renewable Energy Laboratory found that 52-95% of residential customers were willing to pay a modest increase for power from renewable sources. 70% said they were willing to pay an additional $5 per month, 38% would pay $10, and 21% would pay $15 or more.

Some utilities have tried to market green power without much success. When done correctly, the potential for premium green power is enormous. In Colorado, a grassroots marketing campaign for wind energy run by a local environmental group in partnership with the utility managed to sell five times more wind energy than the utility had planned for (see #47).

In February 2001, the Bonneville Power Administration announced plans to purchase 1,000 MW of wind power to address regional power shortages, citing price as a leading reason. "With its gusty Columbia River Gorge and windswept open farm country, the Northwest is a prime location for wind farms. Washington has tens of thousands of megawatts of wind-power potential, easily enough to generate power to supply 3 million homes" (*Seattle Times*, Feb 22nd 2001).

Resources

Austin Energy's Green Power Choice: www.austinenergy.com/greenchoice

Bonneville Power Administration: www.bpa.gov

Center for Resource Solutions: www.green-e.org

Commonwealth Energy Corporation, CA: www.electric.com

Green Energy in CA, PA and NJ: www.edf.org/programs/energy/green_power

Green Power for LA: www.greenla.com

Green Power Network: www.eren.doe.gov/greenpower

SMUD's Greenergy: www.smud.org/green

TVA's Green Power Switch: www.tva.gov/greenpowerswitch

US Green Pricing Activities: www.eren.doe.gov/greenpower/pricing.shtml

Wisconsin Electric's "energy for tomorrow": www.wisconsinelectric.com

> "What degree of proof about the human catastrophe from global climate change do we need, before we are motivated to act to prevent it?"
> — Eric Chivian MD,
> Physicians for Social Responsibility

Set a Goal for Renewable Energy
In response to the public demand for clean energy, Seattle City Council has mandated that municipally owned Seattle City Light will work to obtain 100% of its new energy needs from wind, geothermal, solar, landfill gas, and energy conservation. If this is not enough, it will purchase or organize carbon offsets for the remaining emissions. To achieve the phase-out of CO_2 emissions needed to protect the world's climate, utilities and governments need to set a high target, such as 5% green energy by 2005, reaching 80% by 2025. Seattle has started by selling its 8% stake in a coal-fired power plant, and is inviting bids to develop renewable energy sources.

Top Ten Green Pricing Programs — Most Participants (April 2000)

Utility	Program	Participants
LADWP	Green Power for a Green LA	31,000
PSCo Colorado	Windsource	14,500
Wisconsin Electric	Energy for Tomorrow	12.000
PSCo Colorado	Renewable Energy Trust	12.000
SMUD	Greenergy	6,100
Madison Gas & Electric	Wind Power	5,200
Wisconsin Public Service	Solar Wise for Schools	4,000
Eugene Water & Electric Board	EWEB Windpower	2,700
Hawaiian Electric	Sun Power for School	2,600
Portland General Electric	Salmon-Friendly Power	2,500

Source: National Renewable Energy Laboratory, April 2000

Get your Green Power Certified
How can a customer know if a utility's green power is really green? The world is full of half-truths. Enter the Center for Resource Solutions (CRS), a California-based non-profit dedicated to economic and environmental sustainability that established the Green Pricing Accreditation Initiative known as "Green-e" in 1999 (see #23). CRS has been certifying green power marketers such as GreenMountain.com and the Commonwealth Energy Corporation; in April 2000 it certified its first three utilities: TVA's "Green Power Switch," Wisconsin Electric's "Energy for Tomorrow," and Madison Gas & Electric's Wind Power program. Certification means that a utility has met stringent guidelines regarding the use of renewable resources, appropriate pricing and marketing, and customer education. There is also an annual verification process.

Tell your Customers How their Electricity is Generated
It's not required by all states or provinces, but it's going to be like food labeling: customers will want to know where their electricity comes from, and what emissions it produces (see #63). An annual greenhouse gas emissions statement would also be welcome, to help people see what their emissions are.

Beat the Rush!
In Texas, the state's electric utility restructuring bill requires utilities to generate 2000 MW of energy from renewable resources by 2009, which is causing them to scramble. Texas, Kansas and North Dakota have sufficient wind energy potential between them to meet the needs of the entire US electric grid (3.5 million gigawatt hours). In Arizona, utilities must provide 1.1% of their product from renewable energy by 2007, 50% from solar. In Oregon, new power plants of 25 MW or more must not release more than 0.7 lbs of CO_2 per kilowatt hour, 17% less than the most efficient gas-fired plants. This obliges them to find other ways to reduce CO_2, including buying carbon offsets (see #52).

45

Encourage Residential Energy Efficiency

Why would a utility want its customers to buy less power? Isn't that an odd thing for a profit-oriented business to do? There are three reasons why it makes sense:

First, if a utility can persuade its customers to reduce their peak load power use by 10%, that's equivalent to reducing the entire load capacity by 10%, which can save it from having to increase its generating capacity by 10%. The cheapest way to obtain new energy is to harvest energy that is being wasted. Energy gathered this way costs 0.5 to 3 cents per kWh.

Second, a utility that wants to build loyalty among its customers will service their needs, not sell them energy they don't need. It is only because North American customers have lived in a cocoon of ignorance around energy issues for so long that they continue to pay for an excess of energy without complaining. Customer choice is about to change all that.

Third, a proven reduction in demand from coal- or gas-fired power is a proven reduction in CO_2 emissions, which will help a utility to obtain carbon offset credits and win contracts where proof of CO_2 reduction is required, as it is in Oregon.

Harvesting available electrons by encouraging more energy efficiency is the cheapest and most environmentally sound way to collect power. It has already been generated — you just have to gather it up. In 1997, the Oregon Office of Energy Efficiency estimated that Oregon saved enough energy by its various initiatives to power Portland for a full year. In Sacramento from 1990 – 2000, SMUD saved 240 MW of energy through its various initiatives, as much as a typical power plant produces.

With restructuring looming, many utilities have been cutting their efficiency programs; spending on these has fallen by half. In response, some states have introduced Public Benefit Funds, placing a small surcharge on utility bills to pay for low-income weatherization programs, efficiency programs, and to support renewable energy. In 2000, the American Council for an Energy Efficient Economy did a nationwide survey of utility spending on energy efficiency and found that Washington's utilities came out top with savings of 9.2% of electricity sales, following 20 years of programs. Utilities in only five other states exceeded 4% energy savings: Oregon, Wisconsin, Rhode Island, Minnesota, and Vermont. The national average was 1.74%. Kansas utilities came in last, with zero savings from zero investment. If utilities in every state had harvested the same amount of energy from efficiency initiatives as was achieved in the top five states, US electricity consumption in 1998 would have fallen by 200 billion kilowatt hours, saving 160 million tons of CO_2 from entering the atmosphere (10% of the nation's total from fossil fuels).[1]

What Does New Energy Cost? (2001 US prices)

Saved energy (improved efficiency)	- 0.5 - 3 cents/kWh
Wind turbines	3 - 6 cents/kWh
Combined-cycle gas turbines	3 -- 4 cents/kWh
Conventional gas turbines	3.8 -- 6 cents/kWh
Coal-fired turbines	4.8 -- 5.5 cents/kWh
Gas turbines in tight markets	15 -- 20 cents/kWh
Nuclear energy	15 -- 39 cents/kWh
Solar energy	17 - 50 cents/kWh

Source: Amory Lovins[2]

"America has the technology and resources to meet all its energy needs while safeguarding the earth's climate. The urgent question now is, 'Do we have the will?' At least one city does, and I'm proud to live in it."
— Denis Hayes, Seattle

Woman caulking to seal leaks — one of the easiest and quickest dollar saving techniques.

Ron Coppock Photography, Inc.

Do As They Do In Seattle and Eugene[3]

Seattle City Light is America's third largest public utility, serving some 330,000 residential customers, mostly with hydro. In 1992, it set out to harvest 100 MW of available energy by 2002, using cost-effective conservation. By 1998, it had harvested 66 MW (enough to power 58,000 Seattle homes for a year), and was behind target – but well ahead of most utilities. It currently harvests 6 MW a year. In 2000, Seattle City Council's Energy and Environmental Policy Committee decided to ramp up its efforts, and harvest an additional 100 MW by 2015, saving 850,000 tons of CO_2 from entering the atmosphere.

How do they do it ? By providing or offering:

- extensive information on ways to save energy, and a customized Internet read-out of energy and water usage with specific recommendations for single families;
- a Built Smart program that uses financial incentives to encourage customers to go beyond the code for energy efficiency in new multifamily buildings;
- a region-wide discount on Energy Star light bulbs and fixtures;
- region-wide cash rebates on efficient "WashWise" clothes washers;
- free energy-efficient showerheads and aerators;
- $30 rebates to customers who buy energy-efficient electric water heaters;
- 3% home improvement loans for low- and middle-income households;
- a low-cost weatherization program for landlords of pre-1984 buildings;
- weatherization grants for low-income families;
- a 70% discount on lighting replacements in multifamily building common areas;
- multifamily weatherization grants (70% off lights, 50% off insulation);
- electricity sales through a tiered rate structure (6.3 cents/kWh winter, 4.5/kWh summer);
- community events and displays.

In Eugene, Oregon, the Emerald People's Utility District:

- provides weatherization for electrically heated and mobile homes (40% off, 0% loan);
- subsidizes sealing leaky ducts in mobile homes (50% off);
- buys compact fluorescent light bulbs in bulk and passes on the savings;
- subsidizes new hot water heaters ($30 rebate);
- offers a $75-$100 incentive to replace old fridges with efficient models;
- offers a $100 incentive to buy an efficient clothes washer, $30 for an efficient dishwasher;
- offers free energy-efficient showerheads and water aerators;
- offers a $750 incentive and assistance to install ground-source heat pumps
- offers a "Super Good Cents" $750 incentive to build an energy-efficient house.

In 1989, a pilot program run by the Fitchburg Gas and Electric Light Company in Massachusetts demonstrated that customer participation in an efficient lighting program rose from 4% with a partial rebate to 73% with a full rebate. Even with the utility bearing the full expense, the program was still cost-effective as a strategy to avoid building an expensive new power plant.

Resources

American Council for an Energy Efficient Economy: www.aceee.org

Emerald People's Utility District (Oregon): www.epud.org

PG&E home efficiency: www.pge.com/003_save_energy/003a_res

Seattle City Light:www.ci.seattle.wa.us/light/conserve/resident

Southern California Edison home efficiency: www.sce.com/your_home

46 Encourage Commercial and Industrial Energy Efficiency

All across North America, there are companies that have examined their energy expenditures, invested in energy upgrades, and been astonished at how cost-effective their decisions were. By doing such things as replacing inefficient lights, upgrading buildings, and investing in more efficient turbines, they have realized unprecedented financial returns on investment ranging from 50%–150% (see #31-#40).

For utilities that are approaching restructuring, needing to secure customer loyalty, and experiencing a pressing need to invest in new capacity to meet record-breaking demands for power, the decision to obtain this power by harvesting some of the energy that is being wasted should be a natural.

If only it were so easy! A century of not caring and not understanding how energy is wasted has blunted people's awareness of what needs to be done, often to the point of total ignorance. It is true in the residential sector among architects, builders, and home-owners; it is equally true in the world of commerce and industry, among designers, engineers, facility managers, and corporate finance managers. As a customer, nobody ever asks you if you want to buy more. The power simply flows to your home or office like air towards a vacuum, happy to meet whatever need exists: that is the definition of success in a utility manager's job. You get the bill after you have used it — or wasted it, as is usually the case. It's like providing "pay later" beer to an alcoholic.

How can these megawatts and gigawatts of wasted energy be harvested? Mainly by packaging energy upgrades so that the process becomes as easy as booking a package holiday. This typically involves making a direct approach to businesses, providing energy efficiency education, offering free energy audits and design grants, putting businesses in touch with independent energy service companies (ESCOs), making bulk purchases of commonly used equipment to bring the price down, and spicing the cake with rebates, incentives, low interest loans, and good publicity. It will always make more sense to harvest energy a utility has already generated than to plan, get approval for, and build a new power plant.

"A package of 35 improvements to typical industrial motor systems can save around 50% of their metered energy...with a simple payback under 16 months."
— Amory Lovins

New Lamps for Old

Carol Shook is proprietor of the 68-year-old Five Corners Hardware Store in the St. Anne's neighborhood of Seattle. After hearing about Seattle City Light's small business program, an independent light contractor visited her store and recommended changing her existing (very old) lighting for new, electronically ballasted lamps. The bill came to $3,882, of which Seattle City Light paid $1,780. The new lights save 12,000 kWh a year and reduce Carol's utility bill by $600. Without the rebate, the payback would have been six years. With the rebate, it fell to 3.5 years.

Source: Seattle City Light

"I think when sustainability begins to drive business practices instead of greed, we'll evolve as a species."
— Tom Kay,
President of EcoMall.com

Warren Gretz

Energy-efficient lighting is one of the most cost-effective options available.

Learn from Seattle

Between 1979 and 1998, Seattle City Light's energy conservation initiatives for businesses saved 400,000 MWh of energy, assisting the utility to postpone investing in expensive new capacity. If you operate a business in Seattle, here's what Seattle City Light offers:

- encouragement to join ClimateWise (see #31), which will help you develop and implement a comprehensive plan to save energy, resources, and greenhouse gas emissions;
- cash assistance with building commissioning to ensure that the mechanical and electrical systems operate efficiently;
- an Energy Saving Plan, offering up to 80% of the cost of energy process equipment, variable speed drives, air conditioners, pumps and lights, and refrigeration equipment and controls. A study of five projects under this plan showed a joint yearly saving of $179,000 in energy costs, plus $55,000 in other non-energy benefits;
- an Energy Smart Design Program offering cash incentives and design assistance to encourage efficiency initiatives in new and existing buildings;
- a LaundryWise program, with $50-$150 rebates towards the purchase of energy-efficient equipment;
- a free Operations and Resources Assessment by energy management professionals, which has saved $8 million in business operating costs;
- a Small Business Program offering a rebate to replace inefficient lights, which may be 60% of your business' utility bill;
- involvement in a Neighborhood Power Project. In 1999, the Southeast Seattle project saved 190,500 kWh and 564,000 gallons of water, helped by the efforts of a local steering committee that adopted five goals: helping local businesses and homes to save energy and water; planting trees; picking up litter; helping local schools to save energy and water and plant trees; and helping those most in need.

In May 2000, Pike Place Market and Pepsi joined Seattle's ClimateWise community of businesses. Pike Place Market will save 935,000 kWh by undertaking lighting, waste, and water initiatives, while Pepsi will save 750,000–1 million kWh by installing new lights and controls. In Seattle, every million kWh saved prevents 432 tons of CO_2 from being released into the atmosphere (assuming combined cycle gas turbine generation).

If in Doubt – Bribe Your Own Staff!

Incentive systems work, so why not use them? In his paper *Negawatts: Twelve Transitions, Eight Improvements and One Distraction*, Amory Lovins showed that when one utility started paying a $1 bounty for every kilowatt saved, the volume of savings went up tenfold, while the cost per kilowatt saved fell tenfold.

Resources

Seattle City Light's Business Program:
www.ci.seattle.wa.us/light/conserve/business

Seattle City Light's ClimateWise:
www.ci.seattle.wa.us/light/conserve/business/cv5_cw.htm

EPA's ClimateWise:
www.epa.gov/climatewise

47

Set Up Community Partnerships

One of the most encouraging examples of a utility cooperating with community partners comes from Colorado. It started in 1996, when the Land and Water (LAW) Fund of the Rockies, a non-profit environmental group, lost a series of legal battles to try to force the Public Service Company of Colorado (PSCo) to invest in wind energy. Making a total turnaround, LAW switched its approach and decided to work with PSCo to try to boost customer demand for a new green pricing program through which customers could pay a voluntary surcharge of 2.5 cents per kWh to buy 15% of their energy from PSCo's new Windsource program.

LAW struck an agreement with PSCo, bringing their experience of community organizing to bear on the challenge, plus their credibility as a non-profit society. Using a $70,000 grant from the Department of Energy, they hired a full-time worker and began implementing a Grassroots Campaign focusing on the Aspen area. They reached out to other non-profit societies; they got the Governor and the Office of Energy Conservation involved; they persuaded city and town governments, including Boulder and Denver, to invest in wind power; they created a website (www.cogreenpower.org) where people could sign up; they set up a table at community fairs, shopping malls, and farmers' markets, and direct-mailed 20,000 zip-code targeted customers, receiving an astonishing 6% response. By February 2000, 17,000 residences had signed up. To sign up businesses, they organized a Street Team that went door-to-door among small businesses in the Boulder pedestrian mall; they got chambers of commerce and Rotary Clubs involved; they approached companies that were members of ClimateWise or the Pew Center on Global Climate Change. They also got the media involved, persuading a local TV channel to run spots about the windfarm in the nightly weather spot.

"We need a radical shake-up of the way we use energy, and we need to generate energy in new, sustainable ways. We can't go on damaging the environment as we produce goods. We have to develop new technologies. We all have to 'do our bit' to tackle climate change."
— John Prescott, UK Deputy Prime Minister

Susan Innis.

The Land and Water Fund's grassroots marketing team promoting wind power at a community market.

> **"In the year 2065, on current trends, damage from climate change will exceed global GDP."**
> **— Andrew Dlugolecki,**
> **General Insurance Development, (UK).**

Work Together To Sell Green Energy

By April 1999, LAW had approached 900 businesses and 150 had signed up, including IBM, US West, Rocky Mountain Steel Mills, and Coors Brewing Company. By the end of 1999, their unorthodox marketing had resulted in 25 MW of wind energy coming on-line — five times more than they had been advocating for in the 1996 regulatory proceedings, and five times more than PSCo had expected to sell. 40% of LAW's sign-ups came from Boulder County (where they focused their campaign), which is only 5% of PSCo's customer base, indicating a massive untapped potential. As a result of their success and the media prominence they achieved for Windsource, other utilities started to jump on the bandwagon, including Fort Collins Utilities, where a Sierra Club campaign turned a sluggish response into a sold-out program in just three months.

When it comes to finding solutions to global climate change, there is a huge untapped desire to help. The issue has become the number one global environmental concern, and the concern is shared in homes, schools, and businesses all across North America. People want solutions that make sense, and they are willing to pay for them if it will help protect their children's and their grandchildren's future. The old assumptions about business and environmental lobby groups being traditional enemies can be tossed out the window. By working together, quite unexpected results can be achieved.

Work Together For Bulk Purchase

In 1997, a group of electric utilities, state governments, public interest groups, and industry representatives got together to form the Northwest Energy Efficiency Alliance (NEEA) to promote energy-efficient products and services. By offering a $100 rebate on super-efficient clothes washers, NEEA boosted their sales from 1% to 12% of washers sold in the Pacific Northwest in just 18 months. In another initiative, NEEA worked with Siemens Solar Industries in Vancouver, Washington, to help with furnace design improvements that have reduced the amount of electricity needed to grow single-crystal silicon ingots by 51%, increasing their growth yields by 4% and reducing the amount of argon gas needed by 85%. In 2000, they decided to spend $100 million over five years to seek out energy-saving products and promote their availability. From 1997–2000, they saved 16 MW. By 2010 they expect to save 400 MW, offsetting the need to build two new power plants and preventing 140,000 tons of CO_2 from entering the atmosphere.

Work Together For a Stable Climate

The same kind of co-operative approach can be used by utilities and community partners to bulk-buy efficient lightbulbs and fridges, sell solar PV roofs, or promote street-by-street energy-efficiency upgrades. It can be used by utilities to place bulk orders for micro-turbines, smart meters, and fuel cells. It can be used to lobby together for changes, such as Public Benefits Funds. As soon as we start working together instead of seeing each other as adversaries, we will truly begin to change the world.

Resources

American Public Power Association: www.appanet.org

Climate Challenge's Combined Purchasing & Electric End Use Technology Initiatives: www.eren.doe.gov/climatechallenge/initiatives.htm

Colorado Wind Power: www.cogreenpower.org

Law and Water Fund of the Rockies: www.lawfund.org

Northwest Energy Coalition: www.nwenergy.org

Northwest Energy Efficiency Alliance: www.nwalliance.org

Renewable Northwest Project: www.rnp.org

48 Close Your Coal-Fired Power Plants

"All things are possible, once enough human beings realize that everything is at stake."
— Norman Cousins

In 2000, there were 1,200 electric utility generating units in the U.S., and others in Canada, that were 40 years old or more. Of these, 571 were coal-fired, 426 were gas-fired, and 164 were oil-fired. Fifty percent of the USA's coal-fired plants began operating in 1964 or earlier. The coal-fired plants are particularly inefficient, using a combustion process that wastes almost two-thirds of the energy in the coal. As well as pouring out carbon dioxide, these old plants pour out nitrous oxides, sulphur dioxide, and mercury, three of the nastiest pollutants in North America that are responsible for smog, acid rain, and poisoned lakes and rivers.

According to a study released in May 2000 by the Harvard School of Public Health, the air pollution from just two coal-fired plants in Massachusetts can be linked to 43,000 asthma attacks and an estimated 159 premature deaths each year. In 1999, a study by Abt Associates estimated that in 1997, smog pollution in the eastern USA caused 6 million asthma attacks and sent 150,000 Americans to hospital emergency rooms.[1] The EPA estimates that soot, or fine particulate air pollution, most of which comes from the sulphur dioxide released by coal-fired power plants, causes more than 40,000 people to lose their lives prematurely in the U.S.

Children, senior citizens, and people with respiratory diseases are not the only ones to suffer. Because of the acid rain caused by coal's sulphur dioxide emissions, 25% of the lakes in the mythically beautiful Adirondack Mountains cannot support any plant or animal life, and 6% of Virginia's native brook trout streams are incapable of supporting fish life. Forests throughout the eastern states and up into Ontario and Quebec are suffering and paying the price for the needless use of one of the Earth's dirtiest fuels in some of its oldest equipment.

When it comes to their contribution to global climate change, America's coal-fired power plants release approximately 1 billion tons of CO_2 a year (18% of all U.S. greenhouse gas emissions), about the same as 200 million cars each producing 5 tons of CO_2 per year.[2]

This raises an obvious question: which is technically the easier task, replacing 200 million cars with zero or low CO_2 emissions, or replacing 594 power stations? At one stroke of a legislative pen, the USA and Canada could meet their Kyoto commitments and make a start on the real work, which is an 80% reduction in greenhouse gas emissions.

To see what pollutants power stations are producing in your state, click on "Dirty Power": www.ewg.org/dirtypower/dirty.html

The Ten Most Climate-Damaging Power Plants in the USA

PLANT NAME	OWNER	STATE	CO_2, 1997 (tons)
Gibson	Cinergy Corp.	IN	25,307,140
Cumberland	Tennessee Valley Authority	TN	23,591,360
Scherer	Georgia Power Company	GA	22,367,584
Monroe	The Detroit Edison Company	MI	21,454,630
Rockport	American Electric Power Service	IN	21,404,694
Bowen	Georgia Power Company	GA	21,267,028
James H Miller Jr	Alabama Power Company	AL	21,214,721
W A Parish	Houston Lighting & Power	TX	20,558,269
Martin Lake	TU Electric	TX	19,404,208
Crystal River	Florida Power Corporation	FL	18,618,708

Source: Lethal Loophole: A Comprehensive Report on America's Most Polluting Power Plants and the Loophole that Allows them to Pollute: www.pirg.org/reports/enviro/lethal98

David Parsons

Niagara Mohawk's Dunkirk steam station in New York — soon to be set up for co-firing biomass

How Should They Be Replaced? In a sane world, coal-burning plants would be closed down and replaced with conservation, wind, solar, geothermal, biomass, tidal and hydrogen-generated electricity. Most would probably have been closed already, had they not been grandfathered under the 1970 *Clean Air Act* and given special permission to continue polluting because of their age, thanks to the political maneuvering of their supporters.

The conventional wisdom recommends switching to the most efficient gas-fired technology, the combined cycle gas turbine, which releases 60% fewer emissions than a conventional coal-burning plant. The demand for gas-fired turbines is racing ahead so fast, however, (142,000 MW planned by 2003) that utilities face a four-year waiting list for gas-turbines, and gas prices are rising fast. Gas is a non-renewable resource, like oil, and prices will likely remain high (see p. 43). This raises a major question: if gas prices are rising and wind prices are falling, does it really make sense to invest in new gas-fired plants? Carbon fuels may also soon be made to carry a per-ton carbon levy, while wind will not.

A plant that is planned today will not come on-stream for 3-4 years, by which time wind may be cheaper than gas. Once the plant has been built, it is locked in for 30-40 years, and as the North American natural gas supply runs out, utilities will be forced to import expensive liquefied natural gas. If the gas industry can perfect the art of breaking the carbon-hydrogen bonds in the gas at the wellhead and injecting the carbon dioxide back into the ground (see #53), future gas pipelines may deliver hydrogen for use in fuel cells; but that will not save the gas-fired turbines. Similarly, the coal industry may perfect the art of turning coal into gas and then into hydrogen, but that will not save the coal-fired plants. There are some interim alternatives, such as improving the thermal efficiency of coal-fired plants, or co-firing a coal-plant with sustainably harvested wood-waste, as the Tennessee Valley Authority is planning to do, but these are only stop-gaps.

Phasing Out Coal – The Renewable Way

What is the most cost-effective, climate-friendly way to replace an old coal-, gas-, or oil-fired plant?

- Plan to phase it out within 5-7 years.
- Form a partnership with the state or province, the renewable energy industry, ESCOs, local municipalities, the media, and local environmental groups, and build support to generate an equivalent amount of zero-emissions power.
- Increase the price of power with a 5% Public Benefits Charge and recycle 100% of the income as incentives for conservation and renewable energy.
- Start a huge program of energy conservation, using the policies and techniques described in #24, #32, #45 and #46. Harvesting wasted electrons is a lot cheaper than generating new ones, once you crack the code that releases them.
- Use net metering to encourage the widespread installation of solar and wind energy and gas-fired microturbines within a distributed grid (see #42).
- Issue a Request for Proposals for renewable energy and energy conservation. A request for 750MW will soon get the juices flowing in the new energy industry.

49
Explore New Technologies

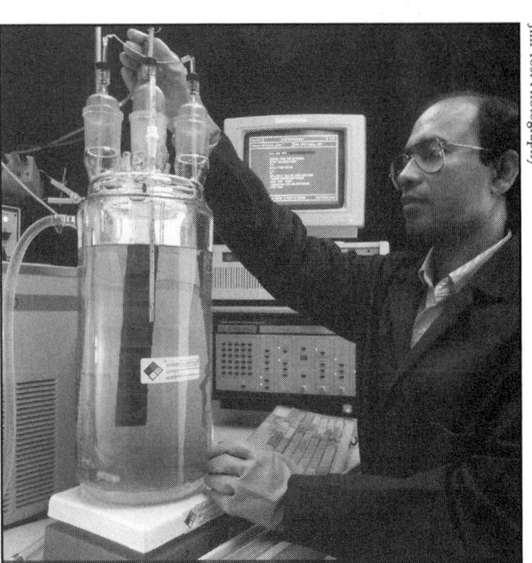
Jim Yost Photography

Superconductive coatings are made with this electrodeposition apparatus.

The world's storehouse of knowledge and technology is advancing faster than at any time in history, and we are only a few years into the Internet revolution, which enables researchers to cooperate across the world and place complex scientific reports in the hands of schoolchildren. In Holland, the government has struck an agreement with Dutch power producers that they will become the most efficient in the world by 2012. They will have to work hard to achieve it, since events are moving so fast.

Superconductivity

Detroit Edison is installing the world's first superconducting cable in 2001 in partnership with the U.S. Department of Energy to supply 14,000 customers in eastside Detroit. Using today's conventional copper cables, approximately 7% of the electricity produced at the power plant is lost before it reaches the customer. One study in Japan estimated that the losses could be as high as 25%.[1] Superconductive cable uses a ceramic substance to transmit electricity three times faster and with twice the efficiency, cutting the 7% loss in half. For techno-philes: "The coated conductors will use the superconducting compound yttrium-barium-copper oxide on a textured substrate technology, referred to as ion beam-assisted deposition. The conductors will use liquid nitrogen as an inexpensive, environmentally friendly coolant." If the whole of the USA's 775,000 MW grid was fitted with the new cables, they could save 27,000 MW, the equivalent of 36 new 750 MW power plants, and save 265,000 tons of CO_2 per year.

Trigeneration

The average coal-fired power plant is only 33% efficient; the rest is lost as waste heat. Cogeneration (known in Europe as combined heat and power) uses a gas-fired turbine process to generate electricity, converting the wasted energy into steam to heat nearby buildings. Trigeneration goes one step further, capturing the waste as steam and chilled water. When the Trigen Corporation installed a trigeneration system in Chicago's McCormick Place Exhibition and Convention Center, it achieved an overall efficiency of 91%, saving the Center $1 million a year in operating costs. The system reduces overall CO_2 emissions by 50% and prevents 24,000 tons of CO_2 a year from escaping into the atmosphere.

> "The future belongs to those who understand that doing more with less is compassionate, prosperous, and enduring, and thus more intelligent, even competitive."
> — Paul Hawken

Resources

Active Power flywheels: www.activepower.com

American Superconductors: www.amsuper.com

Beacon Power flywheels: www.beaconpower.com

California Hydrogen Business Council: www.ch2bc.org

Fuel Cells 2000 Information Center: www.fuelcells.org

Plug Power residential fuel cells: www.plugpower.com

Superconductivity: www.dteenergy.com/environment

Trigen: www.trigen.com

Michele Clements

A Beacon Power Flywheel, before installation underground

Hydrogen Fuel Cells
This technology is moving rapidly off the drawing board and into commercial production. Manufacturers are producing plug-in proton exchange membrane (PEM) fuel cells that convert natural gas into hydrogen and then into electricity, ranging in size from 3–5 to 250 kilowatts, for use in homes and offices. In Chicago, a residential fuel cell being tested by Mosaic LLC could sell for $3,000 and provide enough electricity for a 2,500 sq ft home, plus some heat, for less than the regular price of electricity. As an aspect of the distributed grid, fuel cells will take their place alongside solar, wind, geothermal, and micro-turbine generators, and help utilities avoid expensive investments in new central plants. As long as fuel cells use natural gas as the source for the hydrogen, however, they are still using a fossil fuel. There are several possibilities for generating clean, sustainable hydrogen, including reforming it from ethanol, farming it with algae, splitting it from water using renewable electricity, and separating it from carbon in gas or coal (see #54).

Flywheels
Imagine a cylinder the size of an oil drum. Inside, a 150-pound flywheel spins on magnetic bearings in a near vacuum at 20,000 rpm — an unbelievable 333 revolutions per second. Wire it up, and it can function as a storehouse for solar or wind energy, supply a kilowatt of power for two hours, and provide a back-up against power outages. In June 2000, Beacon Power's flywheels sold for $15,000, a price that is expected to fall rapidly as the market grows. Flywheels will be a boon for on-line businesses, where power interruption can cost $1 million a minute, and for the telecommunications sector in general. Buried underground and monitored remotely over the Internet, they should last for 20 years. San Diego Gas & Electric is installing flywheels instead of batteries as telephone back-up power for 15,000 new homes in Mexico.

Modernized Hydro Technology
In 1991, Pennsylvania Electric and Cleveland Electric upgraded their two Seneca hydroelectric pump turbines, increasing their capacity by 49 MW and their cycle efficiency by 10%. In New York, where Metropolitan Edison is replacing its two 1905 vintage units with a single modern unit, it expects to achieve a 65% increase in capacity and a 20% increase in efficiency.

"Making electricity causes more air pollution than any other industry in the US."
— Julie Blunden, Greenmountain.com

50

Oil Companies – Reduce Your Direct Emissions

Shell, Exxon-Mobil, BP, Chevron, Texaco – these are some of the world's biggest oil companies. Globally, along with the world's other oil companies, they produce and distribute some 72 million barrels of oil per day, pumping or piping it from its resting places deep in the Earth into our cars, planes, trucks, homes and factories. We forget how old the Earth is. It is hard to grasp that these are the molecules of once-living plants and animals, that accumulated over millions of years, gradually turning into oil.

As living organisms, they absorbed carbon dioxide from the atmosphere and stored it in their bodies as carbon. Fast-forward several million years, and here we are digging it up in a frenzy, burning most of it in three or four short generations. A barrel contains 42 gallons of oil, and a gallon produces some 20 pounds of CO_2 when burned. Take out your calculator and you'll find that a year of this behavior releases 9.9 billion tonnes of CO_2 into the atmosphere, 44% of the world's emissions of greenhouse gases from fossil fuels. If you add the production of natural gas, it rises to 65% (the rest comes from coal). So step up to the plate, you big oil companies. What are you going to do about it?

First up is Exxon-Mobil, of *Exxon Valdez* fame (see box) — a good candidate for disinvestment (see #15). It wants us to burn more oil in the belief that oil fuels development, and helps pay to protect the environment. To be fair, it has also formed a partnership with GM, and created a fuel cell processor that uses regular gasoline to create hydrogen, temporarily solving the distribution problem (see #56). (DaimlerChrysler and Shell are working on a similar project.)

Who's next? Texaco. Doing a bit better here. Texaco resigned from the Global Climate Coalition in February 2000, and is pursuing energy efficiency, greenhouse gas emissions management, cleaner technologies such as gasification of fossil fuels and cogeneration (see #49), and tree-planting. In May 2000, Texaco invested $67.3 million to buy a 20% stake in Energy Conversion Devices, a fuel cell company that focuses on making hydrogen safe for fuel cells by using it in solid form rather than in its flammable gas or liquid state.

"There is a lot to be done through employee awareness. I suspect that we can make significant gains in fuel efficiency and emissions control by getting the message home to the people with their hands on the valves."
— A senior technologist at BP[1]

Exxon's Position on Climate Change (June 2000)

We ... understand how the public has become concerned about the wide range of views on the issue and by some upper-end projections that show serious future effects from changing climate. However, such projections are based on completely unproven climate models or more often on sheer speculation, without a reliable scientific basis. We do not believe that the current scientific understanding justifies mandatory restrictions on the use of fossil fuels, and we are certain that large economic harm would result from reducing fuel availability to consumers by the adoption of the Kyoto protocol or other mandatory measures.

(www.exxon.mobil.com/public_policy)

For ExxonMobil's current position on climate change, go to www.exxon.mobil.com/public_policy and click on 'Global Energy'.

David Parsons

The Diamond Shamrock oil refinery, Commerce City, Colorado

Reduce Those Emissions

BP has committed to reduce its emissions to 10% below the 1990 level by 2010. Shell estimates that its 2002 emissions will be 10% below the 1990 level. Both companies are interested in becoming solar/hydrogen energy companies (see #51), and both have quit the Global Climate Coalition. (When an oil company talks about reducing its emissions, it is referring to its direct emissions, not the emissions created by the oil it produces, which are generally ten times greater than its direct emissions.)

In 1998, BP's CEO, John Browne, wrote to 350 of his leading managers seeking their ideas about reducing emissions. They replied with 200 pages of detailed proposals. In response, BP set up an internal trading system between its 90 business units, enabling one unit to meet its target by paying another unit to double its reductions, if this was more cost-effective. At the refinery at Kwinana, near Perth, Australia, CO_2 emissions have been cut by 19% since 1995 by building a gas-powered cogeneration plant to replace coal as a source of power, reducing the flaring of waste gas by 55% and increasing energy efficiency. They are also investing in soil and forest conservation to offset their emissions. In Alaska, BP is introducing "drag reducing" technology along the Alaska pipeline to cut the number of pumping stations needed, eliminating 236,000 tons of CO_2 emissions per year.

Eliminate Venting, Flaring and Leaks

Natural gas is 75-95% methane, with a natural life of 12 years. Over 100 years, it is 23 times more powerful than CO_2. Over 20 years, it is 62 times more powerful. When it leaks or is vented, it escapes as pure methane. When flared, the methane turns into CO_2. Both practices are climate-hostile. Shell has committed to eliminate continuous venting by 2003, and continuous flaring by 2008. In Canada, PanCanadian Petroleum is installing microturbines to generate electricity at its oilfields, using gas that would otherwise be flared. There is a big problem with older pipelines made from cast iron in which the jute bindings leak; estimates vary from 0.05% — 1.4% in North America to 9% in Russia. With the higher leakage rates, gas begins to lose its favored status over oil and coal, but nobody likes to talk about this, and available data is sketchy (see p. 42). The makeshift solution is to fix the joints. The permanent solution, which Consumers Gas is doing in Canada, is to replace the pipes with corrosion-free polyethylene pipes and change the valves and control instruments to a low- or no-bleed configuration. Because of the seriousness of the leakage, this should be legally binding. Besides, the wasted gas represents wasted money.

Resources

Climate Change Solutions—Oil and Gas Production: www.climatechangesolutions.com/english/industry/opportunities/ogproduction

Reducing Greenhouse Gas Emissions from Oil and Gas Production: www.davidsuzuki.org/files/OilandGas.pdf

PetroCanada's Example

PetroCanada has reduced its emissions to within 1% of its 1990 level, despite a 29% increase in the production of gas and crude oil. It established a Global Climate Change Steering Team, and is aiming at a 1% annual improvement in energy efficiency. It is also educating its staff, encouraging its suppliers to join Canada's Voluntary Challenge and Registry program, and supporting various public educational initiatives, including the Climate Calculator (see #1), Destination Conservation (see #13), and the Pembina Institute's Climate Change Solutions program.

51
Become a Solar-Hydrogen Company

"The future of BP is in the sun and hydrogen."
— Peter Knoedel, Deutschen BP

What of the larger picture? For every ton of emissions saved by Shell, BP, or PetroCanada, ten tons are being released by the cars, trucks and planes that burn their oil. It's like polishing the fire extinguisher while the house is burning down. Is it realistic to imagine that a major oil or gas company could become a fire extinguisher for global warming, instead of pumping out more heat?

Most of the world's coal, oil and gas companies are deluding themselves that there will be plenty of time to turn to cleaner fuels — perhaps later this century — but in the meantime, keep on pumping! BP's website says "Any color you want…as long as it's green" — but it then turns around and lobbies furiously to drill for oil in the Arctic National Wildlife Refuge, the birthing ground of the Porcupine caribou herd and the only part of Alaska's Arctic Slope that has not been sacrificed to the oil industry, all for the sake of an additional four to six months of oil, at the rate at which the US wastes energy. "Opening the Arctic Refuge to oil drilling is the equivalent of offering Yellowstone National Park for geothermal drilling, or calling for bids to construct hydropower dams in the Grand Canyon," said Interior Secretary Bruce Babbitt in 1997. This is like a disease — an addiction which has an overpowering grip over BP's management in the Arctic. The fact that it is shared by oil and gas companies all over the world does not make it any less a disease.

If any company has started to look at alternatives, however, it is BP. "Our goal is to play a leading role in meeting world energy needs without damaging the environment", their website says. "Imagine a world where energy is so clean it causes zero pollution, and so simple you hardly know it's there. No sound, no smoke, no CO_2 — just pure power. That world is here now. That world is solar." Through its subsidiary BP Solar, BP has become a major player on the world solar stage: the Athlete's Village at the Sydney Olympics in Australia was one of its many success stories.

For all the publicity BP gets for its solar efforts, 99.95% of its recent investments continue to be spent on fossil fuels. Its solar investments are less than .01% of the company's portfolio. BP spent $100 million, more than double its entire solar investment, just on legal and advisors fees for buying ARCO, the second largest oil company in Alaska. For every $10,000 that BP spent on oil exploration and development in 1998, it spent $16 on solar. It's like building a whopping big oil-fired house, and putting a solar flashlight on the mantelpiece.

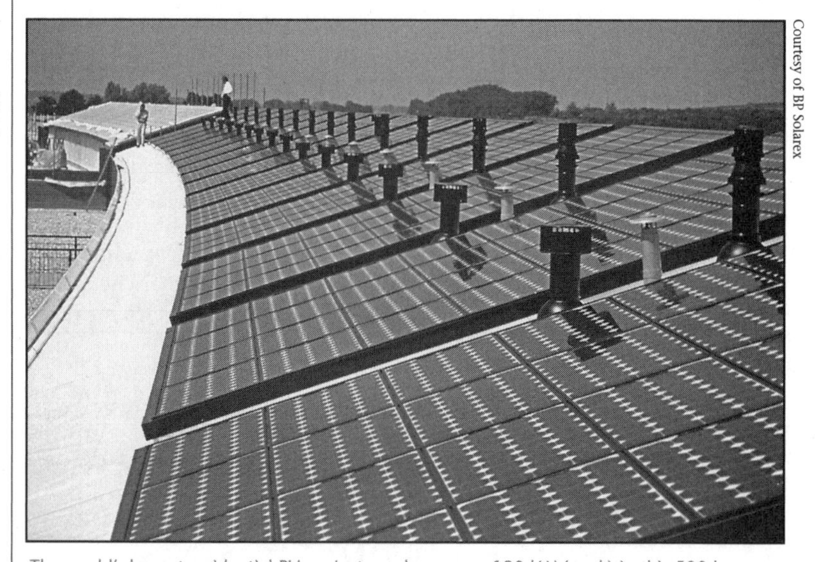
Courtesy of BP Solarex

The world's largest residential PV project produces over 180 kW (peak) in this 500 home development in Amersfoort, Netherlands.

> **"I simply got tired of being on the defensive, and so did our staff. . . that's hardly the way to give people the sense that they are working for a great organization."**
> **— John Browne, CEO of BP**

BP's Solar Future

But what if that solar flashlight on BP's mantelpiece could show us a different future? A future where renewables provided most of the world's energy, where hydrogen was manufactured from renewable sources to power the world's fleet of vehicles, and where desert areas became solar power factories to the world?

A KPMG study has shown that if BP was to build a 500 MW solar plant, efficiencies of scale could reduce the price of PV fourfold to around $1.25 per watt (see p. 138). At that price, every householder and business in the world will be installing PV on their walls and roofs. In April 2000, BP's shareholders were asked to consider a resolution brought forward by activist shareholders to scrap the plans to drill in the Arctic National Wildlife Refuge, and transfer the investment to solar. The motion was defeated, but it won an impressive 13.5% of the vote, enough to persuade BP's directors to consider a massive expansion of the $250 million they had planned to invest in solar up to 2005. BP's first quarter profits in 2000 were $2.7 billion, up by 265% on 1999. The higher the oil prices, the larger the oil companies' profits, providing them with a huge margin of safety to invest in solar and hydrogen, if they want to.

Shell's Hydrogen Future

Shell has its dark side too (think Nigeria, the death of Ken Saro-Wiwa, and his Movement for the Survival of the Ogoni People), but like BP, it is also pursuing a brighter vision of a solar/hydrogen future. The island of Iceland, east of Greenland, derives much of its energy from hydroelectric and geothermal sources, but it runs its vehicles on imported oil, including a huge fleet of fishing vessels. The Icelandic government has become possessed of a dream to convert the island to a hydrogen economy, eliminating greenhouse gas emissions, using hydrogen generated by hydro-power to split water. A company called Icelandic New Energy Ltd has been set up with four partners — Vistorka ('Eco-energy', a public-private Icelandic partnership), Daimler Chrysler, Norske Hydro and Shell Hydrogen. They aim to start by running three fuel-cell buses in the capital, Reykjavik, moving on to the whole bus fleet, then to Iceland's cars, and finally to the fishing fleet by 2030 or 2040.

Shell is also a member of the California Fuel Cell Partnership. It is working with Siemens Westinghouse Power Corporation to develop a Solid Oxide Fuel Cell (SOFC) power plant, and it is researching solid state hydrogen storage techniques as part of its commitment to bring fuel cell vehicles to the market (see #56).

In addition to its hydrogen work, Shell Solar has supplied 10,000 PV panels to Europe's largest solar housing development in Amersfoort, near Utrecht, Holland, and is participating in Germany's 100,000 solar roofs campaign. Another Shell offshoot, Shell Renewables, is part of a partnership that is developing Britain's first off-shore wind-farm at Blyth, Northumberland. Shell is also aiming to capture 10% of the world's biomass energy market by 2010. Overall, however, Shell had invested just 0.6% of its assets in renewable energy in 1999 ($58 million, on a total investment of $9.5 billion[1]).

Resources

BP: www.bp.com

Sane BP: www.sanebp.com

BP Solar: www.bpsolar.com

BP's 1999 Greenwash Award: www.corpwatch.org/trac/greenwash/bp.html

Shell Hydrogen: www.shell.com/hydrogen-en

Shell Renewables: www.shell.com/rw-br

52

Buy Offsets for Your Remaining Emissions

So, you have reduced your company's emissions as much as you can, persuaded your customers and suppliers to become more energy-efficient, and invested in solar, wind, landfill gas and fuel cell projects. You are still producing greenhouse gases, however, and you want to become a carbon-neutral company, or at least meet your CO_2 reduction goals. What else can you do?

A Rational Priority List for Greenhouse Gas Reductions

First: Reduce your own emissions as much as you can.

Second: Encourage your suppliers and customers to reduce their emissions.

Third: Buy carbon reduction offsets in energy efficiency, fuel shifting, landfill gas capture, urban tree-planting, sustainable soil management, or the preservation of forests that would otherwise have been destroyed.

Finally: When you've done all of the above, buy carbon storage offsets to plant new forests.

You can either set up a carbon-emissions reduction project, using one of the growing number of companies that offer this service (see #35), or buy carbon offsets (emissions that have been reduced or sequestered somewhere else and put on the market) from one of the new breed of greenhouse gas trading brokers.

Carbon offsets are still very new, so carbon offset traders are making up the rules as they go along. The goal — supported by some but queried by others — is to establish an international market in carbon offsets, putting a price on the emissions saved by various projects: Solar in a third-world village — $18 a ton. Urban forestry — $36 a ton. The system enables companies to offset their emissions by financing projects that reduce or absorb emissions elsewhere. Once the price system is in place, companies will seek the cheapest offsets first, enabling the world to reduce its emissions in the most efficient manner. The system (also known as "pollution credits" and "cap and trade") was invented for sulfur dioxide emissions in the USA, where it worked very well because SO_2 emissions fall under a common jurisdictional umbrella, and allow for fast and cost-effective reduction.

Carbon dioxide, however, comes from a million different sources that fall under a host of different jurisdictions, making offsets hard to monitor to ensure long-term compliance. There are also two very different types of carbon offsets, that are often treated as one. There are CO_2 reduction projects, which can be realized by methods such as energy efficiency, fuel switching or sustainable soil management; and CO_2 storage projects, where trees are planted to sequester CO_2 that has already been released. The trouble is that it takes a minimum of 40-50 years for a forest to store a useful amount of carbon, and there is no guarantee that it won't burn down or succumb to insect damage caused by the rising heat, releasing its carbon back to the atmosphere.

> **"The times of simply burning oil, one of our most important resources, are coming to an end."**
> **— Dr. Kajo Schommer, Economics Minister of Saxony, Germany.**

Methane Gas Offsets
In Canada, Ontario Power Generation (OPG), with 69 hydroelectric, 6 fossil fuel and 5 nuclear plants, has committed to stabilize its greenhouse gas emissions at the 1990 level, and is engaging in technology upgrades, improved energy efficiency, and the use of cleaner fuels. It has also bought 2.5 million tonnes of CO_2 offsets from the Connecticut-based Zahren Alternative Power Corporation (ZAPCO), which collects methane gas from landfills. This creates a real reduction in emissions — but maybe it should be illegal to operate a landfill without collecting the methane gas in the first place?

Soil Sequestration Offsets
In Iowa, farmer Frank Lewis is using a "zero-till" system of plowing by which he injects his corn and soybean seeds directly into the soil, as a more climate-friendly way of farming. Tilling releases up to 4 tons of CO_2 per acre and weakens soil through the loss of carboniferous organisms (see #36 and #69). He sold his offsets through his crop insurance company (IGF Insurance) and receives $3-$5 an acre from GEMCo, a consortium of Canadian energy companies. IGF estimates that its customers could reduce their CO_2 emissions by 100 million tons a year if they stopped tilling. Since tilling is such a climate-harmful way of farming, however, it might make more sense to ban it, the way we banned DDT and leaded gas.

Climate-Neutral Gas
In Britain, when customers buy gas from Amerada using the Climate Care tariff, the company pays 8% of their tariff plus a $15 donation to Climate Care, which invests the money in energy efficiency, renewable energy projects and tree-planting to offset the CO_2 released.

Tree-Planting Offsets
In April 2000, Texaco committed $900,000 to a 70-year restoration project in the Lower Mississippi Valley, which will see 450,000 tree seedlings planted on 1500 acres of land, sequestering 800,000 tonnes of CO_2. Texaco's 1998 emissions of greenhouse gases were 47 million tonnes, so the forest will sequester 1.7% of its emissions for that year. If Texaco used tree-planting to sequester all of its CO_2 emissions, it would need to plant 58 similar projects every year, requiring 87,000 acres and costing 4.35% of its net annual income ($1.2 billion in 1999). As long as trees are planted in an ecologically and socially sensitive manner, and not as gigantic plantations of eucalyptus that displace local farmers, they are without doubt a good thing. But if we imagine that tree-planting can be the solution to the world's climate problems, we may be making a massive miscalculation. The trees we plant today will not absorb useful carbon for 40-50 years, during which time the greenhouse gas emissions will keep accumulating. Meanwhile, Earth's temperature will continue to rise, and the trees we plant will be at risk of dying from heat, fire or pests, releasing their carbon back into the atmosphere.

Resources (see also #35)

Amerada: www.amerada.co.uk

APPA's Tree Power: www.appanet.org/general/issues/treehome.htm

Climate Care (UK): www.co2.org

Climate ReLeaf: www.americanforests.org

Dutch Electricity Board's FACE project: www.facefoundation.nl

GEMCo (Canada): www.gemco.org

Greenhouse Gas Emission Reduction Trading project (Canada): www.gert.org

Noel Kempff Mercado Climate Action Plan: www.noelkempff.com

The Carbon Trader: www.thecarbontrader.com

Greenhouse Gas Emission Reduction Trading project (Canada): www.gert.org

World Resources Institute: www.wri.org/climate/sequester.html

53

Bury Your Carbon Dioxide

Instead of buying carbon offsets to make up for the CO_2 that a power plant produces, what about capturing it before it escapes?

This is a relatively new idea that has the coal and oil companies very interested. The idea is to capture CO_2 as soon as it is released and store it away, not in forests, but somewhere safe where it can never reach the atmosphere, such as underground, in carbonate rocks, or deep in the ocean. It involves two distinct processes. First, you have to separate the CO_2 from the other escaping gases, and then you have to transport it to its place of storage and lock it away.

Let's Sequestrate Together

In May 2000, BP, Chevron, Norsk Hydro, Shell, Statoil, Suncor and Texaco announced that they would work together on a $20 million project to develop ways to capture CO_2 and store it away in underground geological formations.

Geological Storage

In Norway, the government imposes a $38 per tonne CO_2 tax on offshore carbon dioxide. Norway's Statoil, 240 kilometers off the coast, produces oil and natural gas from the field at Sleipner (with Norsk Hydro, Exxon and Elf). The gas contains 9% CO_2, which is more than the 2.5% its customers will accept, so Statoil uses an on-site chemical plant to extract the excess CO_2, compress it, and pump it 1000 meters beneath the North Sea into a 200-meter thick sandstone reservoir known as the Utsira formation, where it can be stored for thousands of years.[1] Venting would cost $38 million a year in taxes, so even with the capital costs involved, storage makes financial sense. Statoil estimates that it is possible to store 400 years worth of CO_2 emissions from every European power station under the North Sea.

Now that people have seen that it can be done, they are looking at other geological storage options, such as the use of old coal beds, mined-out salt domes, deep saline aquifers, and depleted oil or gas reservoirs, which have the added advantage of enhancing the production of trapped methane. With all these options, the long-term security of the CO_2 has to be assured.

Geological Carbonate Storage

The Los Alamos Vision 21 team wants to store CO_2 in geologically stable mineral carbonates. Nature absorbs CO_2 in the oceans, converting it into the shells of sea-creatures and then into limestone. The Vision 21 team is looking at ways to make a pressurized stream of CO_2 react with magnesium or calcium silicate mineral deposits to form geologically stable mineral carbonates, fixing it permanently with no fear that it could re-emerge to trouble future generations. The main candidates for the process are magnesium-rich ultramafic rocks, primarily peridontites and serpentites, which are widely distributed throughout the world and abundant enough to absorb the CO_2 associated with the world's entire coal reserves.

> **"A true conservationist is a man who knows that the world is not given by his fathers, but borrowed from his children."**
> **— John James Audubon (1785-1851)**

Statoil

Statoil's oil-rig at Sleipner, off the Norwegian coast.

Ocean Storage

Another option that is being researched involves storage or sequestration in the ocean. The oceans are part of the world's natural carbon cycle, storing some 40,000 billion tonnes of CO_2 , compared with 750 gigatonnes in the atmosphere. The question being examined is whether the addition of a few extra billion tonnes in a variety of locations would interfere with the delicate balance of the ocean's ecosystem. If the CO_2 was deposited at a depth of 3,000 meters or more, the ocean would be cold enough to trap it and store it as ice-like CO_2 hydrates. If it was deposited higher up, it would be absorbed into the bodies of phytoplankton and zooplankton. Options that have been suggested for disposal include direct release through a pipeline or from a pipe towed by a ship; dropping the CO_2 overboard as solid blocks of dry ice and letting them fall to the bottom; or pumping it down to natural depressions on the deep seafloor. The big unknown factor is what will happen if the oceans continue to warm, causing the phytoplankton to die or the ice hydrates to release their CO_2. Ocean technologists get excited, but marine ecologists shudder.

CO_2 Separation

Before the CO_2 can be stored, it has first to be separated from other flue gases (CO_2 might be 2%-70% of the flue exhaust). At ABB's Shady Point coal-fired plant in Oklahoma, CO_2 is separated by mixing it into a compound with a solution of dilute monoethanolamine (MEA), heating it to 120°C in a stripping tower to release it, and then compressing, chilling, liquefying, and purifying it, recycling the stripped-off MEA to repeat the process. The CO_2 is then sold for use in freeze-dried chicken. Other methods being researched include the use of molecular sieves, physical solvents, low temperature (cryogenic) processes, high temperature polymer membranes, a "gas-liquid contactor" that creates a whirlwind-like vortex, a tertiary amine scrubbing system, and the use of mats of blue-green algae to absorb the CO_2 through photosynthesis.

Is it Worth the Cost?

The cost of all this — and the energy cost — begins to add up when you consider the full energy cycle, starting with the mining of the coal, plus the transportation, preparation, power generation, particulate removal, flue gas recirculation, CO_2 removal, CO_2 transport, and final CO_2 storage or sequestration. The Department of Energy hopes that it can be done for under $36 per tonne of CO_2 ($10 per tonne of carbon), adding just 0.02 cents per kilowatt hour to the price of electricity. Time will tell.

Resources

Carbon Mitigation Initiative, Princeton: www.princeton.edu/~cmi

CO2 Capture Project: www.co2captureproject.com

DoE carbon sequestration: www.fe.doe.gov/coal_power/sequestration

IEA's CO2 capture & storage: www.ieagreen.org.uk/doc3a.htm

MIT's Carbon Sequestration Initiative: http://sequestration.mit.edu

National Energy Technology Laboratory: www.netl.doe.gov/products/gcc

Statoil's CO2 burial project: www.statoil.com/STATOILCOM/SVG00990.nsf/web/sleipneren

US Office of Science, carbon sequestration: http://cdiac2.esd.ornl.gov

54

Convert Fossil Fuels into Hydrogen

Is a zero-emissions power plant possible? Certainly, if you use solar, wind, or geothermal energy. But what about a zero-emissions fossil-fuel power plant?

The Vision 21 Program

At the Los Alamos National Laboratory near Santa Fe, New Mexico, the US Department of Energy's Vision 21 team is working on a zero-emission process that converts a coal and water slurry into hydrogen and then into electricity using a solid-oxide fuel cell. The process uses a calcium oxide process in an anaerobic (oxygen-free) environment to convert the slurry into hydrogen. In the process, the calcium oxide (CaO — lime) picks up the carbon dioxide (CO_2) and is converted into calcium carbonate ($CaCO_3$ — limestone). Still with us? The calcium carbonate releases a pressurized stream of pure CO_2 that is ready for easy disposal and reverts back into calcium oxide, ready to start over. The solid oxide fuel cell achieves a 50% conversion efficiency, while the other 50%, the waste heat, is re-injected to drive the calcination process. The whole process is done without oxygen and the coal is not burned, so there's no need for a costly oxygen separation system. The nitrogen oxide by-products, which are a leading cause of smog and acid rain, are limited to the nitrogen compounds in the coal itself. The Vision 21 process can be adapted to run on gas or biomass, and generates twice as much energy as a conventional coal-fired power plant.

Shell Hydrogen and Siemens Westinghouse Power Corporation are working on another aspect of the solid-oxide fuel cell process. Solid-oxide fuel cells have a ceramic electrolyte composed of something called yttria-stabilized zirconia (don't you love those names?) that can operate at the incredible temperature of 10,000°C. At this heat, the methane in the natural gas separates naturally into hydrogen and carbon monoxide, fueling the oxidation reaction that produces electricity.

Southern Company, the USA's largest coal-burning electric utility, is also working on a Vision 21 project. In May 2000, it announced that it had successfully tested a process using a "transport reactor" as an advanced pressurized combustor to convert coal into gas for use in a hydrogen fuel cell. Not to be outdone, the Chinese and Italian governments are cooperating on ways to generate hydrogen from China's vast coal deposits.[1]

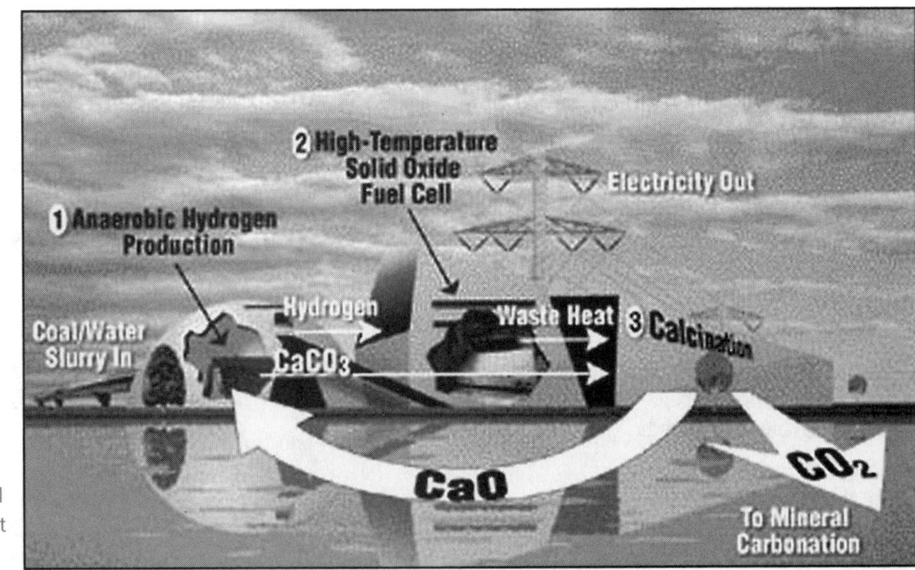

The Vision 21 Zero Emissions Power Plant

> "We feel that hydrogen, generated from a renewable resource such as sunlight, is the only viable long-term solution to providing an absolutely environmentally sound fuel."
> — Henrich Heitmann, BMW Board Member

Use Algae to Produce Hydrogen
Perhaps the most intriguing development in this field concerns green algae, also known as cyanobacteria. In January 2000, five scientists at the University of California at Berkeley and the US Department of Energy discovered a metabolic switch that allows algae to produce hydrogen.[2] It has been known for decades that algae give off a very small amount of hydrogen by means of hydrogenase, a fussy little enzyme that only works in an oxygen-free environment. Healthy algae spend their days doing photosynthesis, consuming carbon dioxide and respiring oxygen to make carbohydrates, so the hydrogenase can only produce its hydrogen during darkness, when the photosynthesis stops and there's no oxygen around. The team's breakthrough was to let the algae grow in the normal manner and then transfer it to a bottle where it was deprived of sulfur, which is essential for photosynthesis. After 24 hours, the algae stopped producing oxygen and switched to "Plan B," which involves turning to its hydrogenase enzyme and using sunlight to feed off its own carbohydrate reserves, splitting water and releasing hydrogen as a by-product. During four days without sulfur, a liter of algae slurry produced 3 milliliters of hydrogen hourly before it had to be fattened up again with its regular diet of sulfur. The process can be repeated endlessly, allowing the continual harvesting of hydrogen. The scientists have patented their discovery and are working on a tenfold improvement in efficiency.

A system for photobiological algal hydrogen production

There is other algae work going on, too. At Tokyo's University of Agriculture and Technology, a group is discovering marine microorganisms with a similar ability. At Montana State University and the University of Memphis, researchers are growing algae as a thin biofilm, or algal mat, and calculating how they could use it capture the CO_2 emissions from a 10 MW power plant as a form of biosequestration.

Now here's a speculation. Combining the two fields of research, could algae feed off a power plant's waste CO_2, turn it into carbohydrates, and then go on a sulfur-free diet to turn the carbohydrates into hydrogen to be fed into fuel cells to produce electricity to power the coming generation of fuel cell cars?

Resources

Department of Energy-Fossil Energy: www.fe.doe.gov

Melis Energy (hydrogen from algae): www.melisenergy.com

Vision 21: www.fe.doe.gov/coal_power/vision21/vision21_sum.html

Zero Emission Coal to Hydrogen: www.lanl.gov/energy/ziock/ziock.html

55 Help the World

"Every generation faces a challenge. In the 1930s, it was the creation of Social Security. In the 1960s, it was putting a man on the moon. In the 1980s, it was ending the Cold War. Our generation's challenge will be addressing global climate change while sustaining a growing global economy."
— Eileen Clausen, Pew Center on Global Climate Change

How do we turn the world around? The Arctic ice is melting before our eyes. Everywhere, temperatures are rising. Among the few, there is a desperate effort to awaken the world to the plight that is before us.

Above all others, it is the energy companies of the world that need to show leadership. In 1997, nearly 80% of the world's emissions from carbon-based fuels were produced by 122 companies; 22% were produced by just 20 private companies.[1] It is our massive use of coal, gas and oil that is responsible for the awful future that lies ahead of us, that will bring climatic chaos to the world and untold suffering to our children and grandchildren, unless we act.

It is not enough to erect a few solar panels, buy a few fuel cells and plant a few trees, while continuing to invest billions in the feverish search for more oil and gas. In the eyes of some, this is so much hypocrisy. All that we ask is that the energy companies change the kind of energy they produce or distribute: become part of the solution, instead of the problem. A world of sustainable technologies and sustainable values is opening up that offers us an alternative to the distress of a runaway greenhouse world.

The British Royal Commission on Environmental Pollution has declared that we need a 60% reduction in CO_2 emissions by 2050. The world's scientists say we need a far greater reduction, far sooner. There are many who believe that if we work together and learn from each other, we can achieve this goal.

Cooperate Regionally

The first place where we can work together is locally, in our states, provinces and regions. Wisconsin owes its success in the area to the fact that its citizens, businesses and utilities have set up several partnerships committed to the development of safe, affordable, sustainable energy: Wisconsin Focus On Energy, Renew Wisconsin, The Midwest Renewable Energy Association, and Customers First!

In the Pacific Northwest, a sense of possibility is developing because Climate Solutions, the Renewable Northwest Project, the Northwest Energy Coalition, the Northwest Council on Climate Change and the Northwest Energy Efficiency Alliance are working together to build partnerships, including utilities, businesses, farmers, churches, municipalities and NGOs. The same is true in Colorado, California, Minnesota, and other regions. Where progress towards sustainable energy has yet to happen, as in large areas of Canada, and states such as Wyoming, the Dakotas, Kansas, Louisiana, Michigan and Arkansas, it is safe to say that it is because these kinds of partnership have not yet been established. (See also #12 & #47)

The Business Council for Sustainable Energy

The BCSE was created in 1992 by senior executives in the natural gas, energy efficiency, electric utility and renewable energy industries in the USA, to advocate for policies that promote a sustainable energy future.

Members include the American Gas Association, Ballard, AWEA, Uni-Solar, the Alliance to Save Energy, Maytag, the Los Angeles Department of Water and Power, SMUD, the Solar Energy Industry Association, Plug Power and Enron Wind. Advisors include the American Council for an Energy Efficient Economy, NRDC, the Union of Concerned Scientists, the Worldwatch Institute, the World Resources Institute, and the Environmental and Energy Study Institute. www.bcse.org

"Our climate is warming at a faster rate than ever before recorded."
— Dr James Baker, NOAA Administrator, April 18th 2000

Ullal, Harin - Central Electronics, Ltd.

A woman at a PV-powered water pump in India

Cooperate Nationally

At the local and regional level, the term "good citizen" carries meaning, because people feel a sense of shared responsibility to the landscape that is their home. At the national level, it becomes fractured, as politicians, businesses and special interest groups draw up battle lines against those who think differently. Two groups that are working to overcome this are the Pew Center on Global Climate Change and the Business Council for Sustainable Energy (see boxes). We need far more of this. Think what we could achieve if we had a broad-based, energetic national alliance combining:

- the passion and grassroots commitment of Greenpeace;
- the knowledge of the Union of Concerned Scientists;
- BP's interest in becoming a worldwide solar energy company;
- the Los Angeles Department of Water and Power's vision of LA as a solar city;
- the Colorado Land and Water Fund's ability to expand sales of green energy using grassroots community marketing;
- Seattle City Light's 20 years practice at energy conservation;
- the Toronto Atmosphere Fund's experience at funding emissions reductions projects;
- the Rocky Mountain Institute's understanding of energy efficiency;
- Shell's interest in building a hydrogen economy;
- Detroit Edison's experience with fuel cells and superconductivity;
- Climate Challenge's know-how about how utilities can reduce their emissions;
- the World Resources Institute's experience in becoming carbon neutral;
- Minnesotans for an Energy-Efficient Economy's vision of a sustainable economy;
- ICLEI's experience of how towns and cities can reduce their emissions;
- the grasp of climate change solutions that organizations such as the National Resources Defense Council, the David Suzuki Foundation and the Pembina Institute have;
- the ability to mobilize grassroots support that groups such as the Sierra Club, Friends of the Earth, PIRG, Ozone Action, Earthday and Climate Solutions have;
- the American Solar Energy Society's commitment to a solar future;
- the American and Canadian Wind Energy Associations' commitment to wind energy;
- ...and so on.

If we can dream it, we can do it.

The Pew Center on Global Climate Change

The Pew Center on Global Climate Change was created in 1998 as a non-profit, non-partisan, independent organization to educate the public and key policy makers about the causes and potential consequences of climate change, and to encourage the domestic and international community to reduce the emission of greenhouse gases. Members include AEP, BP, Enron, Entergy, PG&E, Shell, Toyota, Lockheed Martin, and Whirlpool. www.pewclimate.org.

56

Choose a Sustainable Fuel for the Future

There are 530 million vehicles in the world, and the number is growing by 12 million a year. Almost all of them burn gas, and every gallon burned produces 20 lbs of CO_2. Globally, the impact is enormous. A third of the world's greenhouse gas emissions come from transportation. In 1998, the world's oil companies pumped 74 million barrels of oil out of the ground every day. That's 2.8 billion tonnes of carbon entering the atmosphere every year[1] out of a global total of 6 billion tonnes from fossil fuels.

Demand is soaring, and those oil reserves are going to run out within 50 years, new discoveries in the Caspian region or elsewhere notwithstanding. Somewhere between 2007 and 2020, production will peak and go into a permanent decline, leading to at-the-pump prices of $3, $5, and then $10 a gallon, as demand exceeds supply. While this will create massive profits for the oil companies, it has the potential to cause panic around the world and disaster for countries that depend on imported oil.

We can't afford to wait until this happens. Because of the looming climate crisis, we need to choose a sustainable energy source *now* to fuel the vehicles of the future, and establish an infrastructure to distribute it. Auto companies also need to maximize the efficiency of their existing vehicles (see #57), to build them in a way that does the least damage to the Earth (see #58), to think beyond cars to *mobility* (see #59), and to form new partnerships to realize these goals (see #60).

The fuel of the 21st century is not going to be natural gas; it's going to be hydrogen. Ethanol from agricultural and forest wastes can provide a bridge while providing farmers with a much-needed boost to their incomes. To choose natural gas as a bridge (as the world is doing) is foolish, since it is still a fossil fuel. Gas produces 42% less CO_2 than coal and 16% less than oil,[2] but we need 80% less. Gas also leaks methane, which weakens its advantage (see p. 42). Gas *may* be a fuel of the future when converted into hydrogen by stripping off the carbon, provided the process does not consume too much energy or cost too much, but that is a different story (see #54). Ethanol may also become a primary source of hydrogen for use in the fuel cells.

"Hydrogen is a renewable resource. You can get it from anywhere. It's a wonderful energy carrier...I believe that my grandchildren, at least, will be driving cars that run on hydrogen. So I support the notion that it might be time to bite the bullet and put the hydrogen infrastructure in place."
— Bradford Bates, senior Ford engineer

Resources *(see also p. 40, #49 & #51)*

Alternative fuels data: www.eia.doe.gov/fuelalternate.html
American Hydrogen Association: www.clean-air.org
California Fuel Cell Partnership: www.drivingthefuture.org
Canadian Hydrogen Association: www.h2.ca
Carbohydrate Economy Clearinghouse: www.carbohydrateeconomy.org
Electrifying Times: www.electrifyingtimes.com
Environmentally Friendlier Motoring: www.autonutz.com/eco-car.htm
Forward Drive: The Race to Build Clean Cars for the Future: www.forwarddrive.com
German Hydrogen Association: www.dwv-info.de
Hypercar Center: www.hypercarcenter.org
Hypercar Inc: www.hypercar.com
National Hydrogen Association: www.ttcorp.com/nha
Solar Hydrogen Energy Corporation: www.solar-h.com

*"**Environmentally friendly cars will soon cease to be an option... they will become a necessity."**
— Fujio Cho, President of Toyota*

Some Prototype Hydrogen Vehicles

Opel/GM	HydroGen 1 fuel cell	Liquid hydrogen
	Zafira fuel cell	Liquid hydrogen
	EV1 fuel cell	Methanol + reformer
	Triax	Variable options
DaimlerChrysler	NECAR 2 &5 fuel cell	Methanol + reformer
	NECAR 3 & 4 fuel cell	Liquid hydrogen
BMW	750hL combustion	Liquid hydrogen
Ford	P2000 SUV fuel cell	Methanol + reformer
	P2000 HFC fuel cell	Liquid hydrogen
	TH!NK FC5 fuel cell	Methanol + reformer
Honda	FCX-V3	Compressed Hydrogen
Mazda	Premacy FC-EV	Methanol + reformer
Toyota	RAV4L EV fuel cell	Methanol + reformer
ZEVCO	Millennium Taxi fuel cell	Compressed hydrogen

The most sensible long-term choice for sustainable vehicle propulsion on planet Earth would appear to be electricity, produced by a fuel-cell, using hydrogen that has been obtained from water using renewable energy, or from renewable biomass wastes. This requires sufficient renewable energy and biomass to produce the hydrogen (as well as producing electricity for the grid), and a world-wide hydrogen distribution network.

In the short-term, hydrogen is being produced from methanol (CH_3OH) derived from natural gas, using an onboard reformer. The reforming produces CO_2, carbon monoxide and nitrous oxides when carbon is released from the hydrogen, but 50% less than an equivalent gasoline vehicle. Methanol (which can also be produced from biomass) is similar to gasoline, so it can be distributed through normal service stations, but the need for an onboard reformer makes vehicles heavier.

There is a lot of R&D happening for fuel cell vehicles. The California Fuel Cell Partnership aims to have 50 fuel cell cars and buses on the road by 2003, complete with a fuel distribution system. The partnership includes Ballard, Toyota, Ford, Nissan, Honda, Hyundai, Volkswagen, DaimlerChrysler, Shell Hydrogen, BP, Texaco, the California Air Resources Board, the California Energy Commission, the US Department of Energy, the South Coast Air Quality Management District, and the US Department of Transportation.

In 1999, DaimlerChrysler's fuel cell NECAR 4 (NE = "no-emissions") won an International Engine of the Year award. It operates on liquid hydrogen, and can travel 280 miles on one tank, with a top speed of 90 mph. DaimlerChrysler's goal is to have mass-produced vehicles on the road by 2004, using direct hydrogen fuelling for fleets and onboard methanol reformers for private cars. Honda aims to have a fuel-cell vehicle for sale by 2003, and both GM and DaimlerChrysler are working to produce hydrogen from gasoline, solving the distribution problem until liquid hydrogen distribution can be set up.

DaimlerChrysler anticipates that as the market grows, hydrogen will be derived first from fossil fuels, then from biomass (ethanol) by decentralized local production, then by importation from large remote renewable power stations, and finally by decentralized local production from wind and solar, with a smooth transition from one phase to the next.

The ultimate eco-car will combine ultra-lightweight construction, advanced rigid polymer composite materials, low-drag design, integrated electronics, energy-efficient accessories, and a hybrid-fuel-cell electric engine. It will achieve 75 - 125 mpg, and use 3-5 times less energy than today's vehicles. It may fuel up from a residential hydrogen fuel cell reformer, generating electricity for sale to the grid when not in use. The concept has been developed by staff at the Rocky Mountain Institute, and is known as the "hypercar".

Ballard Power Systems

Using liquid hydrogen, DaimlerChrysler's zero emissions Necar 4, can travel 280 miles on one tank, with a top speed of 90 mph

57
Prioritize Fuel Efficiency

Hydrogen, shmydrogen – and meanwhile, the world's 530 million gas-guzzling vehicles are continuing to pour out 2 billion tonnes of carbon per year. With every gallon of gas that's burned, the world's temperature increases by another fraction of a micro-degree.

Even if we assume the very best use of policy instruments, R&D breakthroughs, smart-vehicle tax-breaks, and grassroots community marketing, it will be 10 years or more before mass-produced methanol-hydrogen fuel cell vehicles begin to dominate the market, and 20 years before mass-produced solar-hydrogen vehicles do so. In the meantime, we urgently need to increase the efficiency of regular gas-powered vehicles and stop producing inefficient ones.

Set Efficiency Goals

Both the know-how and the willpower exist. In 1998, faced with binding legislation they did not want, the European car industry (including GM, Ford, and DaimlerChrysler) signed an "Auto-Kyoto" pact with the European Union agreeing to increase their average fleet efficiency from 29 mpg to 39 mpg by 2008, and reduce CO_2 emissions by 25%.[1] A similar pledge has been made by Japanese and Korean car makers.

If it can be done in Europe and Asia, it can be done in North America. The technology exists: hybrid gas-electric vehicles; regenerative braking; direct injection engines; lean-burn technology; electronically controlled gear-shifting; variable valve-timing; four valves per cylinder instead of two; aluminum, carbon fiber, and plastic composite body parts; increased aerodynamic designs.

In 1996, Greenpeace took Renault's Twingo and adjusted the design to make it twice as fuel-efficient: 66 mpg.[2] The engine is only 360cc, yet its performance equals or exceeds the 1.2 litre Twingo; independent experts have verified that it is just as robust as the normal heavier engines. In summer 2000, Volkswagen's Lupo 3L TDI (3-cylinder turbocharged direct-injection diesel) drove 20,712 miles around the world in 80 days using only 1,000 liters (264 gallons) of fuel. Midway through the journey, the Lupo was averaging 80 mpg.[3]

The average car in the USA achieves 24 mpg (9 liters per 100km). Two decades of fuel efficiency improvements have been wiped out by the increased sales of light trucks and SUVs, encouraged by TV advertising that shows SUVs romping through national parks.

In 1993, GM, Ford, and DaimlerChrysler joined the Partnership for a New Generation of Vehicles with the U.S. government, largely to forestall tighter fuel-economy rules. They agreed to market an 80 mpg family-sized vehicle by 2004 and have chosen to go with direct injection diesel hybrid-electrics in spite of protests that diesel produces noxious air emissions, produces more CO_2 than gasoline, and brings a tenfold increased risk of cancer.[4] Ford plans to produce 10,000–20,000 hybrids by 2003. GM plans to produce 50,000 by 2004. By comparison, 2.8 million SUVs hit the road in the U.S. in 1998, GM is boosting its production of full-sized pick-ups and SUVs to 1.7 million in 2001, and rolling out 20 new big truck models by 2003.

Resources

DaimlerChrysler Research: www.daimlerchrysler.com/research/research_e.htm

Earthsmart Cars: www.nrdc.org/earthsmartcars

FutureTruck SUV challenge: www.futuretruck.org

GM's Future Vehicles: www.gm.com/company/gmability/environment/products/chart/

Greenpeace's Twingo SmILE: www.greenpeace.org/~climate/smile

Partnership for a New Generation of Vehicles: www.ta.doc.gov/pngv

Sierra Club's CAFE Campaign: www.sierraclub.org/globalwarming/cleancars

Toyota Prius: www.prius.toyota.com

"The Big Three sell the dirtiest vehicles and capture the vast majority of U.S. passenger vehicle sales, making them the worst polluters by any measure. DaimlerChrysler, Ford, and General Motors vehicles were the sources of 76% of the CO_2 emitted by vehicles sold in 1998."
— *Pollution Line-Up – An Environmental Ranking of Automakers*,
Union of Concerned Scientists

Guy Dauncey

The Honda Insight, a hybrid gas-electric 2-seater that averages 65 mpg (80 mpg in Canada).

Stop Opposing CAFE and Feebates

The Corporate Average Fuel Efficiency (CAFE) standard was introduced in the USA in 1975 and required that new cars average 27.5 mpg, 20.7 mpg for light trucks and SUVs. Vehicles reached that standard in the 1980s, but since then efficiency has stood still. A revised standard would upgrade the CAFE to 45 mpg for cars, 34 mpg for light trucks and SUVs, but the auto industry has always lobbied to prevent it. If you're sincere about being green, it's time to come clean: stop sabotaging a new CAFE standard. And stop opposing feebate laws, which offer rebates for efficient vehicles while imposing fees on the gas guzzlers (see #66).

Install Fuel Efficiency Indicators

Most drivers have bad driving habits, such as uneven acceleration and braking, that increase their use of fuel. But how can you tell? A fuel efficiency indicator on the dashboard would give you immediate feedback, telling you when your rapid acceleration was burning fuel at 10 mpg and when calm coasting was giving you 40 mpg. Saab offers such a meter in its cars and Honda has one in the Insight. Donella Meadows, as a new Insight driver, wrote that "three weeks of information I never had before have changed 40 years of ingrained driving habits." Every car, truck, and bus should have one. Calling all auto-manufacturers: if you care about the climate, install them now.

Label Your Cars

The European Parliament has passed legislation requiring auto manufacturers to clearly label their vehicles to indicate fuel efficiency and emissions, and to display posters in their showrooms comparing the fuel performance of the 10 most popular models and their CO_2 output. Ford moved quickly to do this, but only in Europe.

Create Intelligent Tire-Valves

Studies in Arizona have shown that 25%–50% of vehicles on the road have at least one tire under-inflated by 4 pounds or more; every 2 pounds of under-inflation creates a 1% loss in fuel efficiency.[5] If 25% of global vehicles have all their tires under-inflated by 1 pound, that's a 0.5% loss of fuel efficiency throughout the world, releasing 37 million tonnes of CO_2. Checking the tire pressures is a fiddly job, however, so the solution is a smart tire valve that gives a digital readout of the tire's pressure or flashes red when a tire needs inflating. With 2 billion wheels needing smart valves, this looks like a good opportunity for a smart engineer.

Present and Future Fuel-Efficient Vehicles[6]

Greenpeace/Renault	Twingo SmILE	Gasoline	64 mpg	1998
Honda	Insight	Hybrid electric	61-70 mpg	1999
Volkswagen	Lupo 3 TDI	Turbo Diesel DI	62-77 mpg	1999
DaimlerChrysler	Smart cdi	Turbo Diesel DI	68 mpg	1999
Toyota	Prius	Hybrid electric	42-50 mpg	2000
Ford	Escape SUV	Hybrid electric	40 mpg (urban)	2003
Ford	P2000	Hybrid electric	63 mpg	2004
Ford	Prodigy	Hybrid electric	72 mpg	2004
DaimlerChrysler	Dodge ESX3	Hybrid electric	72 mpg	2004
GM	Precept	Hybrid electric	80 mpg	2004

58
Become an Earth-Smart Company

Richard Rochon

Aerial rendering of the planned retrofit for Ford's 600 acre Rouge Center complex in Dearborn, Michigan.

With tongue in cheek...

Behold, a typical car hath 15,000 different parts, each of which requireth the use of resources from the Earth and much energy that produceth emissions in the making thereof. Is it possible for a car to enter the Kingdom of Heaven? Verily, I say that for spirits that dwell in the realm of matter, as doth a car, the Kingdom of Heaven is the fullness of the Earth and the protection of the creatures and ecosystems therein. For a car to enter the Kingdom of Heaven, verily, it must use parts that have been reborn from other materials, that can themselves be reborn. In its manufacture it must use energy that cometh from renewable sources, likewise in its running. In its creation it must recycle all emissions that are released, and in its working life it must run clean and pure, with safety. For truly, a car cannot enter the Kingdom of Heaven if it causeth young children to suffer or die a painful death from asthma, or causeth dwellers in the cities to curse at the pollution and noise it produceth, or if its emissions cause the very atmosphere of Heaven to overheat, Earth's coastal realms to drown. What can a poor car do, to clean up its act? I answer you...

Resources

ClimateWise: www.epa.gov/climatewise

Natural Step: www.naturalstep.org

Recycled Paper Coalition: www.papercoalition.org

Reduce Your Greenhouse Gas Emissions

The smartest way to do this is to join ClimateWise, as GM and Ford Motors' Twin Cities Assembly facility have done, and to follow the ten Solutions for Businesses (#31-#40). Ford of Canada, after signing up with Canada's Climate Change Voluntary Register and Challenge, worked with its team of 17,000 people to reduce its emissions by 35% at its assembly facilities, 29% at its engine and engine component facilities, and 16% at its production metal casting facilities. GM Canada has reduced its greenhouse gas emissions by 30% since 1990 by switching to high efficiency lighting, converting from coal and oil burning facilities to natural gas, and other initiatives. GM has also eliminated all ozone-depleting substances, from 750,000 pounds in 1989 to zero in 1994.

Buy Green Energy

At its US headquarters in Torrance and three other California facilities, Toyota is buying 100% renewable energy from GreenMountain.com — one of the largest commercial green energy contracts in California. At Ford's engine plant in Bridgend, South Wales, solar panels provide lighting for 108,000 sq. ft, saving 4,400 tonnes of CO_2 over 30 years. At Ford's Wayne Stamping and Assembly Plants in Michigan, landfill gas powers the plant's boilers and produces 2.4 MW of electricity. An auto plant can command significant leverage with an electric utility, and accelerate green energy production.

"In the end, the question is not 'How do we use nature to serve our interests?' It's 'How can we use humans to serve nature's interest?'"
— William McDonough

Auto Company Environmental Initiatives

- BMW (click on Innovation and Science & Mobility): www.bmwgroup.com
- DaimlerChrysler: www.daimlerchrysler.com/environ/environ_e.htm
- DaimlerChrysler's Operation Green Factory, Germany: www4.daimlerchrysler.com/index_e.htm?/environ/focus3/env1_e.htm
- Ford: www.fordenvirodrive.com
- Ford's Rouge Center: www.mcdonough.com/projects/p_ford.html
- GM: www.gm.com/environment
- Honda: www.hondacorporate.com/environ_tech
- Renault: www.renault.com/gb/pdf/Envgb.pdf
- Toyota: www.toyota.com/times/safety/f_env.html

Maximize Your Recycling
Toyota's vehicles are 85% recyclable. Its goal is 90% recyclable by 2000, which involves simplifying and reducing the number of different parts, switching to more easily recyclable materials, using returnable packaging, developing a Japan-wide collection and recycling system for bumpers, placing identification markers on all parts, sorting shredder dust into 12 categories, publishing a dismantling manual, and developing a 5R program: Refine, Reduce, Reuse, Recycle, and Retrieve Energy. Ford has Recycling Action Teams to increase its recycling rate, following the tradition of Henry Ford, who made vehicle parts from soybean oil and used the wood from packing crates to make floors for the Model T. Plastic soft drinks bottles become grille reinforcements; used computer housings and telephones become grilles; spent battery casings become splash shields. DaimlerChrysler's goal is 85% recyclable cars by 2002. Ford has also joined the Recycled Paper Coalition, making the commitment to purchase paper with at least 30% post-consumer content, and to reduce overall paper consumption.

Establish Green Dealerships
In Umeå, in northern Sweden, the local Ford dealer, Per Carstedt, used an integrated systems approach to build a dealership called the Greenzone, including a car dealership, a McDonalds, and a fuel station. The showroom is heated by thermal solar collectors and a ground-source heat system, with culverts that share excess heat between the buildings. Lantern skylights provide better lighting. Coupled with other design efficiencies, overall energy demand has been reduced by 60%. The remaining energy needs are met by a wind generator, and power from the national grid. The roofs are covered with sedum, perennial grasses, and wildflowers; the interior spaces are purified with living plants; the building materials can all be re-used or recycled. Water needs have been reduced by 90% by collecting rainwater and recirculating water from the carwash and other uses; vacuum toilets use minimal water and treat the wastes locally, composting it into fertilizer. The fuel station offers ethanol as well as gas and diesel, and the whole facility has been certified to the ISO 14001 environmental management standard. The entire staff are educated in environmental matters.

Toyota's "Environmental Action Plan" sets specific targets to achieve by 2005, as part of its Global Earth Charter:

- Reduce electric and natural gas usage by 15% per unit
- Reduce VOCs by 30% per unit
- Reduce hazardous waste disposal at landfills by 95% per unit
- Reduce water usage by 15% per unit

At Ford's huge, 600 acres Rouge Center at Dearborn, Michigan, the architect Bill McDonough is helping Ford undertake a $2 billion retrofit over 20 years that will turn the industrial age mess into an icon of the next industrial revolution, and a paradigm of sustainable manufacturing. The renovated plant will include natural storm drains, porous paving, natural habitat for migrating songbirds, the planting of 1500 trees, a restored waterfront, integrated energy planning, daylighting, fuel cells, solar energy, and the world's largest "living roof".

59

Sell Transport, Not Just Cars

Corbin Motors

The Corbin Sparrow — designed for commuting.

"The automobile business is about to experience the most profound and revolutionary changes it has seen since the Model T first hit the streets."
— William Clay Ford, Chairman of Ford

But what kind of revolution? Will it be just a revolution in engine design, allowing the use of solar/hydrogen fuel, or will it open up a whole new dimension in the realm of transportation? Many things are changing as we enter the 21st century. The fact that the private car has dominated the past 50 years is no guarantee for the future. Generations pass away and new generations appear with different values, different dreams.

It is quite possible, some would say certain, that fuel prices will remain high. If the world supply of oil peaks as early as 2010, the price will skyrocket as the global demand from 530 million motorists outstrips the supply. This will be good news for auto companies that can mass-produce 100 mpg hybrid-electric vehicles. But when oil hits $8 or $10 a gallon, millions of ordinary motorists will start to think differently about the way they want to get from A to B.

If we look further into the future, one thing becomes totally certain: chronically overcrowded megacities such as Jakarta, Calcutta, Beijing or Rio de Janeiro have so many people and so little space that they will never be able to embrace the private automobile as their dominant means of transport. Instead, city governments will turn to mass transit, light rail, jitney-buses, city-wide bicycle routes, rollerblade corridors, car-free neighborhood centers, electronic road-pricing, car-sharing groups, and small electric city-vehicles, rentable by the hour with smart-cards.

In redesigning their cities to embrace these new forms of transport, mayors will create urban civilizations that are much more advanced — socially, environmentally, architecturally, artistically, and intellectually — than the noisy, lonely, car-dominated cities of the late 20th century. Creative people will flock to the redesigned cities and creative businesses will follow them, causing the tax-base of the new cities to thrive.

If you were the CEO of a global auto-company, how would you react to the prospect of such a shift? Would you pin your hopes on the new hybrids and fuel-cell vehicles, and hope that the shift goes away? Would you try to sabotage it by bribing politicians to vote against the necessary planning and policy changes? Or would you see it as an opportunity to widen your company's vision into becoming a provider of transportation and mobility, and not just of cars?

Neighborhood Electric Vehicles

		Range/miles	Top Speed/mpg	Price
Bombardier	NV	30	25	$6,199
Corbin Motors	Sparrow	30-60	70	$14,900
Dynasty	NV	30	25	$10,000
Ford	TH!NK city	53	56	n/a
Ford	TH!NK neighbor	30	25	$6,000
GEM	GEM	20-30	na	$7,000
Nissan	Hypermini	80	60	na
Toyota E-com	Mini EV	60	na	$11,500

"A better world is the ultimate 'must have' product feature. Whoever can deliver it will make friends — or, more accurately, passionate advocates."
— William Clay Ford, Chairman of Ford.

When you leave the world of the regular automobile, you will be astonished at what shows up. It is like entering a parallel universe that has been totally hidden from the public eye. Spend an hour exploring the listed websites, and you'll see what we mean.

Produce Neighborhood Vehicles
50% of the trips we do are under 10 miles, which is where small, nifty electric vehicles like Ford's TH!NK Neighbor and TH!NK City (imported from Norway) come in. A TH!NK City is the official staff car of the Norwegian Embassy in Germany. Ford is also selling TH!NK electric bikes to encourage more people to explore life on two wheels. Neighborhood Vehicles can be organized as local "Station Car" fleets, available for residents or members of a car-share group to rent by the hour, using a smart card. In France, the Praxitele system, being tested in Saint-Quentin-en-Yvelines outside Paris, offers its users access to a fleet of 50 Renault Clios. In Japan, Nissan's Hypermini is being trialled in the city of Ebina in a park-and-ride situation: local residents use them for commuting in the morning and evening, and during the day they become official vehicles at Ebina City Hall. A hundred TH!NK Citys will be used in the new Praesidio development in Los Angeles, where residents will share them for clean-air travel. For the growing number of car-share groups (see #17), neighborhood vehicles are a natural marriage.

Explore the Future
In addition to transit and light rail, some interesting hybrid systems are emerging. In Switzerland, the city of Lausanne is considering an ingenious "Serpentine" electric people-mover that runs on a ground-level MagnétoGlisseur track, next to the road. In Holland, Helmond and Eindhoven are considering an 8-mile TransGlide bicycle wind-tunnel to connect the two towns. In Denmark, trials are underway for a Rapid Urban Flexible (RUF) that allows small electric RUFs or larger MAXI-RUFs to travel directly from the road to a monorail. The RUF has a slot on its underside for the monorail, and carries both road and rail wheels. It can travel at 50 mph on the road, but once on the monorail it can link into a platoon and reach 125 mph, taking its current from the monorail while recharging its batteries, using a smart card to pay for the trip. In Germany, there is a proposed "auto-shuttle" where you drive your car into a see-through container, switch off the engine, and ride next to the highway on a raised Maglev railway.

Join Smart-Card Partnerships
The future of urban and regional travel will almost certainly borrow from Swiss Rail's "Easy-Ride Switzerland" system (see #17). A single smart card will give you access to buses, trains, taxis, bicycle rentals, car-share groups, car rentals and neighborhood vehicles — and pay your parking fines. Throughout the city, electronic boards will let you punch in your destination and show you the easiest way to get there. Heavy shopping will be delivered by neighborhood bicycle carts or electric mini-vans, and urban travel will once again become a pleasant, relaxing, social activity.

Resources

100 Top Transportation Sites: www.100toptransportation.com
Corbin Motors: www.corbinmotors.com
Denmark's Rapid Urban Flexible: www.ruf.dk
Dynasty NVs: www.dynastymotorcar.com
EV Rental, California: www.evrental.com
Ford's TH!NK: www.thinkmobility.com
France's Praxitele: www.epa-sqy.fr/praxib.htm
Germany's Autoshuttle: www.autoshuttle.de
Global Electric Motorcars: www.gemcar.com
Innovative Transportation Technologies: http://faculty.washington.edu/~jbs/itrans
MDI compressed air vehicles: www.zeropollution.com
National Station Car Association: www.stncar.com
North American Car Sharing Network: www.carsharing.net
Poma Group: www.poma.net
Practical Alternatives to the Car: www.bremen.de/info/agenda21/carfree
Skytran: www.skytran.net
Serpentine: www.serpentine.ch
TH!NK City, Norway: www.think.no
Toyota's e-coms: www.toyota.com/html/about/environment/news_awards/crayon.html

60
Form New Partnerships

During the 1990s, Ford, GM, and DaimlerChrysler were all members of the Global Climate Coalition (GCC), along with Amoco, Chevron, Exxon, Texaco, Mobil, and the American Petroleum Institute. The GCC calls itself "a voice for business in the global climate debate," but it has consistently worked to sabotage, obstruct, derail, delay, and diminish the Kyoto Protocol, and to deny that the accepted climate change science is valid.[1] Between 1989 and 1999, GCC's members paid $63 million[2] to the political action committees of members of the US Congress and Senate who have opposed every initiative to reduce greenhouse gas emissions and increase fuel efficiency.

During 1999 and 2000, following a successful campaign by Ozone Action and others (see #20), Ford, DaimlerChrysler and General Motors resigned from the GCC, but the GCC still boasts that its members include "business, agriculture, trade, and manufacturing associations ... that collectively represent more than 4 million businesses, companies and corporations," including the American Highway Users Alliance, which vigorously opposes the Kyoto Protocol and would prefer to see greenhouse gases reduced by making America's roads faster.

The world's auto-manufacturers could urge these trade associations to stop blocking the Kyoto Protocol and other progressive moves to reduce greenhouse gas emissions. They could influence their distributors and suppliers to do the same, and resign from trade associations such as the Coalition for Vehicle Choice, which has similar goals to the GCC.

Instead, they could join progressive partnerships more in keeping with their environmental goals. They could join the EPA's ClimateWise initiative, as General Motors and Ford's Twin Cities Assembly plant have done. They could join the American Council for an Energy Efficient Economy, the Energy Foundation, the Business Council for Sustainable Energy, or Business for Social Responsibility. They could join the Coalition for Environmentally Responsible Economies (CERES), as Ford and GM have done, or the Business Environmental Leadership Council of the Pew Center on Global Climate Change, as Toyota has done. They could join organizations at the front edge of the search for climate change solutions, such as the David Suzuki Foundation, the Pembina Institute and Environmental Defense. They could even form a partnership with Greenpeace, which spent $1.2 million to demonstrate that Renault's Twingo could be 50% more fuel-efficient (see #57). The past is no longer a guide to the future. We need new partnerships for a new century.

Climate-Friendly Legislation that Auto-Companies could Support

- Revised Corporate Average Fuel Efficiency standards (CAFE)
- Distance-based insurance, paid at the pump
- Urban traffic calming initiatives
- State feebate legislation: fees on inefficient vehicles, rebates on efficient ones
- Tax credits for efficient, clean-fueled vehicles
- Carbon taxes and rebates, with revenues going to urban transit and trains
- Fuel efficiency labeling for all vehicles
- Zero emissions laws similar to California's
- A zero-waste 100% recyclable cars bill

"If all of us were renting the Earth, we'd have lost our deposit. At Toyota, we're trying to help restore the planet to the way it was before everyone moved in."
— Toyota Motor Corporation

Form a Green Vehicle Marketing Partnership

Many environmentalists think that corporations are not to be trusted; many corporate managers think that environmentalists are locked into a negative philosophy of blame and criticism. And yes, there's lots of evidence to show that both are right.

But let's leave that aside for the time being, and work out how we can prevent the world from slipping into a climatic disaster that will spell tragedy for all, car-makers and tree-huggers alike. If Ford cares enough about the environment to retrofit its Rouge plant in Dearborn, Michigan so that it creates a net environmental benefit, and if Greenpeace is willing to spend $1.25 million to show that the Twingo SmILE can do 80 mpg, they have more in common than they think.

Auto-companies want customers for their hybrids, electric and fuel-cell vehicles; environmentalists want people to drive more efficient vehicles, and to drive less. When the Land and Water Fund set up a grassroots marketing partnership with the Public Service Company of Colorado (PSCo) to help it sell premium green energy, they sold five times more megawatts than PSCo had anticipated, because people trusted them (see #47).

The American Council for an Energy Efficient Economy is suggesting the need for a Green Vehicle Marketing Partnership, in which governments would establish a fuel efficiency labeling system, set up tax credits and incentives, run a public information program, and order green cars for municipal, state, and federal fleets. Auto-companies would produce the cars and support climate-friendly legislation (see box), and non-profit organizations such as ACEEE, Greenpeace and Climate Solutions would run grassroots campaigns to persuade individuals, companies, municipal governments and taxi-fleets to buy the vehicles.

It's not ridiculous. It's the way we need to work if we are to turn the climate around, and realize the solar/hydrogen revolution.

> **"I want the Ford Motor Company to be a leader in the second industrial revolution — the Clean Revolution."**
> **— William Clay Ford, Chairman of Ford[3]**

Possible Partners for Auto-Companies

American Council for an Energy Efficient Economy: www.aceee.org

Business Council for Sustainable Energy: www.bcse.org

Business for Social Responsibility: www.bsr.org

Clean Cars Campaign: www.cleancarcampaign.org

Climate Solutions: www.climatesolutions.org

Coalition for Environmentally Responsible Economies: www.ceres.org

David Suzuki Foundation: www.davidsuzuki.org

Energy Foundation: www.ef.org

Environmental Defense: www.environmentaldefense.org

EPA's Climate Wise: www.epa.gov/climatewise

Greenpeace: www.greenpeace.org

Pembina Institute: www.pembina.org

Pew Center on Global Climate Change: www.pewclimate.org

Land and Water Fund of the Rockies: www.lawfund.org

The Natural Step: www.naturalstep.org

61
Set Clear Goals

At this stage in the solutions, the game jumps to a higher level. In terms of CO_2 emissions, instead of trying to eliminate 20 tons a year for a typical household or 2 million tons for small city, we are looking at anywhere from 20 to 200 million tons a year for a state or province — or 450 million tons for California.

For most states and provinces, generally speaking, 10% of the emissions come from methane and 90% from fossil fuels, split 50:50 between transportation, and energy use in homes, businesses and industries. By January 2000, 34 states in the USA had completed greenhouse gas inventories, 26 had initiated State Action Plans, and 12 had completed their plans. In Canada, several provinces had begun to develop Action Plans.

In New Jersey, two-thirds of the state is threatened by frequent severe storms, higher seas, and billions of dollars of damage from global warming. In April 2000, the state set a goal of reducing its emissions by 3.5 percent below the 1990 level by 2005; the first state to set a formal goal. In 1990, New Jersey produced 136 million tons of greenhouse gases a year. By 2005, if nothing is done, those emissions will increase to 151 million. The goal is to get them down to 131 million. A third of the reduction is to come from energy-efficiency improvements in homes, businesses and industry; a third from the substitution of fossil fuel energy with wind, solar and biomass energy; and a third from improvements in transportationefficiency and waste management. This may enable New Jersey to reach the Kyoto target of 7% below 1990 by 2008-2012, but it is three times less than is needed to reach the ultimate goal of 80% by 2025, which probably requires a 10% reduction by 2005.

Pursuing such an ambitious goal is a massive undertaking, but it is also exciting, since it is a conscious transition into a new energy era. It is not possible for a state or provincial government to achieve such a transition without the active support and participation of its citizens, businesses, schools, colleges, churches, banks, farms and utilities. A government can only lead where its citizens want to go. In Minnesota, the citizens' organization Minnesotans for an Energy-Efficient Economy works to raise public awareness about the forecasted impacts of climate change, develops policy proposals, and counters those who do not want things to change.

Possible Goals for 2025

- 50% reduction in the amount of solid waste generated
- 95% of all solid waste to be recycled
- 75% of commute trips to be other than by single-occupancy vehicle
- 80% of farms to be managed sustainably
- 90% of single family homes to be roofed with solar shingles
- 100% of businesses to be showing an annual carbon reduction
- 100% of energy produced in or used by the state or province to be renewable

"Every time humankind has switched from an existing fuel to a newer one — from wood to coal, coal to oil, oil to natural gas — the switch has been associated with economic progress. The same will be true for alternative energies, such as wind energy, solar power, cogeneration, and fuel cells."
— Jim Woehrle and Julie Bach, Minnesotans for an Energy-Efficient Economy

"**Global warming means a horrifying future for nature. World leaders must give top priority to reducing levels of carbon dioxide."**
— Jennifer Morgan,
World Wide Fund for Nature

Write a Climate Action Plan
Every state, province, prefecture, region, county, and district in the world faces the same feeling of immensity around this challenge. But while the solutions are fundamentally simple – cut methane emissions, use energy more efficiently and shift to renewable fuels — there are limits to what a state or province can do to change the energy habits of its citizens, businesses and institutions. A governor or premier can order a state or province's vehicles to switch to alternative fuels (as Nebraska did for 25% of its vehicles), or require its office buildings to participate in Energy Star, Green Lights, and other efficiency programs (as Wisconsin is doing), but it cannot walk into someone's home or business and say "become efficient!" or "switch fuels!" The task is a bit like cooking. It involves finding the right mix — 2 oz energy policy, 1/2 lb clean fuel incentives, 1/4 lb tax changes, 1 tbsp educational programs, 1 cup of leading by example — that will persuade people of the merits of the fossil-fuel-free cake, and get them to join in the baking.

The normal starting point for a plan is the establishment of a task force involving scientists, planners, environmentalists, energy specialists, and others who have studied the issues. You should be prepared for an onslaught from the naysayers, who will insist there is no problem and whine to the media that this is yet another example of needless government interference. As authors, our advice is to brush up on the "Counter-Arguments" (see p. 20), and ensure that your task force's terms of reference include the likely impacts of the Runaway Greenhouse Effect, as well as the Enhanced Greenhouse Effect (see pages 14 & 18). It is only proper that residents should be aware of the scale of climatic disturbances that are heading their way — and their grandchildren's way.

When your task force has been established, it might follow an 8-step process:

(1) Publicize the climate change crisis and the proposed plan widely, inviting public input. Keep everyone involved by the Internet and regular emails, and through the media.
(2) Develop a greenhouse gases baseline inventory, so that you know where you stand.
(3) Examine the ecological, social and economic impacts on your state or province for both the enhanced and the runaway climate change scenarios.
(4) Make greenhouse gas projections for "business as usual."
(5) Establish a greenhouse gases reduction goal for 2005 or 2010.
(6) Collect all possible policy options, from A to Z.
(7) Select and rank your recommended policies, and obtain more public input.
(8) Present your proposed Action Plan, along with progress markers, a chart showing who's responsible, a process for monitoring and evaluating, a budget and a leadership structure.

Finally, create a position at the cabinet-table for a Minister or Secretary of State for Climate Action, to push ahead with the policies and solutions that are needed.

Resources

British Columbia's Climate Change Business Plan:www.elp.gov.bc.ca/epd/epdpa/ar/climate

Climate Change—Are You Doing Your Bit? (Canada): www.climatechange.gc.ca

Minnesotans for an Energy-Efficient Future: www.me3.org

New Jersey Future: www.njfuture.org

Public decision-Makers and Climate Change: www.epa.gov/globalwarming/visitor-center/decisionmakers

State Initiatives on Climate Change: www.ccap.org

US State Action Plans: http://yosemite.epa.gov/globalwarming/ghg.nsf/actions/StateActionPlans

62 Show Leadership

The best way for a government to show leadership is in its own buildings, transportation practices, and use of energy. This sets the tone for what is possible, and helps build the market for renewable energy, alternative-fueled vehicles and efficient appliances. With Honda and Toyota marketing hybrid gas-electric vehicles, states and provinces can set purchasing requirements prioritizing the use of vehicles that achieve 50 mpg.

Here are some examples of what different state and provincial governments are doing:

- In Kasumigaseki, Tokyo, the Japanese government has bought 120 bicycles to encourage government employees to cycle between the Diet buildings and other ministry and agency buildings.
- Texas requires all state agencies to purchase alternative fuel vehicles; the Texas Department of Transportation operates more than 5,000 such vehicles. California has purchased 70 fuel-cell powered buses and cars.
- Wisconsin is establishing a comprehensive Transportation Demand Management plan for every state agency, with clear goals for reducing single occupancy vehicle driving. The Wisconsin Energy Initiative was set up in 1990 as a state-wide partnership with the goal of reducing the state's energy consumption by 20% by 2000.
- In Colorado, 90 school districts, city and county governments, colleges, universities, hospitals, and public housing complexes are participating in "Rebuild Colorado". 23 are using performance contracting (see #22) to implement projects, realizing generous energy, cost and CO_2 savings.
- Maryland has installed a large photovoltaic system on the University of Maryland campus, and is supporting solar cost-sharing programs for schools, farms, residential rooftops, and local government buildings.
- The governors of Colorado and Nebraska have issued executive orders encouraging state agencies to buy renewable energy. Five Nebraska state agencies and the governor's mansions in both states are now purchasing wind energy from local utilities.
- In British Columbia, the government's "Green Building BC" program requires that new government buildings must beat the national energy efficiency standards by at least 50%, and existing buildings are to be upgraded for energy, water and resource efficiency, saving money and reducing greenhouse gas emissions.

"State and local governments [in the US] have 10 billion square feet of public space and spend $16 billion annually on energy — equal to the gross domestic product of Costa Rica.
The energy used in these public spaces is responsible for 134 million tons of carbon emissions, but 30% of that energy is wasted because of inefficiency. Your efforts will make or break our nation's efforts to address climate change. Your efforts are where the rubber hits the road."
— Jerry Clifford, EPA

Resources

Clean Cities Alternative Fuel Vehicle Fleet Buyers Guide: www.fleets.doe.gov

Clean Government -- Options for Governments to Buy Renewable Energy: www.repp.org

Energy Star for governments: http://yosemite1.epa.gov/estar/business.nsf/webmenus/Government

Energy Star Purchasing Toolkit: www.epa.gov/nrgystar/purchasing

Green Buildings BC: www.greenbuildingsbc.com

Public Employees for Environmental Responsibility: www.peer.org

Renew America: www.solstice.crest.org/environment/renew_america

Rebuild Colorado: www.state.co.us/oemc/rebuildco

"If global warming is not contained, the West will face the choice of a refugee crisis of unimaginable proportions, or direct complicity in crimes against humanity."
— George Monbiot[1]

Establish Green Purchasing Guidelines

State and provincial budgets run into the billions, including state-ordered purchases of a huge amount of energy, paper, and other materials, all of which carry an embodied carbon dioxide cost. By creating a green purchasing policy, states and provinces can help build the market for green energy, energy efficiency, and recycled materials while showing responsibility and reducing CO_2 emissions.

Green Purchasing in Action

- Arizona uses a list of prohibited tropical wood products.
- In 1995, Missouri purchased $11.3 million of products made from recovered materials.
- Minnesota's recycling program for state buildings has resulted in the development of lease language for rented facilities that requires janitorial recycling services.
- Illinois has an overall waste reduction strategy and the state printer routinely questions agencies submitting printing orders that do not require dual-side printing.
- In 1992, New Jersey, Vermont, and Maine jointly purchased glass beads with 95% recycled content for reflective highway stripping. The contract price was almost 50% less than the states paid individually for the same product.
- Missouri law requires all contract printers working for the state to use soy-based ink and recycled paper if the cost is no more than 10% greater than comparable inks.
- Utah runs the Utah Energy Efficient Procurement Program to ensure that the state procures energy-efficient products.
- Washington requires all communities building new schools to complete an Energy Conservation Report detailing energy-using systems that have the lowest life-cycle cost. Designers must either recommend a design option within 10% of the lowest life-cycle cost or have the school board vote on an alternative design.
- California evaluates bids for office copiers on a life-cycle cost basis, including energy consumption.

A Chicago Transit Authority bus powered by Ballard fuel cells and XCELLSIS fuel cell engines.

Green Purchasing Resources

Environmental Purchasing Guide
www.pprc.org/pprc/pubs/topics/envpurch.html

Forty Ways to Make Government Purchasing Green:
www.gpp.org/40ways/40ways.html

Government Purchasing Project:
www.gpp.org

63

Build an Energy Democracy

In a healthy democracy, citizens have access to information about public decisions, free choice in making decisions, protection against the manipulation of power, and a guarantee that elections are not unduly influenced by vested interests. Democracy is not a new institution. It has developed and been bitterly fought for during hundreds of years when societies were ruled by whoever assumed power. We say that our countries are democratic, but the process of building a healthy democracy with equal participation by rich and poor, empowered and disempowered, has hardly begun.

There is a lot of effort being expended today to build a better energy democracy. People are waking up to the shocking reality that the way we produce energy is giving the entire Earth a fever — melting the glaciers and the polar ice — and they are realizing that unless we get Earth's temperature down, we will face hundreds of years of climate-induced tragedy. We urgently need a healthy energy democracy. Here are seven initiatives and reforms that can help build one.

1. Invest in Public Education
To most people, climate change is bewildering enough without the remaining scientific uncertainties, and the disinformation that is put out by the fossil fuel industry. The public urgently needs climate change and energy information in films, TV shows, posters, websites and books. Without the foundation of a properly informed public, politicians and other leaders cannot advance the policy initiatives needed to wean society off fossil fuels, and usher in the solar/hydrogen era.

2. Support Regional Partnerships
Regional non-profit partnerships focused on climate, energy and sprawl —such as New Jersey Future, Climate Solutions (Pacific Northwest), and Minnesotans for an Energy-Efficient Economy — are today's equivalent of the suffragettes or the civil rights groups, which made such important progress for democracy. They are building a grassroots response to global climate change. If your province, state or region doesn't have a climate action partnership yet, it's time to build one.

3. Make Sure Everyone Has Access to Energy Information
What is the fuel source of your electricity? What are its CO_2 emissions? This is information that should be disclosed. By August 2000, 15 states (but no Canadian provinces) had disclosure laws. In Illinois, they use color-coded pie charts to show the information on utility bills. Household appliances, cars, and other equipment should all carry energy labeling. The electricity meter on your wall could do with a jump-start out of the past — a digital smart meter could give real-time, room-by-room information on energy-use and CO_2 emissions (see #42). With personal CO_2 calculators (see #1), individuals, households and businesses could do an annual CO_2 assessment alongside their taxes, planning next year's CO_2 reduction strategies accordingly.

> **"What SMUD has done to put solar technology into the hands of its customers... could be easily replicated in community after community."**
> **— Ed Smeloff, President, Sacramento Public Utility District (SMUD)**

Educational Resources

Climate Change Outreach Kit (slides, CD): www.epa.gov/globalwarming/publications/outreach

EPA's Global Warming page: www.epa.gov/globalwarming

The Heat is On-Line: www.heatisonline.org

Turning Down the Heat (National Film Board of Canada): www.davidsuzuki.org/energy

Union of Concerned Scientists: www.ucsusa.org

Warren Gretz

Photovoltaic system located next to the closed-down nuclear power operation at Sacramento Municipal Utility District's Rancho Seco Facility.

4. Give Communities the Right to Aggregate (Community Choice)
Many states in the US are in the middle of electricity restructuring, which can either weaken consumer rights or provide important benefits. One of the rights that is fundamental to an energy democracy is the ability to aggregate as communities and co-operatives. This enables common citizens to join forces and gain bargaining power with energy producers, putting them on a level playing field with the large institutional power customers. This is crucial in the emerging retail power marketplace, where larger customers are busy striking deals with the power providers. In the restructuring plans passed by several states, communities have lost the right to aggregate. In Massachusetts, activists won the right for towns and cities to aggregate their demand, and form energy buying clubs. The Local Power Project is leading the drive for this reform. (See also #23)

5. Set up Net Metering
Net metering gives everyone the right to generate their own energy — whether by solar, wind, fuel cell or microturbine — and to feed it into the grid instead of storing it in a battery. The meter spins backwards when there is a surplus, giving you a credit (see #42). It is being used with great success in Denmark and Germany. By March 2001, 31 states in the US as well as Manitoba and part of Ontario in Canada had introduced it. The best legislation is in Iowa and Ohio, which impose no limit on the energy a producer can generate, and Washington, where an energy surplus can be calculated over a full year, allowing a summer solar surplus to offset a winter deficit. Net metering is a critical component of the distributed grid.

6. Allow Customer Choice
Customer choice is essential for an energy democracy. Alberta and 22 US states give consumers the right to choose the kind of energy they want. By choosing green energy, they encourage utilities to invest in more solar, wind, geothermal and biomass projects.

7. Legislate Campaign Finance Reform
It is essential to weaken the grip of those who influence politicians by making hefty donations to their political action committees. In North Dakota, where there is the greatest potential for wind power in the US, but where the existing power comes from four huge coal-powered utilities, for some strange reason almost no wind turbines are being installed. The states of Maine, Vermont, Massachusetts and Arizona have all chosen Clean Money reforms for their state elections. In Holland, only registered voters are allowed to contribute to political campaign funds, and every dollar is matched by the government. No corporations or other organizations are allowed to donate. Now there's an idea...

Resources

American Local Power Project: www.local.org

Common Cause: www.commoncause.org

Net Metering: www.eren.doe.gov/greenpower/netmetering

Public Campaign Finance Reform: www.publiccampaign.org

Public Citizen Campaign Finance Reform: www.citizen.org/congress/reform/refhome.html

Wind & Net Metering: www.awea.org/policy/index.html#netbill

64 Produce Cleaner and Greener Power

In 1998, 51.7% of the electricity consumed in the U.S. came from plants that burned coal (anthracite, coke, lignite wastes); 18.6% from nuclear energy; 15% from natural gas; 9% from hydroelectric; 3.5% from oil; and 2% from renewables – wind, geothermal, biomass, and solar energy. In all, 70% of electricity came from burning fossil fuels, releasing carbon dioxide and methane. In Canada, 29% came from fossil fuels, 56% from hydro, and 14% from nuclear.

For states and provinces, going clean and green is a clear winner. A Wisconsin study showed that generating 750 MW of renewable energy would produce $3.1 billion more in gross state product and 65,000 more job-years than using fossil fuels.[1] Here are some steps to make your state or province a clean energy winner.

Make Them Clean Up Their Act

One of the dilemmas in U.S. energy politics is that the oldest, dirtiest coal-burning power plants, built before 1977, were given exemptions from the 1990 Clean Air Act. This allowed them to go on producing cheap, filthy energy. If they were forced to clean up their act, they would have to increase their prices considerably or go out of business. In May 2000, following mounting political and grassroots pressure, Massachusetts became the first state to require its coal-burning plants to exceed the federal standards and become as clean and efficient as new plants within three years or face hefty penalties. Hopefully, this will open the door for other states to follow.

Create a Mandatory CO_2 Standard for Energy

In 1997, Oregon passed a landmark law that establishes a "carbon dioxide standard" for all new power plants of 25 MW or more, requiring that they achieve CO_2 emissions 17% below the most efficient gas-fired plant currently operating in the U.S. The standard can be met by investments in efficiency, cogeneration, renewable energy, or off-site carbon-offsets, which can be arranged by paying a fee to the Oregon Climate Trust to make carbon-offset investments on their behalf. The company building the new Klamath Falls gas-fired cogeneration plant is reforesting 6,250 acres in western Oregon, expanding the geothermal heating system in Klamath Falls, generating 32 MW from waste methane, and supporting the installation of 182,000 PV systems in rural villages in India, China, and Sri Lanka. Using a similar approach, Wisconsin requires its electric utilities to add a cost of $15 per ton of CO_2 emitted when comparing and selecting technologies in the state's advance planning process.

Don't Shift To Natural Gas

Coal produces 100 units of CO_2 for every Btu of energy produced, but gas releases only 55 units, which is why the use of natural gas combined cycle turbines is seen by some as a move in the right direction. Ninety-six percent of new energy production in the U.S. is planned to come from gas. But gas still releases CO_2, and it also releases methane, which is 62 times more powerful than CO_2 as a greenhouse gas over 20 years (see p. 42). If natural gas production peaks between 2010 and 2020, as some energy commentators think it may, we will see even higher price increases. With fossil fuel prices heading upwards, renewable energy prices heading downwards, and carbon taxes on the horizon, renewables are a more intelligent long-term investment.

Resources

Clean Power Surge - Ranking the States: www.ucsusa.org

Database of State Incentives for Renewable Energy: www.dcs.ncsu.edu/solar/dsire/dsire.cfm

Interstate Renewable Energy Council: www.irecusa.org

Long Range Energy Alternatives Planning: www.seib.org/leap

Oregon's Climate Trust: www.climatetrust.org

Oregon's CO_2 legislation: www.leg.state.or.us/97reg/measures/hb3200.dir/hb3283.a.html

Public Benefit Funds: www.aceee.org/briefs/mktabl.htm

ReInState: http://gem.crest.org/sbs

Renewables Portfolio Standard: A Practical Guide: www.naruc.org

Vermont's Wind Resource: www.state.vt.us/psd/ee/wind/ee-wind.htm

Sandia

Wind turbines at Bushland, Texas.

Set Up A Public Benefit Fund

Public Benefit Funds (also known as System Benefits Charges) are built up from a small fee collected as a surcharge on electricity bills, and used to encourage energy efficiency and renewable energy. This is the policy tool that Germany is using to build its solar market. By August 2000, 21 US states had set up such funds, led by California, Massachusetts, Connecticut, New Jersey, Oregon, and Minnesota. By 2010, they will have channeled $1.7 billion into the development of renewable energy. California's fund is the largest, at $496 million over the period 1998–2002.

Create A Renewables Portfolio Standard (RPS)

An RPS is a state policy that mandates that a percentage of all electricity must come from renewable sources. By August 2000, 13 states had made RPS commitments. Massachusetts leads the way, requiring 7% of total electricity sales to come from new renewables by 2012, followed by Connecticut (6%), and Minnesota (4.7%). Texas has committed to 2000 MW of wind energy (2.2%), energizing the Texas wind industry. Arizona has committed to 1.1% by 2007, including 50% from solar energy. Maine has committed to 30%, but since it already produces 45% of its electricity from biomass and electricity, this is in fact a step backwards. In order to reduce CO_2 emissions by 80% by 2025, RPS commitments need to be around 35% in all states by 2012.

Introduce Electricity Feed Laws, Tax Incentives, and Other Policies

In Denmark and Germany, where wind energy is having enormous success, electricity feed laws are used to oblige electric utilities to purchase renewable energy at a guaranteed price equal to 90% (Germany) or 85% (Denmark) of the retail price. In Denmark, wind energy producers also receive eco-tax relief worth 9 cents/kWh and electric utilities receive a 1.5 cents/kWh subsidy for producing wind energy. Oregon offers a tax credit of up to $1,500 for residential solar systems, and businesses can write off 35% of the cost of renewable energy investments over 5 years. Maryland has introduced tax incentives to encourage the use of solar power, fuel cells, and biomass fuels. California pays consumers a 1.5 cent/kWh subsidy for choosing green energy.

Other policy tools include barrier removal, the use of investment tax credits, property tax reductions, subsidies and loans (e.g. California's low-interest loans and 50% cost subsidies for solar panels or backyard wind generators), direct production incentives, and R&D assistance. States and provinces can publish wind energy maps, as Oregon and Vermont have done, and pass net metering legislation, ensuring that anyone producing renewable energy is able to export the surplus to the grid (see #42).

Stop Licensing Further Oil and Gas Exploration

What sense can there be in searching for more fossil fuels, when we need to phase them out?

How Much Renewable Energy Is Needed?

Megawatts of electricity capacity of U.S. electricity in 1996: 775,872

Megawatts needed in 2025 if efficiency reduces demand by 50%: 400,000

Megawatts of U.S. renewable energy capacity in 2000: 3,600

Megawatts of renewable energy capacity planned for 2012: 9,770

Megawatts of non-CO_2 hydro and nuclear capacity in 2000: 209,500

Megawatts of renewable energy needed for 80% non-CO_2 emitting by 2025: 152,000

Megawatts of potential wind energy capacity in Texas: 525,000

Sources: Union of Concerned Scientists; U.S. Energy Information Agency; Texas Sustainable Energy Development Council

65

Build an Energy-Efficient State or Province

Residential and commercial buildings account for 35% of US CO_2 emissions. Building owners can achieve a 10-50% savings on their energy bills by investing in energy-efficiency retrofits, realizing an excellent return on their investment. Why is it, then, that so few people do it? From a climate change perspective, it is an enigma that needs cracking urgently, to unlock the formula that will cause the owner of every home, office, factory and school in North America to retrofit it to the highest level of efficiency.

From a purely economic perspective, it makes sense. Every dollar not spent purchasing energy for an inefficient building is a dollar that can be spent elsewhere. In one year recently, Wisconsin people spent $6 billion of their wealth importing energy, a bigger chunk of their state economy than the forest products and agriculture sectors combined. So what is the key that can unlock this enigma?

The Cheaper it is, the More they Burn

Every so often, the US Energy Information Agency publishes figures that show the price of personal energy consumption in each state. It bears close analysis, because it demonstrates that the cheaper energy is, the more people burn. *In the five states with the cheapest energy, people consume the most* (see box). The cheap energy does not save them money, however, for these same five states come top of the list for yearly expenditures on energy per person. By contrast, out of the ten states where energy is the most expensive, eight are states where people consume the least, and spend less overall.

The moral of the story is clear: cheap energy gets squandered, while expensive energy encourages efficiency. There could hardly be a stronger argument for raising rates by placing a surcharge on electricity, gas and home-heating oil, and recycling it back to the consumer in the form of efficiency rebates and incentives.

The Cheaper It Is, The More They Burn

Cheapest Energy	Price per million Btu	Million Btu per person/yr	Cost/person 1997
Louisiana	$5.81	920.3	$3,473
North Dakota	$6.25	547.6	$2,651
Wyoming	$6.51	881.6	$3,903
Alaska	$6.69	1,151.8	$3,575
Texas	$6.94	590.8	$2,841
Most Expensive Energy			
Hawaii	$13.34	204.6	$1,920
Arizona	$11.75	251.4	$1,883
New Hampshire	$11.58	260.5	$2,154
Connecticut	$12.56	252.3	$2,218
District of Columbia	$12.84	329.0	$2,518
US average	**$8.82**	**352.2**	**$2,119**

Source: Energy Information Administration, State Energy Price and Expenditure Report 1997: www.eia.doe.gov/emeu/seper/contents.html (click on 'State Rankings').

FACT: **If every person in America consumed the same average energy per year as residents in the ten states with the highest energy prices, total consumption and CO_2 emissions from energy would fall by 27%, while annual energy expenditures per person would fall by $282.**

Source: Energy Information Administration data, combined with US census population data

Jon Cosner

San Francisco's Commercial Energy Conservation Ordinance becomes enforceable on existing buildings whenever a building's title is transferred, an addition increases the heated space by more than 10%, or a renovation worth more than $50,000 occurs.

Unlock the Efficiency Enigma

The key that can unlock the 'Efficiency Enigma' has five notches to it, answering the five excuses that people give for not investing in more efficient buildings. Use them right, and at least the buildings in your state or province will become energy efficient. For industry, there are similar approaches that can be developed.

Excuse No 1: I am not required to make my building more efficient.

Solution: Upgrade your state energy codes to 50% above the International Energy Conservation Code and make the code mandatory, as Oregon does. Take a leaf from San Francisco's book, and pass legislation (or enabling legislation that municipalities can use) that requires an audit to demonstrate conformance to the code whenever a house is sold, and on other occasions (see #24). Require all public buildings to meet the code, or suffer budget penalties.

Excuse No 2: I don't know where to find out about energy upgrades.

Solution: Establish a 1-800 hotline, as Minneapolis-St. Paul did.

Excuse No 3: I can't afford an energy audit.

Solution: Make the cost of energy audits free, financed through a Public Benefit Fund derived from a surcharge on utility bills.

Excuse No 4: I don't trust the various businesses to do a good job.

Solution: Ask the industry to set up a system of accreditation for energy service companies and other operators.

Excuse No 5: I can't afford the cost.

Solutions:

(a) Use tax incentives to encourage the ESCO sector to expand, enabling them to use performance-based contracting more widely.

(b) Pass legislation that prohibits mortgages for buildings that do not meet the code and makes energy-efficient mortgages and mortgage extensions the norm.

(c) Pass legislation that allows the State Energy Office to sell general obligation bonds to establish energy-efficiency loan funds, as Oregon does.

(d) Pass legislation that allows construction bonds to be issued only for building work that conforms to the code (thanks to David Morris[1] for this one).

(e) Establish tax credits to cover the cost of an efficiency upgrade, including efficient appliances and solar installations, as Oregon and Maryland[2] do, and the construction of efficient new buildings, as New York does.

(f) Set up a partnership with banks and credit unions enabling them to issue retrofit loans that are repaid through subsequent savings.

(g) Use money from a Public Benefit Fund to issue efficiency loans to low-income households.

(h) Grant tax relief on money invested in private sector energy efficiency and renewable energy.

Resources

Advanced Buildings, 90 technologies and practices (Canada): www.advancedbuildings.org

American Council for an Energy Efficient Economy: www.aceee.org

Alliance to Save Energy: www.ase.org

Building Codes Assistance Project: www.crest.org/efficiency/bcap

Canada's Office of Energy Efficiency: www.oee.nrcan.gc.ca

Energy Efficient Mortgages, National Home Energy & Resources Organization: www.national-hero.com

EPA's Energy Star Buildings & Homes: www.energystar.gov

New York's Green Building Tax Credit: www.nrdc.org/cities/building/nnytax.asp

Oregon Office of Energy: www.energy.state.or.us

Oregon's tax credit legislation: www.energy.state.or.us/res/tax/finalrtc.htm

Rebuild America: www.eren.doe.gov/buildings/rebuild

US Building Standards & Guidelines Program: www.eren.doe.gov/buildings/codes_standards/buildings

Vermont Energy Investment Corporation: http://homepages.together.net/~veic

66

Build a Sustainable Transport System

and Stop Sprawl

The way we travel is a huge factor in the global climate change crisis. Transportation accounts for 31% of the USA's and 27% of Canada's greenhouse gas emissions, and the miles driven and the gasoline consumed are both increasing. The "big picture" solutions lie in switching to alternative fuels and more efficient vehicles; switching from driving to cycling, walking, ride-sharing, transit, rail, and working at home; stopping the spread of suburban sprawl; and building a fast, nation-wide inter-city rail service. What policies and initiatives can a state or province adopt to accelerate these solutions?

Encourage Fuel-Switching

Maryland's Clean Energy Incentive Act reduces the state's titling tax by up to $2,000 for electric and fuel-cell vehicles, and up to $1,500 for gas-electric hybrids. Oregon's Residential Energy Tax Credit provides a $750 credit for all electric and alternative fuel vehicles, and towards the cost of conversion or installing a home charging or fueling system. In California, 10% of new cars and light trucks must release little or no pollution by 2003 — 4% must be zero-emissions and 6% must be near-zero emissions; Maine, Massachusetts, New York and Vermont are following California's lead. California's Consumer Assistance Program pays people $1,000 to retire old gas-guzzlers, and helps pay for the cost of repairs to cars that don't meet the state's emissions standards. Arizona offers a $10,000 tax credit for the purchase of electric vehicles.

Aerobus International Inc.

Aerobus — see www.aerobus.com

Encourage Fuel Efficiency

In Canada, Ontario's Tax for Fuel Conservation imposes a fee on fuel-inefficient vehicles, while giving a $100 rebate to vehicles which use less than 6 liters per 100 km (38mpg). Twenty states in the US have tried to introduce a similar "feebate", but the government has argued that the federal Corporate Average Fuel Efficiency (CAFE) statute makes it illegal for states to enact laws relating to fuel efficiency. The absurdity is that the CAFE standards are so low (see #76). Oregon is lobbying the federal government to update the CAFE standards as part of its Climate Change Action Plan. Maryland has introduced a tax incentive to encourage the use of hybrid vehicles.

Voters Support More Sensible Transportation

- 78% say the US should increase its use of electric vehicles
- 80% favor raising the fuel-economy standards for cars and SUVs
- 57% oppose drilling for oil in areas such as the Arctic National Wildlife Refuge
- 76% say the environment and global warming will influence how they vote

Source: 1998 Sustainable Energy Coalition poll

"If you build it, they will come. Cities that build freeways will sprawl. Cities that emphasize transit will not."
— Peter Newman, co-author of *Sustainability and Cities: Overcoming Automobile Dependence*

Encourage Sustainable Land-Use
One of the main components of Oregon's Climate Change Action Plan is the Transportation Planning Rule, which says that local land-use and transportation plans must show how they are going to reduce parking, and achieve a 10% reduction in vehicle miles traveled by 2015, 20% by 2025 (compared to 1995). In return, the Department of Transportation provides grants for transportation growth management. Oregon's 1973 state law requires municipalities to develop comprehensive land-use plans and establish growth boundaries around their perimeters.

In Maryland, the *Smart Codes Act* creates incentives for local governments to adopt model development codes that encourage infill development and smart growth in appropriate areas. Wisconsin offers smart growth dividends to local governments to encourage more efficient development patterns. State and provincial grants to municipalities need to be made conditional on the existence of smart growth and sustainable transportation plans, and CO_2 reduction plans.

Sustainable transportation legislation can level the playing field and ensure that transportation dollars support cycling, transit and rail, not just roads and highways. In Minnesota, the 2000 transportation funding plan diverted revenues from the motor vehicle excise tax into transportation and mass transit. In the US, federal TEA21 transport grants can be used to advance cycling, walking and transit.

Pennsylvania has completed a top-to-bottom review of all state programs, policies and regulations to see where they encourage sprawl, and all state agencies are required to support smart land use planning and promote the preservation of farmland, open space, greenways, natural areas and historic areas. California, Oregon, and Washington are using a set of land-use and energy analysis tools called "Planning for Community Energy, Environmental and Economic Sustainability" (PLACE^{3}S) to help communities save energy and reduce CO_2 emissions through land-use planning.

Between 1990 and 1996, Cincinnati's population grew by 2%, but its land area grew by 12%, and its drivers spent 200% more time in gridlock. The Sierra Club of Ohio's Transportation and Livable Communities Campaign is working to establish greenbelts around Ohio's urban centers, to purchase and preserve open space, and link Ohio's major cities with a light rail system, encouraging the public to get involved through the Citizen Advisory Committees that serve Ohio's Metropolitan Planning Organizations.

Encourage Cycling
Maryland's Bicycle and Pedestrian Access Act 2001 created a Director of Bicycle and Pedestrian Access, with a brief to develop a 20 year Statewide Bicycle-Pedestrian Master Plan, help Maryland's counties develop more facilities for cycling and walking, and ensure that the best engineering practices are used. In British Columbia, the Cycling Network Program provides 50:50 provincial funding for cycling investments by local governments. For every $1 invested in cycling, a city saves $2 in transportation costs, so it makes financial sense to establish a goal of 10% cycling for city trips and allocate 20% of the transport budget to achieve the goal.

Resources *(see also #27, #28, #29 & #77)*

Once There Were Greenfields: How Urban Sprawl is Undermining America's Environment, Economy and Social Fabric: www.nrdc.org.publications

Smart Growth: Myth and Fact (Urban Land Institute) www.uli.org

Taking Its Toll: The Hidden Costs of Sprawl in Washington State: www.climatesolutions.org

BC's Cycling Network Program: www.tfa.gov.bc.ca/grant/cycling.asp

Maryland's bicycling legislation: www.ohbike.org/mbpac

National Association of Railroad Passengers: www.narprail.org

Ontario's Tax for Fuel Conservation: http://iisd.ca/greenbud/taxfuel.htm

PLACE^{3}S: www.energy.ca.gov/places

Sierra Club Report on Sprawl: www.sierraclub.org/sprawl/50statesurvey

Stopping Sprawl in Ohio: www.ohio.sierraclub.org/sprawl

Transport Equity Act: www.tea21.org

Vermont Forum on Sprawl: www.vtsprawl.org

Victoria Transport Policy Institute: www.vtpi.org

67 Shift Your Taxes

"Carbon or energy taxes are not supposed to create a shock effect but instead only to restructure the economy and fiscal system. They are mainly effective in the long run."
— Kai Schlegelmilch, Wuppertal Institute, Germany

If there's one thing that states and provinces are good at, it's collecting taxes. Nobody likes it, but we hold our noses and do it because that's the way our world works.

But what if it worked differently? What if, instead of there being taxes on "goods," such as income and jobs, there were taxes on "bads," things we wanted to discourage, such as pollution, traffic, and carbon emissions? This is what tax shifting is all about. It's not about charging more taxes — it's about shifting them from one thing to another. Environmental taxes have been used in northern European countries for over a decade (see box) and now they're coming to North America.

Under Vermont's current perverse tax-regime, property taxes fund the cost of road construction and maintenance and provide a hidden subsidy to motorists, while the sales tax provides an exemption on the use of energy, chemical fertilizers, and pesticides for farming. By way of reform, Vermont's legislators have been looking at the following proposals:

- a carbon tax on fossil fuels with the income going as a yearly refund to residents and businesses;
- reducing motor vehicle license fees while increasing fuel taxes;
- a motor fuel tax of 4 cents per gallon, with the income going to reduce property taxes;
- a pesticide and fertilizer tax with the income going as tax credits to farmers and to help them move towards organic and low-impact farming;
- a solid waste tax with the revenues going to subsidize recycling, composting, and waste reduction.

The Vermont Department of Public Service's tax-shift report estimates that the first two proposals could cut fossil fuel use by 16.2% within a year. Energy used for transportation would fall by 30%, and there would be a 38.7% increase in the use of renewable energy. A poll of 500 voters showed that 86% supported tax reform that would promote well-paying jobs; 93% supported tax reform that would protect clean air and water; and 88% supported tax reform that would control sprawl style development.

Tax-shifting is also being considered by British Columbia, Maine, Michigan, Minnesota, and Oregon, where a report by Northwest Environment Watch (*Tax Shift*) estimated that a tax on pollution, carbon, and traffic could reduce business and income taxes from 46% of state revenues to 14%, and eliminate property taxes.

Good Taxes on "Bads"

- Sweden: Carbon dioxide emissions, lead, diesel fuel, chemical fertilizers
- France: Water pollution, sewage, carbon
- Denmark, Britain, France: Non-recycled waste
- Germany: Disposable fast-food packaging, energy
- British Columbia: Tires, car batteries, paint
- Denmark: Pesticides, herbicides, fungicides, energy
- Switzerland: Volatile organic compounds, fuel
- Holland: Electricity, fuel

"The supreme reality of our time is... the vulnerability of our planet."
— John F. Kennedy, 1963

Carbon Credits and Rebates

Imagine there was a website that enabled you and every household, business, or public institution in your state or province to:

(1) analyze your annual climate impact (see Carbon Calculators, #1);

(2) establish a CO_2 reduction goal for next year;

(3) plan a CO_2 reduction strategy; and

(4) purchase carbon offsets to absorb your outstanding CO_2 emissions.

This is all about the power of information to help you take responsibility for your share of the climate crisis. Most people and businesses haven't a clue how much CO_2 they produce, so they're not motivated to think about it, let alone plan a reduction strategy.

Let's take the idea a step further. Imagine a government tax-shift that gave you a tax-credit towards the purchase of renewable energy (as Utah does), an energy efficiency upgrade (as New York does) or a fuel-efficient vehicle (as Oregon and Maryland do), and for buying a bicycle or for planting trees. Imagine the government granting tax-free status to investments which reduce greenhouse gas emissions and introduce clean energy, as the Dutch government has been doing since 1996.

Now imagine completing an annual carbon assessment, alongside your tax return, when you plan next year's CO_2 reduction strategy and seek to maximize your tax-credits. If the government introduced carbon taxes and rebates (see #73), you would use the same system to claim your rebate. Put it all together, and the process could make carbon reduction as familiar as recycling.

With carbon calculators and tax-incentives, a government could make its grants and subsidies conditional on a proven track record of carbon reduction, as a form of persuasion. If you knew in advance that you could only obtain a grant if you had a track record in carbon reduction, you would soon start participating. Municipal governments and school boards could follow suit. It would not be long before everyone was doing it.

Resources

British Columbia's Tax Shift paper: www.gov.bc.ca/ges/key/taxshift.htm

Citizens Guide to Environmental Tax Shifting: www.foe.org/envirotax/taxbooklet

Environmental Taxation Worldwide: www.greentaxes.org

Environmental Tax Reform in the States: www.sustainableeconomy.org

Green Budget Reform: http://iisd.ca/greenbud

Green Tax Shift: www.progress.org/banneker/shift.html

Minnesota's Ecological Tax Reform Project: www.me3.org/projects/greentax

Tax News Update (Center for a Sustainable Economy): www.sustainableeconomy.org

Tax Shift, by Alan Durning: www.northwestwatch.org

This Place on Earth: Guide to a Sustainable Northwest, Chapter 5, Green the Tax Code: www.northwestwatch.org

Vermont Fair Tax Coalition: www.foe.org/envirotax/vermonttax report.html

68
Capture Your Methane

Capstone Turbine Corp. www.microturbine.com

Capstone MicroTurbine™ generators burning landfill gas at a southern California landfill.

It's time to shift the focus away from carbon dioxide and onto methane, which is the leading contributor to global warming from human activities after CO_2. Methane is the stuff that creates that distinctive fragrance when you've eaten too many beans. It is an atmospheric trace gas with a global warming potential over a 100-year period that is 23 times more powerful than carbon dioxide. Methane only lives for 12 years in the atmosphere, however, compared to up to 200 years for carbon dioxide. The IPCC has calculated that over 20 years, it has a global warming potential that is 62 times more powerful than CO_2. Because of the IPCC's choice of 100 years as the time span for measuring global warming potentials, the potency of methane is being overlooked, and the importance of eliminating it is being under-estimated.

Methane escapes from three major sources – landfills, agriculture, and the energy industry (see box). Methane represents 9% of US greenhouse gas emissions, and 13% of Canada's. Since 1800, methane concentrations in the atmosphere have increased by 145%, from 700 to 1,850 parts per million.[1] Since 1992, the rate of global methane emissions has slowed.

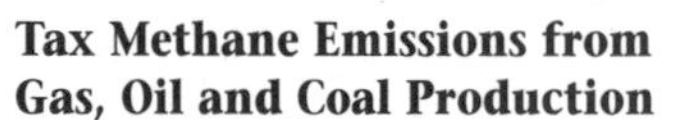

Tax Methane Emissions from Gas, Oil and Coal Production

Methane is a natural component of natural gas, and is also found in conjunction with coal and oil. It escapes during the production, processing, storage, transmission and distribution stages, and can be captured by a variety of technologies. 30% of the methane can be captured on a cost-effective basis, based on the value of the saved methane (EPA estimates), so this is a natural place to impose a state or provincial methane tax to increase the financial incentive. Since 1990, US emissions in this sector have fallen by 6.3%, due to a fourfold increase in methane recovery from coal mines.

Reduce Methane Emissions from Livestock

The problem here is twofold: cows, sheep and goats produce methane as part of their normal digestive process (known as "enteric fermentation"), which can be reduced by changes in feeding practices. Methane is also released when the manure from pigs and dairy cows is treated in a liquid anaerobic system, as opposed to a dry solid waste system (see #36). Alternative farming practices would reduce both sources (see #69). Since 1990, US emissions from livestock-have risen by 6.9% due to an increased number of hog farms, and larger sized cattle.

Sources of Methane

	% of US methane	% of US GHGs	% of Canadian Methane	% of Canadian GHGs
Landfills	34%	3%	23%	3%
Energy Industries	35%	3.2%	43%	5.6%
Cattle (digestion)	18%	1.6%	20%	2.6%
Manure	10%	0.9%	4.6%	0.6%
Other	3%	0.27%	9.4%	1.2%

Source: US Energy Information Administration[2] & Environment Canada[3]

"I know of no restorative of heart, body, and soul that is more effective against hopelessness than the restoration of the Earth."
— Barry Lopez

CFS Alternative Fuels

CFS Alternative Fuels Inc. demonstration plant, producing liquid natural gas and marketable CO_2 from landfill gas, Victoria, B.C., using CryoFuel Systems technology. The landfill could produce enough LNG to run Victoria's city bus fleet.

Reduce Methane Emissions from Landfills

When landfill waste decomposes, it gives off a mixture that is 50% methane and 45% CO_2. Methane can be captured and burned as fuel, which reduces its global warming potential to that of CO_2. This also reduces odours, which are often a local problem. Methane gas can be used to generate electricity, heat, or steam; as a fuel in school buses or taxis; in fuel cells; and to heat greenhouses. Out of 6,000 landfills in the US, only 270 are engaged in landfill-gas-to-energy projects. The EPA estimates that another 700 could be converting methane to fuel in a cost-effective manner. With an appropriate methane tax and a phased ban on methane releases, all landfills could - and should — be capturing their methane in both the US and Canada. Since 1990, US emissions from landfills have fallen by 11.3%.

In the US, the EPA's Landfill Methane Outreach program helps landfill operators form local partnerships and conduct project feasibility studies. One million tons of landfill waste generate 300 cubic feet of methane per minute, which can generate 7 million kWh of electricity and prevent 7,000 tons of CO_2 from entering the atmosphere per year. In 1998, Americans threw 150 million tons of waste into their landfills.

At Pattonville High School, in Maryland Heights, Missouri, the school's Ecology Club came up with the idea of using the gas from a nearby landfill to heat their school. The landfill operator donated the gas and the school paid $175,000 to run a 3,600 foot pipeline to the school's boilers. The school is now saving $40,000 a year and will recoup its investment in five years.

In Groton, Connecticut, the ONSI corporation is using landfill gas in a demonstration project to power a fuel cell, extracting the hydrogen from the gas and mixing it with oxygen to produce 140 kilowatts of electricity, enough to power 100 homes. The landfill saves $500,000 by not having to install a flare to burn off the gas. Portland, Oregon, operates a 200 kilowatt fuel cell at its Columbia Wastewater Treatment Plant, using methane from the plant to produce electricity.

Landfill gas can also be reduced at source. In the US, 24 million tons of leaves and grass clippings are landfilled unnecessarily every year. Lawns can be cut with mulch-mowers, and leaves make the perfect compost, so a landfill ban on leaves and lawn-clippings is quite feasible. In Nova Scotia, all food wastes are banned from landfills.

Resources

Annual Estimates of Global Anthropogenic Methane Emissions: http://cdiac.esd.ornl.gov/trends/meth/ch4.htm

Biopower: www.eren.doe.gov/biopower

CryoFuel Systems Inc.: www.cryofuelsystems.com

EPA's Landfill Methane Outreach Project: www.epa.gov/lmop

Methane and other Greenhouse Gases: www.epa.gov/ghginfo

ONSI Corporation: www.onsicorp.com

Opportunities to Reduce Methane Emissions in the US: www.epa.gov/ghginfo/opportoc.htm

Pattonville High School's initiative: www.mcs.net/~rogers/globe/trash.html

Reducing Methane Emissions: www.eia.doe.gov/oiaf/1605/vrrpt/chapter5.html

Reducing Methane Emissions from Livestock: www.ciesin.org/TG/AG/livestck.html

Tends in Global Methane Emissions: http://cdiac.esd.ornl.gov/trends/meth/methane_intr.htm

69
Support Sustainable Farming and Forestry

Farming
With farming representing 8% of US and Canadian greenhouse gas emissions, it is sensible to ask what states and provinces can do to reduce their farming emissions (see also #46, #78). Farming adds to the problems of climate change through the use of nitrogen fertilizers, the loss of soil carbon, the way cattle are fed, the way manure from cattle and pigs is treated, and the use of fossil fuel energy. Farmers can contribute to a solution by reducing these emissions, by generating wind energy from their land, and by using biomass to produce energy or manufacture ethanol.

Here are some steps that states and provinces can take to decrease the nitrous oxide and methane emissions, and increase the sequestration of carbon in soil:

- Ensure that agricultural colleges and advisory agencies give soil conservation and greenhouse gas (GHG) reduction advice.
- Develop a soil carbon measurement test and establish soil carbon goals.
- Establish a GHG measurement procedure and GHG reduction goals.
- Phase out the liquid and anaerobic management of manures.
- Place a tax on the sale of chemical fertilizers and pesticides; recycling it as farm grants for organic transition, soil conservation, tree-planting, and GHG reduction.
- Make farm subsidies conditional on soil conservation and GHG reduction.
- Use net metering, wind maps, and tax incentives to encourage farmers to install wind turbines on their land.
- Use tax incentives to encourage the use of biomass wastes for ethanol production.
- Use tax incentives to encourage tree-planting on farmland.
- Use tax incentives to encourage conservation tillage, as South Carolina does.
- Encourage farmers to go organic by supporting training and marketing initiatives.

"We desperately need to recognize that we are the guests not the masters of nature, and to adopt a new paradigm for development based on the costs and benefits to all people bound by the limits of nature herself, rather than the limits of technology and consumerism."
— Mikhail Gorbachev, President of Green Cross International

- Give organic farmers a five-year tax-break to help them get established.
- Sponsor an Organic Harvest Week, as Minnesota does every September.
- Enact legislation to control industrial mega-farm operations.
- Enact model legislation to encourage the sale of farm development rights.

The Impact of Agriculture[1]

	Methane		Nitrous oxide		Carbon dioxide			TOTAL	% of GHGs
All data MMTCE	Cows' stomachs	Liquid manure	Soil	Animal manure	Soil loss	Biomass burning	Fossil fuels		
Canada	4.9	1.4	8.2	1.3	0.3	n/k	n/k	16.1	8.7%
USA	33.7	22.9	83.9	4	n/k	n/k	n/k	144.5	7.9%

Frances Hunter

Wildwood Forest, on Vancouver Island, managed ecologically since 1938 by Merv Wilkinson. It has produced 140% of its initial volume in timber, while the whole forest is still standing.

Forestry

A climate-friendly forest policy would seek to maximize the carbon content in the forest by encouraging harvesting methods that cause minimum disturbance to the forest floor during harvesting, while maximizing the accumulation of carbon in the forest (see also #37, #79).

A typical coastal Douglas fir forest in the Pacific Northwest grows by 2 - 4% per year until it reaches an old growth condition. Industrial forestry clearcuts the timber in one area and then leaves the forest to regrow from scratch. Clearcutting has been banned in Switzerland and parts of Germany because soil is lost, along with its carbon, and the sustainability of the forest cannot be guaranteed beyond two or three cuts.

Ecoforestry, by contrast, views the forest as an integrated ecosystem and uses single tree selection to harvest the annual growth, leaving the forest standing. In the 94,000 acre Almanor forest in Northern California, the Collins Companies have been following this method since 1943, keeping careful records. The forest has been logged continuously for five decades and has yielded more than 1.7 billion board feet of timber. Yet today it has a higher inventory of timber than it did in 1943, much of it in mature trees. Ecoforestry methods reap the forest's interest and protect the principal, while industrial forestry methods reap the principal and hope to hell that it will regrow. The loss of carbon in the soil during clearcutting results in reduced carbon in the new trees, and causes overall yields to be lower. Industrial forestry may maximize short-term profit, but it minimizes sequestered carbon and weakens long-term profit, while undermining the ecological integrity of the forest, along with its habitat and recreational values.

Certify all State and Provincial Forests

The Forest Stewardship Council's system of certification guarantees that a forest is managed in a climate-friendly manner (see #37). In Minnesota, the Department of Natural Resources has certified 550,000 acres in Aitkin County. Pennsylvania has certified 2,200,000 acres of state forest. New York has 700,000 acres that are about to be certified.

Introduce Forest Product Feebates

Non-certified forest products could be charged a "non-sustainability fee" with the revenue used as a rebate on certified products. This would reflect the loss of long-term forest sustainability that industrial forestry dismisses as an accounting externality, and help to balance the ecological and the carbon books.

Ban Wooden Pallets

Wooden pallets represent an extraordinary waste of resources (see box). With durable pallets made from recycled plastic available, it is time for states and provinces to ban the use of wooden pallets by a certain date. This would send a signal to importing countries that they must use plastic pallets or face a penalty. In Germany, some pallets are bar-coded, with the original owner receiving a royalty each time a pallet is used and being charged each time it needs fixing.

The "Unpalletable" Numbers

- % of total lumber cut in US used for wooden pallets: **11%**
- % of total hardwood cut in US used for wooden pallets: **40%**
- Total number of pallets in the US: **1.5 billion**
- Number of new pallets made every year: **400 million**
- Number of houses that could be framed each year from discarded pallets: **300,000**
- Annual cost to New York businesses to dispose of pallets: **$130 million**

Source: *Natural Capitalism*

70

Build a Green Economy

Warren Gretz

A 9,000 liter fermenter at an Ethanol Process Development Unit.

Think big! Think big! That is the only reasonable response, for instance, to the information that our way of living has caused the natural wealth of the Earth to decline by 33% since 1970.[1] What are we doing? We think it good for a nation's economy to grow, but do we realize that with each percentage increase in our economic wealth, we are shrinking our world's ecological wealth? Our culture encourages us to consume at a level that, if the rest of the world were to do the same, would require two more planets to provide the necessary raw materials and absorb the CO_2 emissions.[2]

This road has an end to it, as surely as pasture will die if too many cattle graze on it. The warming of the atmosphere is one indicator that the end is approaching; the loss of topsoil and freshwater is a second. The extinction of so many species is a third indicator; for these creatures, the road has already ended. If we continue much further in this direction, the polar bears will join them, starving to death as the Arctic ice melts. How are we going to explain this to our grandchildren?

A state or a province is a powerful entity, where leadership and initiative can make a big difference. Some, such as Alaska, Ontario, and Quebec, are twice as big as Europe's largest nation (France). Others, such as California, New York, and Texas, have more people than most European nations. But what does it mean to change course so fundamentally that your state or province's economy heads in a sustainable direction, and no longer lives off the world's ecological capital?

This is the question many people are beginning to ask in Oregon, where Governor Kitzhaber passed an Executive Order in May 2000 directing that "The State of Oregon shall develop and promote policies and programs that will assist Oregon to meet a goal of sustainability within one generation -- by 2025." This is what we mean by "Think big." Anything less will not achieve the changes needed before the road runs out. The Oregonians who are taking the lead know that the challenge is complex and that many steps will be involved. They also know that the process will require the participation of all Oregonians. We live in a political world where not everyone agrees with the need for sustainability; indeed, many go out of their way to oppose it. A government can only move into new territory if it has support from the people who voted it into office. British Columbia,[3] Minnesota, Wisconsin, Connecticut, Florida, and North Carolina have also begun to explore ways in which they can adopt sustainable practices.

> **"What is the use of a house, if you haven't got a tolerable planet to put it on?"**
> **— Henry David Thoreau**

Resources

British Columbia's Green Economy Initiative: www.gov.bc.ca/ges

Community Indicators: www.rprogress.org/resources/cip/links/cips_web.html

Minnesota's Alliance for Sustainability: www.mtn.org/iasa

Oregon Solutions for a Sustainable Future: www.OregonSolutions.net

Redefining Progress: www.rprogress.org

Sustainability Indicators: www.sustainablemeasures.com/Indicators

The New Rules Project: www.newrules.org

"I know of no restorative of heart, body, and soul that is more effective against hopelessness than the restoration of the Earth."
— Barry Lopez

What Does "Sustainable Development" Mean?

There are many definitions, so it is good to be clear. Our preferred definition modifies the United Nations' version by adding the words in italics:

"Sustainable development meets the needs of the present without compromising the ability of future generations *of humans and other species* to meet their own needs, *while restoring local and global ecosystems."*

The challenge may be big, but you have to start somewhere. This is the essence of the thinking that underlies *The Natural Step* (see #40), a cohesive framework that helps businesses, institutions, and governments change course from business-as-usual to sustainability. Think big, keep your eyes on the vision, and take it one step at a time.

Oregon's first step has been to focus on the government's own operations, issuing an order that all state agencies and employees are expected to promote sustainable practices, from waste reduction and energy efficiency to building maintenance and vehicle use. To help things along, they have:

- set up a Sustainable Supplier Council to develop appropriate purchasing policies;
- adopted sustainable facilities standards and guidelines;
- purchased renewable energy;
- launched a website www.OregonSolutions.net, which showcases a host of sustainable projects and initiatives;
- awarded a monthly Sustainability Award to a person or team within the state government;
- set up a Sustainability Learning Network to promote mutual learning;
- required state investments in facilities, equipment, and goods to reflect the highest feasible efficiency and lowest life-cycle cost, lowering the barrier that blocks efficiency investments;
- asked the Oregon Progress Board to assess the best indicators to measure the state's progress towards sustainability;
- established a wide-ranging Sustainability Work Group to develop proposals for additional steps the state can take;
- undertaken a project to develop an environmental management system for use within the government, using sustainability principles;
- developed an initiative to form 15 locally developed business, citizen, and government partnerships structured around sustainability principles.

Can Sustainability Protect Jobs?

One of the fears that causes some people to oppose sustainability initiatives is the belief that the initiatives will destroy jobs. When the spotted owl habitat in the forests of Washington and Oregon was protected, many loggers lost their jobs. When coal mines close, coal miners will lose their jobs. Overall, however, layoffs due to environmental regulations in the US have been less than one tenth of 1% of all layoffs; most mining and logging job-losses have come from increasing mechanization, not new environmental laws. From 1980-1999, US coal extraction rose by 32%, but employment in the industry fell by 66%, from 242,000 to 83,000 workers.[4]

We all have to stop and retrain at some point in our lives; the important question is not whether particular jobs will be safe, but whether the overall transition to sustainability will produce a net job loss or job gain. Building a green economy has generated 14 million new jobs worldwide, and is going to generate many more. There will be jobs in energy efficiency, solar energy, wind energy, recycling, remanufacturing, designing for durability, ecotourism, pollution control technologies, ecoforestry, organic farming, ecovillage development, etc. Overall, investing in these activities will generate many more jobs than continuing to invest in extractive industries and fossil fuels.[5]

71

Launch a New Apollo Project

In Holland, where there is concern about rising sea levels and the possible shut-down of the Gulf Stream, the Dutch government is planning to reduce greenhouse gas emissions by 80% by 2050. In Britain, where the government is aiming to reduce emissions by 21.5% by 2010, a Royal Commission has recommended a 60% reduction by 2050. In Denmark, they are planning to produce 50% of their electricity from renewable energy by 2030.

Yet none of these goals will reduce CO_2 emissions sufficiently to stabilize the climate and protect Earth's ecosystems from the climatic disorders that are heading our way. It has been the authors' contention throughout this book that we need to reduce our emissions by 80% by 2025 if we are to escape the devastating impacts of a permanently raised level of greenhouse gases in the Earth's climate. If everyone pulls together, we can do it.

We need a New Apollo Project, but this time as a Mission to Planet Earth, not to the Moon. Apollo is the Greek God of the Sun, so his is an appropriate image. We need to get motivated, to dream, and become excited about moving into an age of ecological restoration powered by sunlight and hydrogen. A New Apollo Project should aim to make clean energy cheaper than fossil fuels by 2005, and to generate 50% of the world's energy by 2015, 80% by 2025. It is an immense task — but so was the idea of going to the moon in 1961.

There is no shortage of solutions. We need to pull together, the way we did to defeat Hitler. We all burn fossil fuels, so we must all be part of the solution. We need an alliance of people and governments that reaches around the world. This is not a project to be feared, as some economists would have us believe. It is an exciting project and a natural successor to Apollo I, which prepared the way for solar panels and fuel cells to become commercial technologies.

A New Apollo Project will need a huge educational program to explain the scientific realities to an often-confused public, and to show how we can all reduce our greenhouse gas emissions. It will need personal incentives and local efforts to get streets, neighborhoods, and businesses involved. It will need new energy policies, new taxation practices, and an end to fossil fuel subsidies. It will need new policies and practices for farming, forestry, and transportation. It will need new agreements between industry and government to achieve an emissions-free economy, as they do in Holland.

Above all, it will need a willingness to step back from the material wastefulness of the past 50 years and find ways to consume fewer of the planet's resources, while living more simply and efficiently, but with more fulfillment.

Resources

Good Practice Guidance and Uncertainty Management in National Greenhouse Gas Inventories: www.ipcc-nggip.iges.or.jp/public/gp/gpgaum.htm

National Climate Change Plans:

- Belgium: www.environment.fgov.be/Root/tasks/atmosphere/klim/set_en.htm
- Britain: www.environment.detr.gov.uk/climatechange
  - UK Royal Commission on Energy & Climate: www.rcep.org.uk/newenergy.html
- Canada: www.climatechange.nrcan.gc.ca
- Denmark: www.ens.dk/Publikationer/2000/Climate2012
- Europe: www.europa.eu.int/comm/environment/climat/home_en.htm
- France: www.effet-de-serre.gouv.fr
- Germany: www.bmu.de/english/fset800.htm
- Holland: www.minvrom.nl/minvrom/pagina.html?id=1&goto=1314
- Switzerland: www.buwal.ch/e/themen/umwelt/klima
- USA: www.state.gov/www/global/oes/97climate_report
  - US 1997 Climate Action Report: www.epa.gov/globalwarming/actions/national

"We are going to need an international effort pursued with the same urgency as the Manhattan Project or the Apollo space program. The roles of governments and market entrepreneurs...need to be considered comprehensively. It is our hope that the potential adverse effects of humanity on Earth's climate will stimulate new industries in the 21st century, as the Second World War and the Cold War did in (the 20th) century."
— Martin Hoffert, physicist, New York University, and 10 other scientists, *Nature* Oct 29, 1998

In every country where people have met to seek solutions to the problems of climate change, one issue keeps recurring: that the primary reason why we dump our wastes into the atmosphere is because it is free. No one has to pay for it.

At the core of the New Apollo Project, therefore, in addition to all the other policies and programs, there needs to be a simple carbon tax on all CO_2, methane, and the other greenhouse gases that enter the atmosphere. This will mean more expensive electricity from fossil fuels, more expensive beef, and more expensive gasoline, but it will encourage people to invest in more efficient appliances, renewable energy, and more energy-efficient homes, vehicles, businesses, and industries. Where government taxes on gasoline are already high, as they are in Europe, a portion of the existing tax could be re-assigned to become a carbon tax.

To balance the cost, and to get everyone involved, every householder and business should be invited to complete an annual CO_2 inventory, on the honour system, like a tax return. For each ton of CO_2 saved, compared to the previous lowest year, there should be a carbon rebate (see #73), valued at twice the amount of the original carbon tax. Coupled with tax credits for energy efficient appliances, vehicles, retrofits, and renewable energy, households and businesses that reduce their emissions may find themselves in pocket at the end of the year.

The income from the carbon tax should accumulate in a New Apollo Trust Fund, separate from the government's general revenues, and be used exclusively for actions to accelerate the clean energy revolution:

- to cover the tax credits and carbon rebates;
- for research to develop and commercialize clean energy technologies;
- for energy efficient and renewable energy loans and grants;
- to retrain displaced workers (e.g., coal workers);
- to assist low income households to do retrofits;
- for investments in transit, rail, and smart transportation initiatives;
- for community climate change solutions grants;
- for a country-wide climate change educational program.

Jim Hamm

Wind Turbines in Holland

A new global treaty is not a precondition for a New Apollo Project; the project can stand on its own since it will create economic rejuvenation in its own right, as well as bringing health and environmental benefits. The pundits and doomsters who predict that a carbon tax will cause economic life to grind to a halt are simply displaying their ignorance of what is possible, and their attachment to old-fashioned theories from a passing age.

Goals for an 80% Reduction in CO_2 Emissions by 2025

	Increased Efficiency	Renewable Electricity	Renewable Auto-Fuel
2005	.5%	.3%	.2%
2010	.25%	.15%	.8%
2015	.50%	.35%	.28%
2020	.65%	.50%	.54%
2025	.80%	.80%	.80%

72
Level the Playing Field

> "Taxing emissions makes little sense if governments simultaneously continue to subsidize fossil fuels."
> — Douglas Koplow and Aaron Martin, *Fueling Global Warming*

Fifty to sixty years ago, during the 1930s depression and the war years, a number of subsidies were put in place for the coal, oil, gas (and later the nuclear) industries to encourage economic development. Today, many of these subsidies still exist.

The oil companies are enormously rich. Between 1968 and 1972, the seven major US oil companies accumulated net profits of $144 billion, but paid less than $2 billion in federal taxes: an effective tax rate of 5%. In 1973, at the start of the first oil crisis, western governments allowed them to keep the profits from increased prices in the belief that they would re-invest them to secure the future energy supply. Their worldwide earnings increased by 71% and their net profits were $6.7 billion, but they paid only $642 million in taxes. As the profits poured in, they were invested in shopping malls, grocery chains, timber products, minerals, and coal companies. By the end of the 1970s, the oil companies owned fourteen of the largest coal reserves, two of the three top uranium producers, and three of the largest uranium reserves.[1] This was their "re-investment."

The tax breaks, unfair wealth, poverty, and disasters that are heading our way from global climate change are intimately connected. Subsidies and tax loopholes keep the price of fossil fuels artificially low, making it hard for renewable energy to compete.

The fossil fuel companies give big-time campaign donations and political action committee contributions to encourage politicians to keep the subsidies in place and the special program subsidies flowing. From 1993-1999, Southern Company, Chevron, General Electric, ARCO, Texaco, and 121 other companies that benefited from federal energy subsidy programs gave $39 million in campaign donations and received $7.3 billion in government grants over seven years: a 186:1 return on their investments.[2]

The solutions are clear:

(1) Eliminate all tax loopholes and subsidies for coal, oil, and gas.
(2) Implement a carbon tax to cover the true cost of fossil fuels.
(3) Transfer the subsidies to renewable energy to speed up the changeover.
(4) Transfer all energy research funding to renewables and energy efficiency.
(5) Reform the campaign finance laws (see #63) to end the corrupting influence of corporate donations.

Resources

Fuelling Global Warming: www.greenpeace.org/~climate/oil/fdsub.html

Gas-Guzzler Loophole Report: www.foe.org/gasguzzler

Give Green Power a Chance!: www.EcoAction.ca

Paying for Pollution: www.foe.org/eco/payingforpollution

Subsidizing Big Oil: www.ucsusa.org/transportation/pipeline.html

Taxpayers for Common Sense: www.taxpayer.net

The Subsidy Scandal (Europe): www.greenpeace.org/~comms/97/climate/eusub.html

Removing the Subsidies[3]

China: Subsidies cut from $24.5 to $10 billion, 1990-1997. 26,000 coal mines closed.

Russia: 50% cut since 1990. 30% fall in carbon emissions.

UK: Coal subsidies cut from $7 billion to $0, 1989-1997. 62% increase in use of natural gas.

Not Removing the Subsidies[4]

Canada: Average federal spending on fossil fuels, 1970-1999: $1.3 billion per year. On renewable energy: $12 million per year.[5]

Japan: Coal price supports: 88% of the price (1993)

USA: $18.3 billion per year in fossil fuel subsidies

Global: Total subsidies $235-$350 billion per year. Reduction in CO_2 emissions if removed: 4-18% (IPCC).

Tax Loopholes

A variety of tax loopholes benefit oil companies in the US, allowing them to:

- depreciate machinery, equipment, and buildings faster than their service lives;
- deduct more than their investment in oil properties from their taxes;
- count a portion of their foreign tax payments as a credit instead of a deduction;
- delay the payment of taxes on certain foreign earnings;
- deduct expenses related to multi-year oil well assets in the current year, rather than capitalizing them;
- get a 15% income tax credit for "enhanced oil recovery" methods;
- deduct 70% of their intangible drilling costs immediately, even though these amount to 75%-90% of the cost of getting an oil or gas well into production;
- get a production tax credit for fuels produced from non-conventional sources, such as oil from tar-sands (shale) and synthetic fuels from coal and gas;
- get a 15% write-off on the cost of a drilling operation as a "depletion allowance;"
- pay lower state taxes, since state taxable income is based on federal tax returns.[6]

The average effective federal tax rate paid by oil companies in the US fell from 21.9% in 1981 to 11.9% in 1995; and it's not just the oil companies. Coal companies are required to reclaim mining areas after a mine is depleted, but they can deduct the cost before they do the work. They can also treat income from royalties as capital gains, paying a reduced rate of tax. Taken together, the fossil fuel and nuclear producers gain $14.7 billion a year from various tax loopholes, which contributed to a net profit in 1997 of $29.8 billion.[7]

US Fossil Fuel Subsidies

Tax subsidies and loopholes:	$14.7 billion per year ($40 million a day)
Federal programs	$2.36 billion per year
Military protection	$10.5 to $23.3 billion per year
Strategic Oil Reserve	$1.6 to $5.4 billion per year
Sub-total	**$29 to $46 billion per year**
Cost per US citizen	$105 to $167 per person per year
Environmental health costs	$39 to $182 billion per year
Grand total	**$68 to $228 billion per year**
Cost per US citizen	**$247 to $829 per person per year**

Auto Tax Loopholes

The auto-manufacturers avoided $10.2 billion in taxes on 1999 models of SUVs, pick-up trucks, vans, and minivans because of a loophole in the CAFE law establishing fuel-efficiency standards. General Motors and Ford Motor Company benefited the most, each avoiding over $3.3 billion in taxes in 1999.[8]

Program Subsidies

Between 1948 and 1998, the US federal government spent $111.5 billion on research and development for the energy industry. About $66 billion went to nuclear energy and $26 billion to fossil fuel energy, but only $12 billion to renewables and $8 billion to energy efficiency.

Military Protection Subsidies

The US doesn't need an army to protect its wind supply, but it does to protect its Persian Gulf oil shipments. Estimates vary from $10.5 to $50 billion a year. The Strategic Petroleum Reserve Subsidy, used to stabilize the supply of oil in case of a disruption to imports, costs an additional $1.6 to $5.4 billion a year.

Environmental, Social and Health Subsidies

Dirty air and dirty water result in increased health care costs, loss of wages due to sickness and premature death, reduced agricultural output, and damage to buildings. A 1995 report estimated the annual cost of illness and early death from air pollution (almost all of which results from fossil fuels) at $182 billion; another estimate came in at $39 billion. In Austria, France, and Switzerland, where air pollution accounts for 6% of deaths each year, the health costs of vehicle pollution come to 1.7% of their gross domestic product.[9] These figures do not include the cost of present and future climate change impacts, or various additional subsidies to motorists (see #76).

73

Introduce a Carbon Tax and Rebate

A carbon tax is a tax on carbon dioxide. Its purpose is to encourage us to reduce our emissions and stop treating the sky as a free dumping ground for pollution. By early 2000, nine countries had raised their taxes on environmental harm, using the revenue to reduce income and other taxes.[1]

Finland was the first country to introduce a simple carbon tax on fuel inputs. It worked very well until they joined the European Union, when their industries suffered competitively. They shifted to a more complicated carbon tax on electricity ($19 tonne/$CO_2$), with a 39% exemption for industry and greenhouse cultivators. Finland is now pressing for an energy tax on all 15 members of the European Union.

Germany has increased its energy taxes on gasoline, diesel, heating oil, natural gas, and electricity, and is using the revenue to lower old-age pension premiums.

France is planning a carbon tax in 2001, starting at $84-$114 tonne/CO_2, rising to $279 by 2010.[2]

Britain has a fuel tax that increases by 6% every year and is introducing a climate change levy in April 2001. They recycle the revenue as reduced employment taxes, with an 80% rebate for industries that undertake energy saving.

Norway has a CO_2 tax on offshore oil and gas production that caused gasoline prices to rise by 10%, heating oil by 15%, and achieved a 3-4% reduction in CO_2 emissions. They are planning a more general carbon tax of $49 tonne/$CO_2$ and a landfill disposal tax to reduce methane emissions.

Switzerland passed a law on the reduction of CO_2 in 1997, and set a goal of 10% below the 1990 level by 2010. They have a mileage tax on heavy vehicles and a surcharge on motor vehicle fuel, and use the revenue for public transportation. Their law invites industry and the public sector to achieve voluntary reductions, but if this doesn't work by 2004, a carbon tax of $142 tonne/$CO_2$ will be introduced, with exemptions for industrial sectors that achieve voluntary reductions. The Swiss Association of Trade and Industry supports the threat of a high CO_2 tax because it gives industry a strong incentive to make reductions, but the Swiss Parliament is having doubts about the mandatory 2004 date.

Holland's green taxes account for 14% of national taxes. A study has estimated that their fuel tax has saved 1.7 million tonnes of CO_2 since 1994.

Denmark's energy tax has been in place since 1966; residential fuel consumption fell by 15% from 1966 to 1991.

Sweden's Green Tax Commission estimates that their CO_2 tax caused an 11% reduction in CO_2 emissions from 1987-1994. Most European countries are allowing key industries to be exempt when they face cross-border competition from countries without such taxes.[3]

In *Japan*, the government's Environment Agency is recommending a 1 cent per liter carbon tax, with the revenue being used to pay down the government's fiscal deficit.

"The Kyoto Protocol will delay warming by less than ten years."
— Bob Watson, Chairman of the Intergovernmental Panel on Climate Change.

Resources

Center for a Sustainable Economy Tax News Update: www.sustainableeconomy.org

Environmental Tax Shifts Multiplying, by David Roodman, Vital Signs 2000 (Worldwatch Institute): www.worldwatch.org

Green Budget Reform in Europe: Countries at the Forefront, by Kai Schlegelmilch (Wuppertal Institute/Springer): www.wupperinst.org

Taking Stock of Green Tax Reform Initiatives, by Carola Hanisch: www.pubs.acs.org/hotartcl/est/98/dec/hanish.html

The Natural Wealth of Nations, by David Roodman (Norton, 1998)

"Our old system (of carbon taxes on fuel inputs) had a very beautiful and clear structure. There were no exemptions. It promoted renewables and helped the replacement of coal by gas. The tax was very easy to raise year after year, and you have a predictable perspective for industry in the long run." — Markku Nurmi, Director General, Finnish Ministry of Environment

Can carbon taxes work in North America? For sure they can, except that Canadians and Americans have been so spoiled by subsidized fuel they think it is a right.

The first step towards a New Apollo Project has to be a massive campaign of grassroots activism and public education to open people's eyes, and end the policy logjam that the fossil fuel companies have set up by cosying up to the legislators.

The next step must be to eliminate the various subsidies and tax loopholes that the fossil fuel industries have wangled for themselves (see #72). If just the tax loopholes were removed, the $14.7 billion of taxpayers' money that is being kept by the oil companies would give every individual in the USA a $53 annual bonus. The Worldwatch Institute estimates that if *all* of the various subsidies that industry in the USA gets for environmentally destructive activities were removed, there would be a yearly bonus of $500 per person, or $2,000 for a family of four.[4]

In 1998, a poll conducted in the US for Friends of the Earth showed that 71.5% of the respondents supported a shift in taxes from payrolls and income to fossil fuels, right across party lines.[5] By establishing a $50 ton/$CO_2$ carbon tax with a $75 rebate for every ton of CO_2 saved[6], and by requiring everyone to fill in a yearly personal CO_2 assessment, managing the money through an independent New Apollo Trust Fund, people all across the nation would start to feel involved. A New Apollo Project must belong to everyone, and not become just another government project.

What About a Sky Trust?

In Alaska, a portion of the revenues from oil are shared out equally through the Alaska Permanent Fund, which gives every man, woman, and child a yearly royalty check reflecting their share of the oil income. A national US Sky Trust is being proposed by the Corporation for Enterprise Development, whereby the sky would be held to be a common resource, with companies paying for the right to dispose of their CO_2 in it.[7] If a carbon emissions permit cost $100 per ton, the yearly dividend would come to $400 per person. For details, see www.skytrust.cfed.org

One drawback is that this would encourage people to look forward to their yearly payment instead of thinking of ways to reduce it; pollution of the sky should not be treated as a windfall benefit. Charging $100 per ton would increase the price of gas by $1 a gallon.[8] The average motorist consumes 460 gallons a year (for 11,500 miles), so the charge would cost an additional $460 per year on the price of gas.

The Impact of a Carbon Tax ($50/ ton) and Carbon Rebate ($75/ton)

Average electricity:	1.45 lbs of CO_2/kWh (1 ton per 1,380kWh). Tax = 3.6 cents/kWh
Inefficient household:	11,000 kWh @ 8 cents = $880 + $396 carbon taxes = $1276
Efficient household:	5,000 kWh @ 8 cents = $400 + $180 carbon taxes = $580
Efficient household + carbon rebate:	@ $75/ton for 4.3 tons CO_2 reduced: $580 - $322 = $258
Gasoline:	20lbs of CO_2/gallon (1 ton/100 gallons). Tax = 50 cents/gallon, 13 cents/litre
Inefficient car, 25 mpg:	460 gallons @ $2 = $920 + $230 carbon taxes = $1150
Efficient car, 45 mpg:	255 gallons @ $2 = $510 + $127 carbon taxes = $637
Efficient car + carbon rebate:	@ $75/ton for 2 tons CO_2 reduced: $637 - $150 = $487

74

Establish an Energy Efficiency Strategy

The great puzzle in the climate change debate is why, when energy efficiency can reduce emissions with such easy, profitable returns, do less than 2% of new homes achieve a high level of efficiency, and why are most businesses equally inefficient?

There's no doubt that a $50 ton/$CO_2$ carbon tax would help. In Japan, where electricity costs 20 cents/kWh, homes and businesses are far more efficient than they are in North America. Even in the USA, when you compare the states where people pay more for electricity with those where they pay less (see #65), it is apparent that the higher the utility rate, the lower the overall fuel bill. Higher fuel prices save money, while lower fuel prices encourage people to be wasteful.

In the realm of electricity, however, carbon taxes alone will not do the job. A host of everyday barriers stands in the way of harvesting the inefficiencies that are lying around. Homeowners, business owners, landlords, tenants, architects, developers, plant managers, and finance managers either know nothing about energy efficiency, or say "Why bother?" and keep on buying inefficient equipment that is more costly in the long run. For nation-states, these barriers need tackling from the top.

Appoint a Cabinet-level Chief for Energy Efficiency. Set a goal of a 30% improvement in efficiency over ten years and tie the salaries of the chief and staff to their achievements, while encouraging them to be creative. If you were that chief, what do you do? First get advice from the groups that know the barriers inside-out such as the Rocky Mountain Institute, the American Council for an Energy Efficient Economy, and the Alliance to Save Energy, and then move into action.[1]

Step 1. Legislate an Energy Efficient Equipment Act
After the normal negotiations, legislate that every piece of energy equipment — from toasters to industrial motors — must carry a clearly displayed energy rating from 0 to 4 stars based on its efficiency, where 0 stars means "lowest 10%" and 4 stars means "best 10%." Give all 4-star equipment and industrial retrofit costs a 100% tax credit (50% for 3 stars) and announce that equipment with two stars or less will not be allowed to be sold after a set date.

"There is a huge body of business evidence now showing that energy savings give better service at lower cost with higher profit. We have to tear down barriers to successful markets and we have to create incentives to enter them."
— President Bill Clinton, *National Geographic*, October 1997

Becoming Efficient in the USA

A proposed Energy Efficiency Buildings Incentive Act (S. 2718), as introduced to the 106th US Congress in June 2000, would provide tax deductions for:

- efficient new residential buildings which save 30% or 50% of the owner's energy costs compared to national model codes (higher incentives for higher savings);
- efficient heating, cooling, and water heating equipment that reduces consumer energy costs by 20% (lower incentives) and 30%-50% (higher incentives) compared to national standards;
- new and existing commercial buildings with 50% reduced energy costs;
- solar hot water and photovoltaic systems.

For a progress report, see www.floridagreenbuilding.org/s2718

D&R Int., Ltd.

Energy Star-labeled windows can reduce overall home energy use by 15%, saving several hundred dollars each year.
www.energystar.gov/products/windows

Step 2. Legislate an Energy Efficient Buildings Act
Write an improved Building Code and legislate that every building that is for sale or in a change of rental or lease must be rated for energy efficiency, meet the Code, and display an Energy Rating from 0-5 stars (5 stars for zero energy buildings). Give all upgrade costs a 100% tax credit and stipulate that rental contracts must allow landlords and tenants to share in the upgrade benefits. Announce that all government buildings must increase their efficiency by 50% by 2010 (the US has set a goal of 35%) and negotiate for a percentage of 5-star zero-energy buildings to be built each year as a demonstration that it is possible.

Step 3. Establish a Universal Public Benefit Charge
Legislate that all electric utilities levy a Public Benefit Charge on each unit of electricity sold, and provide matching federal resources. Distribute the income through local non-profit Energy Saving Trusts, using public auctions to grant it to whoever can deliver the greatest efficiency savings for the lowest cost. As soon as there is a market for efficiency savings, businesses will step forward to deliver them. (See ESCOs, #22). Require every utility to provide its customers with smart electronic meters that show real-time energy costs and give CO_2 readouts, and offer "feebate" contracts to new customers, charging a fee on heavy usage and rewarding efficient customers with a rebate. Allow utilities' profits to be based on their ability to cut customers' bills, not on the amount of power they sell.

Step 4. Tie New Generating Capacity to Proven Energy Savings
Legislate that no new licenses for generating plants will be granted until the would-be licensee has delivered an equivalent amount of energy through efficiency initiatives. They'll quickly find that it's cheaper to give away energy-efficient light-bulbs (Southern California Edison Company gave away over a million), or to subsidize lamp manufacturers to produce them more cheaply, than it is to build a new plant. During a water shortage in Goleta, CA, a similar initiative stipulated that a developer who wanted to build must first deliver twice the amount of water required through local savings. This immediately put a price on saved water, reducing Goleta's water use by 30% in one year with no loss of service quality, and deferring indefinitely a planned multi-million dollar sewage treatment works expansion.[2] Exactly the same could happen for energy.

If these measures are combined with an annual CO_2 assessment, a carbon tax-and-rebate, the abolition of tax loopholes, and the measures described in Solutions #24, #32, #45, $46 and #65, energy use will soon begin to fall, carrying CO_2 emissions with it.

Becoming Efficient in New Zealand

In July 2000, New Zealand passed the Energy Efficiency and Conservation Act, proposed by the Green Party. The Act requires the government to develop a national strategy to pursue energy efficiency, energy conservation, and the use of renewable energy, with clear goals and targets, giving the government the power to set higher energy efficiency standards. The existing Energy Efficiency and Conservation Authority has become a permanent independent authority with a wide range of responsibilities for reducing energy consumption.[3]

For details, see www.energywise.co.nz/content/ew_government/Intro.doc

75
Establish a Renewable Energy Strategy

What kind of policies should a government adopt to achieve the fastest possible transition to renewable energy? In the US, more than 80% of the public favors tax supports to increase green energy.[1] The policies suggested have been assembled from the best in Germany, Holland, Denmark, and the USA.

The first step should be to change the ministry or department of energy into a ministry or department of Sustainable Energy, changing its vision and goals. Next, create a Renewable Energy Act with 12 components. The first three — End Fossil Fuel Subsidies, Establish Carbon Taxes, Introduce Individual Carbon Assessments and Carbon Rebates (#71-#73) — establish the financial environment in which renewables can take off. The remaining nine enable a government to deliver on its commitment.

Resources

America's Global Warming Solutions, Tellus Institute: www.tellus.org

Global Climate Change—Kyoto Protocol Implementation: Legal Frameworks For Implementing Clean Energy Solutions, by Richard L. Ottinger & Mindy Jayne: www.solutions-site.org/special_reports/sr_global_climate_change.htm

PIRG Campaign for R.E.A.L. Energy: www.pirg.org/enviro/energy

Renewable Energy Policy Outside the United States, by Curtis Moore and Jack Ihle, REPP: www.repp.org

4. Support Research and Development
Wind energy has reached relative maturity, but solar, geothermal, biomass, tidal, and hydrogen need continuing R&D and demonstration projects. The US Department of Energy deserves credit for the work it does on renewables. Just think what it could do if it could use the 84% of government R&D funds that go to nuclear and fossil fuels.

5. Organize Full Grid Integration
People need easy guaranteed access to the grid, and an end to administrative and regulatory barriers, especially for photovoltaic (PV) and wind energy. This involves the nation-wide adoption of net metering (see #63), certification of green energy, agreed interconnection and installation standards, nation-wide wind maps, and a requirement for states and provinces to establish wind-zones with energy targets (as Denmark and Germany do) with model ordinances and regulations.

6. Encourage Community Buy-in
The public needs to understand the damage that fossil fuel energy does, and the many benefits of renewable energy. As well as nation-wide education campaigns (preferably run by independent non-profit societies), energy products must be clearly labeled to show where the energy comes from and what their emissions are. The reason why wind energy has been so successful in Denmark is that it started with people in rural areas forming their own wind energy cooperatives, and having to deal with everything that was thrown at them (see #20). When farmers, shop-owners, doctors, vets, and musicians invest in their own wind turbines, strengthening the local economy, there is community support. Winning community buy-in involves establishing the legal right for people to form renewable energy cooperatives, guaranteeing people's right to choose green energy, supporting community marketing campaigns (see #47), and ensuring that renewable energy has majority representation on the various energy boards that advise the government.

> **"Every generation faces a challenge. In the 1930s, it was the creation of Social Security. In the 1960s, it was putting a man on the moon. In the 1980s, it was ending the Cold War. Our generation's challenge will be addressing global climate change while sustaining a growing global economy."**
> **— Eileen Claussen, President, Pew Center on Climate Change**

Warren Gretz

Heliostats reflect the sun's energy onto a receiver atop a centrally located 300-foot tower at Solar Two, the world's most advanced central-receiver power plant, located in the Mojave Desert.

7. Set Mandatory Goals

The best policy is a secure, long term Renewables Portfolio Standard (RPS) (see #64) by which utilities must provide 20% of their electricity from renewable sources by 2010; 80% by 2025. The US has currently set a voluntary goal to achieve 7.5% of its electricity from renewables by 2010 and to triple its use of biomass energy. Canada has no such goals.[2]

8. Establish an Electricity Feed Law

Denmark combined its RPS with an Electricity Feed Law by which utilities must pay independent producers 85% of the ultimate sale price of their energy. In Germany, where utilities must pay 90% under the wonderfully named *Stromeinspeisungsgesetz* law, the policy has allowed an incredible growth in wind and solar energy.

9. Establish Subsidies (see box)

Denmark uses 15 different kinds of subsidy to encourage renewable energy. A full package (financed out of carbon tax revenues) is needed to create a steady, long-term investment environment to give investors and homeowners the assurance they need to make a serious commitment.

10. Phase Out Non-Sequestered Fossil Fuels

The writing must be placed clearly on the wall for all fossil fuel energy that does not sequester its CO_2. Denmark has simply banned the construction of new coal-fired power plants. One policy involves placing an annual cap on permitted CO_2 emissions per MW of produced energy (as Denmark and Oregon do, see #64) and scheduling it to reach zero within 35 years, coupled with an emissions trading system that allows utilities to buy CO_2 reductions elsewhere to help them achieve their reductions. This will send a clear signal to the natural gas industry, as well as to coal- and oil-fired utilities.

11. Lead Globally

Stop supporting fossil fuel projects in developing countries through export credits, the World Bank, and other means. Transfer this support to renewable energy projects.

12. Lead by Example

National governments buy an enormous quantity of energy to fuel their various initiatives, which they can use to help lever renewable energy into the market. The US Department of Energy has ordered that 7.5% of its energy must come from renewable sources by 2010 and that its greenhouse gas emissions must fall to 30% below the 1990 level by 2010. The EPA laboratory in Richmond, CA, is buying renewable energy from the Sacramento Municipal Utility District (see #43); the US Post Office is buying it for 1,000 facilities in California from Go-Green.com. See #92 for a proposal that would have a revolutionary impact on the price of solar cells for everyone on the planet.

Renewable Energy Subsidies and Supports

- tax credits for the purchase of green power (Germany pays 3 cents per kWh);
- guaranteed prices for the sale of net-metered energy (Germany guarantees 46 cents per kWh for 20 years);
- low-interest loans for solar mortgages (Germany's national banks offer 2%, 10-year loans);
- capital grants and subsidies (Japan pays $2,560 per kW for solar installations; Denmark paid 30% of the cost of wind turbines and for biomass);
- production subsidies (Denmark pays 3.8 cents per kWh);
- corporate investment subsidies (Germany pays $96 per kW; Holland gives 40-52% tax-relief; the US pays 1.5 cents per kWh on wind and biomass);
- personal tax relief on interest from renewable energy investments (Norway and Holland allow 100%);
- accelerated depreciation of renewables equipment (as Holland allows);
- buy-down grants for solar, wind, and earth energy (California pays $3 per kW for PV installations);
- mortgage breaks: eg a 30% break for 5-Star homes, 20% for 4-Star homes (see #74); (Holland offers 20% mortgage breaks for sustainably built houses).

76
Establish a Zero Emissions Vehicles Strategy

The enormous volume and low efficiency of the world's vehicles, coupled with the growth in vehicle ownership and the urgency of the climate change crisis, make it essential that governments use every means in the book to encourage, cajole, and require their vehicle-manufacturers to smarten up.

Some manufacturers say they are ready to assist, but even while they are proclaiming how green they are, their lobbyists in Washington, Paris, and Berlin are busy trying to sabotage progressive legislation that could make a difference, and are financing auto-friendly politicians. We need an annual report card on the conduct of the auto companies as a whole, as well as on the greenest vehicles (see #3), so that people can see how green the companies really are.

To send a clear message, government transportation departments should upgrade their visions and goals and re-emerge as Departments of Sustainable Transportation, with a commitment to provide "mobility without emissions." Here are eight policy initiatives for the vehicle-based part of a Sustainable Transportation strategy (see also #77).

1. Set National ZEV Goals
The day after California announced its Zero Emissions Vehicles program in 1990, requiring that 4% of all new vehicles produce zero emissions by 2003, investments started pouring into the hydrogen fuel cell companies. One of the most effective ways that governments can reduce greenhouse emissions from vehicles is by setting national ZEV standards for new vehicles. We suggest 5% by 2010; 25% by 2015; 50% by 2020 and 100% by 2025.

2. Upgrade the Corporate Average Fuel Efficiency (CAFE) Standard
In the US, the auto industry has stopped the CAFE standard from being upgraded since 1975, with the result that it remains stuck at 27.5 mpg for cars and 20.7 for vans and pickups, with SUVs being cleverly classified as pickups, not cars. A new CAFE standard must raise the required efficiency to 45 mpg by 2010, 65 mpg by 2018, and 80 mpg by 2025, with equivalent standards for heavy trucks and buses. In 2000, Congress voted to study the matter.

3. Establish a CO_2 Emissions Standard
In reality, it's emissions that matter, not efficiency. If a car has zero emissions, it can be as inefficient as its owner wants it to be. A Sustainable Transportation Act should bring in mandatory fleet standards for CO_2 and other emissions, starting with a 25% reduction by 2010, leading to 50% by 2018, 80% by 2025, and 100% by 2040. In Europe, faced with mandatory legislation, the auto companies voluntarily agreed to a 25% reduction by 2010, meaning that their cars will need to achieve 39 mpg (see #57); Japanese and Korean auto manufacturers have also agreed to these standards. This system, coupled with the elimination of subsidies and the use of a carbon tax-and-rebate, will radically increase consumer demand for fuel-efficient vehicles and drive the shift to biofuels and hydrogen. If the industry prefers a voluntary agreement, it should be coupled with serious consequences in the event of failure, such as a doubled CO_2 tax on vehicle fuels.

> **"Six billion people share this planet, and the days when the auto industry could just crank out cars without concern for their impact on the environment are over."**
> **—Toshiaki Taguchi, President of Toyota Motor North America**

Ballard Power Systems

Ford's hydrogen fuel-cell TH!NK in front of Ballard Power.

4. Establish a National Feebate and Clunker Program

A feebate is a fee on inefficient vehicles matched by a rebate on efficient ones. California (1990) and Maryland (1992) attempted to bring in such a program, only to have it quashed because the CAFE law has a cunning clause that prohibits state legislation of this kind. California has a successful $1000 bounty program that encourages people to turn in their old clunkers. Both programs should run nationally. A rebate for the most fuel efficient and low-CO_2 vehicles is a sure way to attract buyers and increase manufacturers' willingness to increase production.

5. Legislate Clear Labeling and Fuel Efficiency Indicators

All cars should be clearly labeled with a 1-5 star system, similar to that for appliances and housing (see #74), so consumers know what they are buying. In Europe, the EU has agreed that all cars must display a clear fuel efficiency and pollution label, all showrooms must display a poster comparing the ten most popular models, and every country must publish an efficiency guide to every model on the market. In addition, every car should be obliged to have an onboard fuel efficiency indicator (see #57), just as it has seatbelts.

6. Create a Hydrogen Development Strategy

We need a Canadian/US/European equivalent of the California Fuel Cell Partnership (see #56) to speed up development of a hydrogen economy. Clean hydrogen producers need R&D grants and production subsidies, hydrogen filling stations need to be given investment credits, and hydrogen investors need tax incentives.

7. Legislate for Distance-Based Insurance

Under the current system, a person who drives 2,000 miles a year pays the same insurance as someone who drives 42,000 miles, penalizing low-mileage drivers while subsidizing high-mileage drivers who produce the most emissions. The solution lies with distance-based insurance, with a core payment and an additional payment collected at the pump or based on an odometer audit. The Progressive Auto Insurance Company in Texas offers just such a system using Global Positioning Satellite technology to track when, where, and how much its customers drive, and charging them accordingly. About 1,100 people have signed up for a test run.

8. Lead by Example

Every year, the US government buys 56,000 new cars from Detroit for use by its various departments. If governments around the world were to announce that from 2003 their purchases were going to prioritize hybrid-electric, hydrogen fuel-cell, or alternative fueled vehicles, this would give manufacturers a head start on their production numbers, bringing down the price for everyone.

Resources

California's Mobile Sources Program: http://arbis.arb.ca.gov/msprog/msprog.htm

California's Zero Emissions Vehicles Rules: http://arbis.arb.ca.gov/msprog/zevprog/zevprog.htm

Driving the Future—Zero Emissions Vehicles: www.DrivingTheFuture.com

Progressive Auto Insurance: www.progressive.com

Sierra Club's Clean Cars Campaign: www.sierraclub.org/globalwarming/cleancars

77

Develop a Sustainable Transportation Strategy

Eliminate the Hidden Subsidies
Private automobile drivers are cushioned and coddled to the tune of $1,700 a year[1] in hidden subsidies for costs not covered through vehicle and gas taxes. This includes highway improvements, policing costs, ambulance and hospital costs, subsidized or free parking costs, and health care costs due to air pollution. For a typical driver (11,500 miles a year), that's a 15 cent subsidy for every mile driven, a $3.70 subsidy for every gallon. Our transport system produces a dangerous pollutant (CO_2) and we are subsidize it to the hilt with cheers from the auto companies who know that if the subsidies were taken away, millions of people would demand a more affordable system of transport. More efficient vehicles is half the solution: more efficient transportation is the other half.

Build the Political Will
There are 101 ways to move people and goods around affordably without building more roads (see #27, #28, #29 and #66). The problem is not a lack of solutions: it's a lack of political will. People who have been spoiled by years of subsidized gas prices are easily swayed by politicians who want to kill proposed gas taxes and rail and transit projects. Those who walk, bike or use transit need to be much more vocal. In a 1999 Canadian survey, 50% of the respondents said that expanding transit was the single greatest improvement that could be made to local transportation (up from 41% in 1994). 19% wanted road widening, and 7% wanted new roads.[2]

Build Bicycle Greenways
Bicyclists ride a very elegant machine that eliminates pollution and contributes to our general health and well-being, but for their pains, they are subjected to a daily threat to life and limb. It is time for every country to build a nationwide network of bicycle paths, separate from other traffic. In Britain, Sustrans is building a 10,000-mile National Cycle Network that will run through the middle of most major towns and cities, with support from landowners, governments, and thousands of members. When complete in 2005, it will put 20 million people within a ten-minute cycle ride of their nearest route and carry an estimated 100 million trips a year, 60% utility trips and 40% leisure. In Canada, the Trans Canada Trail Foundation is working on a 16,000 km cross-country trail; in the US, the East Coast Greenway is a planned 2,600 mile traffic-free path from Key West, Florida, to the Maine-Canada border, passing through

Bikeways

National Bicycle Greenway: www.bikeroute.com
Rails to Trails Conservancy: www.railtrails.org
Sustrans (UK): www.sustrans.co.uk
Trans Canada Trail: www.tctrail.ca

Busways

American Public Transport Assn: www.apta.com
Busways: www.busways.org
Canadian Urban Transit Assn: www.cutaactu.on.ca
Future Busways in LA:
www.busways.org/pages/busways/la/renderings.html

Railways

California-Nevada Maglev project:
www.ci.las-vegas.nv.us/maglevproject
Gulf Coast Maglev: www.gulfcoastmaglev.com
High Speed Ground Transportation Assn: www.hsgt.org
Magnetically Levitated Trains: www.maglevinc.com
National Assn of Railroad Passengers: www.narprail.org
Northeast Corridor High-Speed Rail Program:
http://project1.parsons.com/ptgnechsr

Other ways

Aerobus International: www.aerobus.com
Transport 2000 Canada: www.transport2000.ca
Transport Demand Management: www.vtpi.org/tdm

"Once every great while, the little person, the nobody, and the common man see a great opportunity...they see their chance to merge their dreams with a grand shift in consciousness. The Transportation Equity Act (TEA-21), which is providing untold support to bicyclists across the Nation, as well as to walkers, joggers, skateboarders, wheelchair athletes and in-line skaters, is that rare opportunity."
— Col. Dan Petkunas, National Bicycle Greenway

Washington, New York and Boston. For the US as a whole, Martin Kreig and the National Bicycle Greenway are dreaming of a coast-to-coast network of multi-use transportation and recreational bicycle trails.

For 100 years, almost all national transportation funding has gone into building highways and roads. It's time to reverse the tide, and use our national resources to build a more sustainable, planet-friendly transportation system.

Fund More Transit

In 1999, Americans took more than 9 billion trips on public transportation and ridership increased by 4.5% (the highest in 40 years), compared to a 3% increase in air travel and a 2% increase in motor vehicle travel. In small towns, there was a 12% increase in people using the bus. The US government has allocated $4.2 billion to transit over five years through its TEA-21 program, but given $175 billion to highways, a subsidy of $250 per vehicle per year.[3] In Canada, transit needs $9.2 billion over the next 5 years, but has been given only $1 billion. It's time to reverse the priorities and give transit a 10-fold increase in funding, saving the land and forest to store carbon dioxide.

Fund More High Speed Trains

For intercity travel, high speed trains are the future. Europe is full of high-speed trains, and the Japanese, who already have their shinkansen (bullet train), are designing an aerotrain that runs on solar and wind energy using four horizontal wings, that is scheduled to "fly" in 2004. The next big excitement is Maglev (magnetic levitation) trains. Amtrak hopes to build high speed train corridors in California, New York, the Northeast, the Pacific Northwest, the Southeast (through Virginia, the Carolinas and Georgia), the Gulf Coast and the Chicago hub, but it all needs funding, just as highways are funded. The British government is going to spend $270 billion over the next ten years on non-automobile transport, with $43 billion going to railways. The US, with 4-5 times the population, has earmarked just $2 billion to railways, over five years. The next time around it should be $200 billion, and $20 billion from the Canadian government.

Tie Federal Payments to CO_2 Reduction

Whenever the federal government makes payments to states, provinces or municipalities, it has the leverage to require adherence to CO_2 reduction goals, half of which should come from transportation emissions. Reduce your carbon emissions, and you can stash the cash.

Lead by Example

Free parking encourages federal employees to drive to work. When the Canadian government increased the parking rates for its employees in Ottawa, 16% started using public transport and 23% stopped driving to work. Set up a Trip Reduction Incentives Program (see #27), and tell every department to charge the market rate for parking.

Use Taxes Creatively

- Give tax credits for the purchase of bicycles, and for monthly or yearly transit and rail passes.
- Give a $1,700 tax credit to car-free individuals and families to compensate for the hidden subsidy that cars receive.
- Give a $1,000 tax-break to people who telecommute.
- Allow full tax exemption for employer-provided transit benefits (as the US does).
- Subsidize lower interest rates on location-efficient mortgages (see #29).
- Remove all hidden parking subsidies.
- Remove vehicle taxes on vehicles used for car sharing and car-pooling.
- Give tax incentives for investments in transit and high speed trains.

78

Encourage Sustainable Farming

When we approach farming from a national perspective, some major issues arise. Global climate change is going to bring an increase in the frequency of droughts, torrential downpours, pest invasions, and late and early frosts; all will hinder the ability of farmers to produce the export surpluses on which both farmers and food-importing countries depend. Crop yields may rise in some areas as a result of increasing CO_2 levels, while they may fall in other areas because of unseasonable frosts, floods, pest-damage, or lack of winter snow cover.

Farming's business-as-usual agenda involves a call for increased government subsidies to help farmers compete with their European competitors, who receive larger government subsidies. Beneath this call for more handouts, however, lies a far more complex picture. Over the last 50 years, farming in North America has become increasingly industrialized and chemically dependent. But when farmers do their annual accounting, they do not list the topsoil that has been lost and the soil-carbon that has broken down and been released as carbon dioxide. They do not list the harm that has been done to rivers, lakes, streams, and groundwater by toxic run-off from chemical herbicides and fertilizers, or by open-air manure lagoons that have burst, leaked, or overflowed. They do not list the loss of biodiversity caused by their overuse of pesticides, or the public health costs of treatment for cancers that have been caused at least in part by those pesticides.

A "Climate-Safe" agenda for farming needs to eliminate the practices that release nitrous oxide, methane, and CO_2 that makes up 8% of US and Canadian climate change emissions; restore the lost soil; make a country-wide transition to sustainable organic practices that can better withstand the decades of troublesome climate impacts that lie ahead; eliminate the further loss of farmland to sprawl; and bolster farmers' incomes by encouraging the use of biomass and wind energy on their land.

The task is encouraged by the growing demand for organic food and consumer rejection of genetically engineered and chemically treated food. Being a step ahead of North America, Europe is blazing the way. In a report prepared for the European Union, agriculture specialist Dr. Nic Lampkin found that while there were 6,300 organic farms in the European Union in 1985, by 1998, there were over 100,000.[1] By 2005, 10% of all agricultural land in western Europe will be organic; Austria has already passed the 10% mark.

Carolyn Herriot

"You know, there are farmers who have to wear a mask when they go to their fields, when they should be breathing the good, clean air."
— Henri Paque, Belgian organic farmer

"A true conservationist is a man who knows that the world is not given by his fathers but borrowed from his children."
— John James Audobon

Legislate a Sustainable Agriculture Act

Countries need wide-ranging sustainable agriculture legislation that would:

- lay down standards and goals for sustainable farming;
- establish a timetable to phase out farming-related greenhouse gas emissions;
- eliminate subsidies that support unsustainable farming practices;
- eliminate livestock production subsidies;
- control animal factory farm pollution;
- establish a tax on chemical pesticides, herbicides, and fertilizers, using the revenue to help farms convert to organic methods or integrated pest management systems;
- establish a tax on methane emissions and use the revenue to encourage dry manure composting systems and methane-to-biogas projects;
- provide tax credits for the production of sustainable energy;
- set up a 50:50 Farmland Fund to help states and provinces protect their farmland;
- change the focus in agricultural colleges and advisory agencies towards climate-safe farming.

Set up Organic Conversion Grants

In Europe, EU legislation sets common standards across the 15-nation bloc, and provides government support to help farmers break their dependence on artificial fertilizers and pesticides, and convert to organic methods. In Denmark, where conversion grants are administered by a wide-ranging Council for Organic Agriculture, a farmer can receive $180 per hectare in the first year, $96 in the second, and $36 in the third. The conversion must cover the entire farm, and the farm must continue as organic for at least two years.

Carolyn Herriot

The Power of Consumers

Consumers can help by buying organically grown food and by lobbying local farming organizations, cancer patient support agencies, and NGOs such as the Natural Resources Defense Council, World Resources Institute, American Farmland Trust, Environmental Defense Fund, Sierra Club, World Wide Fund for Nature, The Land Institute, and FarmFolk/CityFolk to press for a Sustainable Agriculture Act.

Resources *(See also #36)*

Alternative Farming Systems Information Center: www.nal.usda.gov/afsic

Community Alliance with Family Farmers: www.caff.org

Denmark's Organic Food Sector: www.ecoweb.dk/english

Ecological Agriculture Projects: www.eap.mcgill.ca

FarmFolk/CityFolk: www.ffcf.bc.ca

Organic Farming in Europe:www.irs.aber.ac.uk/research/Organics/europe

Sustainable Agriculture Network: www.sare.org

The Land Institute: www.landinstitute.org

World Resources Institute Sustainable Agriculture: www.igc.org/wri/sustag

79

Encourage Sustainable Forestry

Before white settlers came to North America, most of the continent was either prairie or forest. The forests were huge, the trees were huge, and they stored a huge amount of carbon, playing an ancient role in the planet's natural carbon cycle.

Today, America has 747 million acres of forestland (29% less than in 1630[1]) and Canada has 1,032 million acres.[2] The planet's boreal forests, which reach across Canada and Russia, store 90 billion tons of carbon in their trunks, branches, and leaves, and a further 470 billion tons in the soil.[3]

As the world's temperature rises, the forests are becoming stressed by heat. They are being attacked by new insects and are suffering an increase in forest fires. The number of disturbances from pests and fire has doubled in Canada since 1970; between 1980 and 1997, Canadian forests experienced five out of the seven worst fire seasons in recorded history — coincidental with the warmest years. Due to increased age, fires, logging, and insect attack, Canadian forests are no longer a net carbon sink of greenhouse gases. Prior to 1980, they sequestered 188 million tons of carbon per year. Today, they *release* carbon.[4]

As an ecosystem type, forests move in response to gradual climate change. When the planet warmed up after the last ice age, forests moved north into what used to be tundra, and tundra moved north into what used to be ice. But forests today cannot move fast enough for the speed of warming that is taking place. Individual trees that cannot take the heat will die, releasing their carbon, and in many places, trees will find their passage north blocked by cities or farmland.

Legislate a Sustainable Forestry Act

A Sustainable Forestry Act would require Forest Stewardship Council certification as a condition of all timber operations, whether on private or publicly owned land, and place a carbon tax on all non-certified imported forest products to account for the loss of carbon that occurs during harvesting.

Reduce Demand

We can also take action to reduce demand, so that there is less reason to interfere with the forests. We need legislation requiring all newsprint to have 50% recycled content, for instance, and a nation-wide phase-out of the use of wooden pallets (see #69).

End Perverse Subsidies

In Canada, perverse subsidies amount to an estimated US$2-2.7 billions per year. In the Province of British Columbia, alone, subsidies to the industry in 1997 totaled US$2 billion, of which half was estimated to be contributing to loss of old growth forests. In the US, timber programs in National Forests have lost over $2 billion between 1992-97 according to government analyses, including a continuing subsidy for commercial logging in Alaska's Tongass National Forest.[5]

"The warming appears to be stressing the forests by speeding moisture loss and subjecting them to more frequent insect attacks."
— Gordon Jacoby, Lamont-Doherty Earth Observatory, on trees in northern and central Alaska

Resources *(see also #37)*

Canada's Forests at the Cutting Edge (Sierra Club): www.sierraclub.ca/national/forests

EcoForestry Institute: www.ecoforestry.ca

Forest Conservation Archives (12,000 articles): www.forests.org

Greenpeace Ancient Forests Campaign: www.greenpeace.org/~forests

Perverse Habits: The G8 and Subsidies that Harm the Forests & Economies: www.wri.org/forests/g8.html

Society of American Foresters: www.safnet.org

World Resources Institute Forests program: www.wri.org/forests

Is Your Furniture Destroying the Amazon?

In 2000, Greenpeace completed a two-year investigation tracking timber that was illegally exported by transnational loggers from the Amazon to prestigious institutions such as the British Museum and Heals Furniture in Great Britain. The Brazilian government reckons 80% of logging in the Amazon is illegal, but is unable to control the trade because of the huge profits involved. The official environmental inspectors, IBAMA, have just one inspector for each area of jungle the size of Switzerland. The trade in illegal wood is masterminded by Asian companies that have been repeatedly fined by the Brazilian government for illegal logging, but it continues with the aid of death threats, bribery of officials, links to the drugs trade, and the unwitting support of British companies who are told the wood comes from legal sources. Until all wood products have independent certification of their origin and the sustainability of their management practices, anyone could be contributing to the demise of the planet's ancient forests by buying any wood product.

Source: Greenpeace

Reduce Logging — End Clearcutting — Shut Down the US Forest Service

We need to end active forestry wherever wildlife and wilderness need protecting, and embrace ecologically certified forestry practices everywhere else. We should phase out forestry wherever we can and change our methods of forestry where we can't. Certified forests grow more carbon than industrial forests, and do not lose it through soil destruction.

So what do we need to do? End all logging in US national forests, as the proposed National Forest Protection and Restoration Act would do, including the entire 17 million acres of the Tongass rainforest in SE coastal Alaska. Convert the US Forest Service into a new Forest Ecological Restoration Service. The national forests produce less than 4% of the US total timber consumption, but from 1992-97, they cost the US taxpayer $2 billion in subsidies. End road-building through roadless areas. End logging in the Great Bear Rainforest, the Stoltmann Wilderness, and other threatened rainforests of British Columbia. End all oil and gas exploration in national parks and forests. End clearcutting as a method of forestry throughout Canada and the USA.

Protect Tropical Forests

In the last 50 years, almost half of the world's tropical rainforests, which have stood for thousands or millions of years, have been cleared, producing around 12% of the greenhouse gas emissions from human activities. For every mahogany tree in the tropics that is found and cut down, a bulldozer smashes its way through 60 other trees.[6] We need to end the import of mahogany and other threatened tropical woods. We need to write legislation to prevent US and Canadian corporations from logging outside North America in forests that have not been certified, or tax the import of such forest products so heavily that consumers cease to buy them. We need to rewrite the world's trade rules so that laws of this kind are not ruled illegal by the World Trade Organization.

The richer northern countries should assist the poorer countries of the south to pursue development paths that preserve their forests, instead of subsidizing their destruction by giving export credit loan guarantees. We need to do all we can to protect what remains of the world's forest carbon store.

Ornamentum

Ornamentum Furniture products are manufactured only from wood from forests certified by the Forest Stewardship Council. www.ornamentum.bc.ca

80
Steer Towards Sustainability

Our ancestors did not intend to lose the topsoils of Kansas and Ohio in the dust storms of the 1930s. Nor did they intend, going back many generations, to turn the lush agricultural lands of Mesopotamia into a desert. These things happened out of ignorance about how soils worked, and insensitivity by the landowners who chose to ignore the warning signs.

Our ancestors did not intend to unleash the storms of climate change, either, when they first started burning fossil fuels and cutting down the world's forests. We're getting better at managing soil, but now we're repeating the pattern with the atmosphere. We're ignorant about how it works, we're greedy to go on using it as a dump for our emissions, and we're choosing to ignore the warning signs. As a result, we're heading for a similar disaster, except this time it will be global.

As a species, we are courageous and innovative. We travel with a suitcase full of self-confidence but only a smidgen of ecological knowledge about the territories we move into. On a deep level, we believe that we deserve it all — the best homes, the best furnishings, the best holidays. We don't *intend* to ruin the rainforests, or disturb the atmosphere, but deep down, most of us still don't accept that the planet has limits. Our philosophy is embarrassingly simple: if we can buy it, it must be alright, whether it's a Ford Excursion SUV that does 10 mpg in the city or a table made from teak cut illegally from the Amazon rainforest.

The problem is that we don't receive the warning signals until it is almost too late, and then we waste time going through denial, and dealing with the confusions spread by the fossil fuel producers. We are heading in the wrong direction.

The government's task is to steer, but in a democracy, a government can only steer where the people want to go, so it's the *people* who have to receive the signals, as well as the government. For the economy, we have monthly data on GDP, growth rates, unemployment, vacancy rates, prices, currencies, investment rates, and so on, that we use to set the interest rates, the rudder of the economy. For the ecosystem, on which the whole system depends, we have almost nothing, just emergency reports from the front that tell us whenever we have hit an iceberg. If we are to steer away from danger and towards sustainability, we must have the signals to guide us there.

"Why is our [US] government the only one in the civilized world with a stupid, short-term energy policy? Why do our elected officials consider a European or Japanese-type energy tax not only unpassable but undiscussable?"
— Donella Meadows

Five More Policies for National Governments

- Develop domestic carbon emissions trading before international trading starts in 2008.
- Set a timetable for the rapid phase-out of methane, N_2O, HCFCs, HFCs, PFCs and SF_6 emissions, faster than required by the Kyoto Protocol. The Danish government is showing the way by banning the use of HFCs after 2003, taxing CFCs, HCFCs and industrial greenhouse gases, and moving towards a complete phase-out of these gases
- Stop financing and underwriting fossil fuel and forest destruction projects in developing nations.
- Require a K-12 climate change education program in all schools.
- Develop global partnerships to share sustainable technologies (see #92).

The Genuine Progress Indicator includes social, health and environmental losses and gains, as well as economic data.

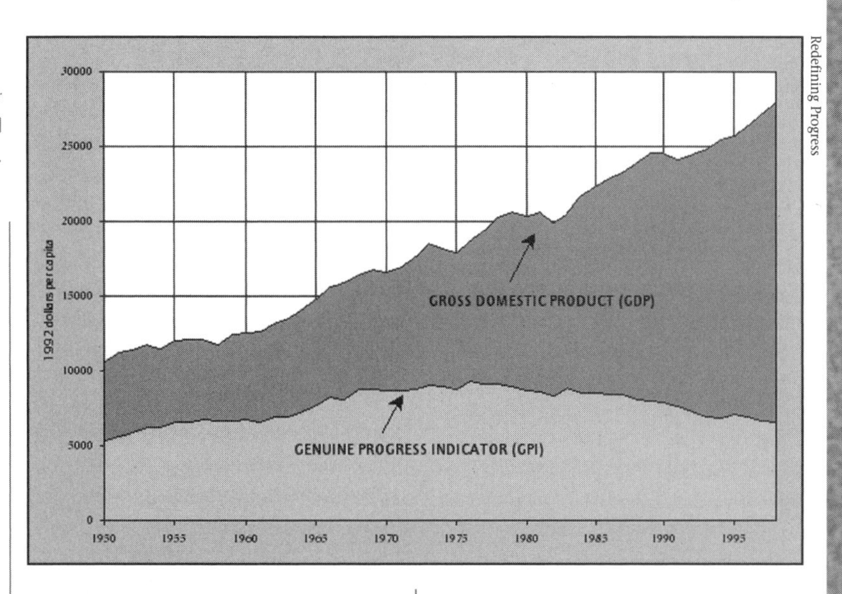

Redefining Progress

Establish Genuine Progress Indicators

Are we heading in the right direction? Who knows? The current system of measuring the nation's wealth as GDP sends no signals about the loss of social wealth (unemployment, crime, overwork, stress, suicides, road accidents, divorces) or environmental capital (forests and wetlands lost, topsoil lost, immune systems weakening, atmospheric stability gone). Under the current system, if you have a car accident, divorce your partner and then blow up half the town in an act of revenge, each expenditure is counted as a plus in the nation's GDP, and the economists celebrate. Traditional economists who continue to rely on this system have surely to be some of the most blinkered professionals in the world. There is a better system developed by the folks at Redefining Progress called the Genuine Progress Indicator (GPI) that includes data for social and environmental losses and gains and shows if we are making genuine progress or not. The GPI has been falling since 1976, true to many people's experience. The GDP should be coupled with regularly published GPI figures so that politicians and ordinary people might wake up and stop fooling themselves that the picket fence is white, the lawn is green, and everything is fine.

Replace Sales Taxes with Ecological Taxes

In the world of consumer purchases, there are two signals that matter: certification (carbon-neutral products, eco-certified timber, organically labeled food) and prices. For prices, carbon taxes and rebates (see #73) are a good beginning. A Dutch company called PRi Consultants has developed software called ECO-it that can analyze the environmental impact of any product within minutes, using a database of over 100 eco-indicators. Using software of this kind, every product could be required to display a 0-5 star status and be taxed according to its ecological impact, using its barcode to replace the sales tax (GST in Canada, VAT in Europe). This would give consumers an extra reason to buy less harmful products, and encourage manufacturers to reduce the ecological and climate change footprint of their products.

Resources

Denmark's ban on industrial greenhouse gases: www.mst.dk/news/04020000.htm

Denmark's tax on CFCs and industrial greenhouse gases: www.skm.dk/printvenlig.php3?SlutFilId=1464

New Economics Foundation: www.neweconomics.org

PRi Consultants (Holland): www.pre.nl

Redefining Progress, Genuine Progress Indicator: www.rprogress.org

Vital Signs – Environmental Trends that are Shaping Our Future (Worldwatch Institute): www.worldwatch.org

81
Follow the Sustainable Path

For years, politicians who are close to the coal and oil companies have complained that climate change is not a proven science, and that reducing carbon emissions will slow economic growth and destroy jobs. Now they are saying that before they sign the Kyoto Protocol, the developing nations must agree to sign too.

To many leaders in the developing world, it is galling that a super-wealthy country such as the USA, which produces far more emissions per person than any developing nation, refuses to reduce its own emissions unless the developing nations do so too. "This is totally unjust," they say. "Our cities are bursting apart at the seams. We have terrible poverty. Our fresh water supplies are disappearing. There's a growing specter of AIDS. We have to deal with ethnic and tribal conflicts, and an increase in natural disasters. We are losing our forests and natural habitat. Amid all this, we are forced to repay our debts to the west, and meet the structural adjustment terms of the IMF. And now you are demanding greenhouse gas reductions?"

"We're sorry," the USA seems to be saying, "but we've filled Earth's atmosphere with our emissions, and there's no room left for yours. You'll have to reduce them." To many a nationalist politician, this is fuel to the fire. Throughout the developing world, politicians, economists, and millions of ordinary people believe that building more fossil fuel power plants is the way to achieve a western standard of living, and there are consultants, CEOs, engineers and bankers from the west lining up to help them. Some developing nation leaders even believe that climate change is a western conspiracy, invented to stop them from catching up.[1]

"The next period will be defined by three revolutions: the continued globalization of the world economy, the changes in national governance occasioned by the related revolution in civil society, and ... a "clean revolution" -- the widespread and continuing development and adoption of ever less polluting and more resource-efficient products, processes and services. "
— Dr. Supachai Panitchapakdi, Deputy Prime Minister of Thailand, 1998

If the developing nations carry on burning fossil fuels the way they are, however, they will be producing more CO_2 emissions than the developed world by 2020. As your authors, we want to shout *Stop!* Carbon reduction is a *blessing,* not a curse. Combined with sustainable transportation, renewable energy, and the more efficient use of energy, carbon reduction will build stronger economies, reduce air pollution, sidestep the riots and upheavals that will accompany rising oil and gas prices, and put developing nations on the path to genuine, peaceful progress. The Philippines plans to complete half of its remaining national electrification exclusively with renewables. The renewable energy revolution is the gateway to a prosperous, sustainable 21st century.

The 20th Century Industrial Path	vs.	The 21st Century Sustainable Path
Dirty fossil fuels	vs.	Clean renewable energy
Firewood, kerosene	vs.	Clean renewable energy
Much wastage of energy	vs.	Energy efficiency
Central power generation	vs.	Distributed power generation
Clearcut, industrial forestry	vs.	Certified, sustainable forestry
Chemically-based farming	vs.	Organic, sustainable farming
Trickle-down economic development	vs.	Community-based economic development
Roads, highways, cars, pollution	vs.	Trains, transit, bikes, community cars
Garbage disposal in dumps	vs.	Zero waste redesign and recycling

Ramakrishna Mission staff trainees installing a PV module in Gosaba, Sundarbans Region, West Bengal, India

Harin Ullal

The Climate Impacts Are Going To Be Huge

Recent years have seen devastating floods throughout the developing world, from Central America to China. Hurricane Mitch has devastated Honduras, and an unprecedentedly powerful cyclone has inundated Orissa state in India. In some cases, the setback to development is measured in decades. Sub-Saharan Africa is becoming drier: in East Africa, 40-year droughts are now arriving every four to five years. In India, the Ganges, Brahmaputra, and Indus rivers, fed by the Himalayan glaciers, may dry up by 2040, devastating the valley lands where so much food is grown. In Bangladesh, rising sea-levels will create 20 million refugees.[2] There will be more typhoons, more torrential downpours, more mudslides, more floods, more droughts, more searing heat, more forest fires, more refugees. Small island states threatened by rising sea levels have joined to demand global action to avert catastrophic climate change - but the entire developing world is similarly threatened. Developing nations tend to be on the tropical storm track, and far more than developed nations, their economies are dependent on farming and natural resource production that is vulnerable to climatic disruption. Nor do developing nations have the disaster response and recovery capacities of developed nations.

Oil and Gas Prices Will Continue to Rise

The world's oil supply is expected to peak somewhere between 2007 and 2020 (see p. 7). As the supply diminishes, prices will rise sharply, bringing balance of trade difficulties, higher food prices, unemployment, heightened ethnic tensions,and civil unrest. Why build a future around a fuel that is rising in price, poisoning people's lungs, and destabilizing the planet's atmosphere?

Leapfrog Over the Dirty Industrial Age

Between 2000 and 2020, developing nations will invest $68 billion a year in new power capacity.[3] A typical investment lasts 30 - 40 years, so it does not make sense to invest in expensive, dirty, old-fashioned power plants when cleaner, more efficient technologies are available. By-pass the centralized power systems and invest in a million sources of solar, wind, biomass, geothermal, and tidal energy, linked by a distributed grid (see #42). Instead of a few central power plants feeding electricity through a costly grid of power lines, millions of micropower plants fueled by the sun, wind and hydrogen can supply individual buildings, or whole villages and neighborhoods. For the two billion people who are not yet on the grid, distributed generation represents the fastest and least costly way to gain electrical services. A parallel exists in telecommunications, where nations such as China are moving directly to wireless telephony; China is now the world's largest cellular phone market. Villages without power can move straight to decentralized solar systems (see #82); their children can study for degrees over the Internet. Towns and cities can embrace bus and bikeway systems (see #83). Factories can embrace zero-emissions goals. The technical solutions to the 20th century's environmental mess already exist. For the rest — the vision, solidarity, political will, and strategy[4] — it's up to us.

> **"Few scientists now doubt that climate change will be among the most pervasive environmental threats of the ... century."**
> **— Red Cross[5]**

82

Create Solar Villages

Delight in Sukiki Village, Solomon Islands, on delivery of the new solar system.

Solar Electric Light Fund

Karma Dorjee lives in a small village in the Phobjikha Valley of Bhutan with his mother and three sisters. Light comes from candles and a smoking kerosene lamp, along with fumes and the risk of fire. Power for their radio comes from a dry battery, and their cooking fuel comes from the firewood that his mother gathers in the surrounding forest. When Karma grows up, he wants to move to Thimphu, the capital of Bhutan, where there are bright lights and a chance to get ahead. He may not know it, but he shares a similar life with two billion people who live in rural villages in the developing world.

One day, a young man arrives in the village carrying a large black case. He meets with the elders and, as Karma and the other children cluster around, he opens his case and takes out a solar panel, assembling it to show them how it works. When the sun shines on the panel, the light comes on. When he shades it, the light goes off. He takes out a laptop computer and a small satellite receiver, plugs it into the solar panel, logs onto the Internet, and invites the kids to explore. Karma is enthralled. He has heard about this "Internet," but he never imagined he would try it for himself in his own village. His thoughts about moving to Thimphu fade a little.

Karma's mother is persuaded, and two months later they have bought a 50W solar home system that powers a fluorescent light, a radio, and a television. The system cost $450 and was financed through a revolving loan fund that costs them only slightly more than they were paying for kerosene, candles, and dry-cell batteries. With persistence, the village school may get a computer and a satellite hook-up.

In Bhutan, the solar energy systems are being introduced by the non-profit Solar Electric Light Fund (SELF) in partnership with Tshungmed Solar Inc. and the Bhutan Development Finance Corporation. Around the world, SELF has helped villagers install reliable systems in China, South Africa, Vietnam, Sri Lanka, Uganda, Indonesia, India, Brazil, Tanzania, and the Solomon Islands.

In South Africa, SELF worked with the KwaZulu Finance Corporation to set up a Women's Solar Cooperative in Maphephethe, KwaZulu-Natal, and to install a system at Myeka High School, where the 830 students now have a computer center with Internet access. In Vietnam, SELF worked with the Vietnam Women's Union to train solar "motivators" who visit villages, sign up families, and collect their down payments. In Sri Lanka, SELF worked with the Sarvodaya rural development movement. In Uganda, the banks are prepared to lend money for the purchase of solar energy systems. The International Solar Energy Society is organizing a Utility Initiative for Africa, with the goal of providing power to every African family by 2020.

The Many Advantages of Solar Light

- Reduced air pollution. Inhaling kerosene fumes is equivalent to smoking two packs of cigarettes a day.
- Reduced emissions. A 50W PV system saves 6 tons of CO_2 over a 20 year life.
- Reduced ground pollution from discarded lead-acid batteries.
- Improved literacy; evening schoolwork.
- Increased evening work-time for home craft industries.
- Reduced need for urban migration because villagers no longer feel so isolated from the world.
- Reduced birth-rate, with more to do in the evenings.
- Reduced risk of fire and burns from kerosene.
- With credit, monthly payments are the same as for kerosene and batteries.

"The solar equipment has had a huge impact on the culture of learning and teaching in our school. The excitement has rubbed off on the learners...the drop-out rate has declined considerably over the past two years. You will never understand how much difference the intervention of SELF has made in the education of an African child."
— Melusi Zwane, Principal of Myeka High school

Sustainable Villages
What does it take to turn a village from a place of hardship, poverty, occasional famine and frequent exploitation into a just, sustainable community? SELF has part of the answer. The Sarvodaya movement has part of the answer, working in 12,000 in Sri Lankan villages with a Buddhist attitude of compassionate awareness to build schools, sanitary systems, and solar village centers. The Grameen Bank has another part of the answer, encouraging Bangladeshi women to become solar agents in the 35,000 villages where they have micro-lending programs, enabling the villagers to hook up to light, telephones and the Internet.

Technology is part of the answer. The world's largest solar cooker, at Taleti, Mt Abu, India, can cook 33,800 meals and boil 3,000 liters of water in a day, using 84 parabolic mirrors to generate steam (saving 400 liters of diesel). There are biogas stoves that use methane from animal wastes to create cooking fuel; there are solar dryers that can protect crops from spoiling; there are solar distillation and desalination systems that can provide fresh water from polluted or salty water.

What Will it Take to Make Every Village Solar by 2020?
Around the world, 400 million households use kerosene for lighting and firewood for cooking. The villagers want power, but linking to the grid is expensive, and will increase the burning of fossil fuels.

SELF has shown that many villagers are able to buy their own solar systems, provided there is a partnership with a local organization. Globally, 5% of the 400 million households could pay in cash, and 25% could pay with short-term credit. 25% could pay with long-term credit, and the remaining 45% would need subsidies.[1] If all 400 million households were to have a 50W system by 2020, this would require 20,000 MW of new solar capacity, averaging 1,000 MW a year. The world's total PV production in 1999 was 201 MW.

Solar technology is improving all the time. The key to mass production lies with better policies, increased purchasing commitments from governments, corporations, municipalities, utilities and home-owners, and reduced prices. Taken together, these could release a solar cascade.

The Kyoto Clean Development Mechanism might be able to help. A 50W system costs $300, and prevents 6 tonnes of CO_2 from being released by kerosene over 20 years. If carbon offset credits sold for $25 per tonne of CO_2, the solar panels' offsets could provide a $40 subsidy. A "Debt for Solar" swap could increase the subsidy. It's just a matter of determination, public support, and political will.

Resources

Grameen Bank: www.grameen-info.org

Greenstar Solar Internet villages: www.e-greenstar.com

International Solar Energy Society: www.ises.org

Microfinance Gateway: http://nt1.ids.ac.uk/cgap

Renewables for Sustainable Village Power: www.rsvp.nrel.gov

Sarvodaya Movement of Sri Lanka: www.sarvodaya.org

Solar Electric Light Fund (SELF): www.self.org

Solar Cooking International: www.solarcooking.org

Village Power 2000 Millennium Program: www.villagepower2000.org

Women and Sustainable Development: www.energia.org

Villages on the Internet

Comunidad Indígena Asháninka, Peru: http://ekeko.rcpip.net/ashaninka

Myeka High School, South Africa: www.myeka.co.za

Otavalo, Ecuador: http://abyayala.nativeweb.org/ecuador/otavalo

83

Build Ecological Cities

"There is no endeavor more noble than the attempt to achieve a collective dream. When a city accepts as a mandate its quality of life, when it respects the people who live in it, when it respects the environment, when it prepares for future generations, the people share the responsibility for that mandate. This shared cause is the only way to achieve that collective dream."
— Jaime Lerner,
past Mayor of Curitiba

We often call our cities "lively", "colorful", or "bustling"; we rarely call them "ecological". All over the world, cities suck in resources and spew out wastes. In the developing world, most cities are a chaotic jumble of noisy vehicles, polluted rivers, and crowded shanty-towns, with a minimum of green space.

Life changes, however. A hundred and fifty years ago, many of Europe's most beautiful cities were a fetid mass of pollution with no drains or clean water. A hundred years from now, many developing world cities will be well-managed urban paradises, with clean-flowing rivers, parks, bikeways, pedestrian areas, fast and efficient bus transportation, thriving green economies with zero emissions, and millions of trees.

The ecological cities of tomorrow's carbon-neutral world will be far more attractive than today's jumbled urban areas in both the North and the South. The process of getting there will be step-by-step; the cities that make the fastest progress will attract the most businesses and strengthen their tax-base.

Brazil's Ecological Capital

The residents of Curitiba, in south-eastern Brazil, think they live in the world's best city — and they make a good case for it, as Curitiba's ecological and social innovations are emulated by cities around the world. The capital of Parana state, with a population of 1.6 million, Curitiba is known as Brazil's ecological capital because of the many initiatives it has taken. Business is flourishing and major manufacturers are moving in, yet CO_2 emissions from transportation are 30% below average. What is Curitiba doing that the rest of the world could learn from?

The story started in 1965, when the city planned two major arterial routes out of the city to take pressure off the downtown. When Jaime Lerner was appointed Mayor in 1971, he turned the arterials into high density transit corridors. In 1982, he added three more transit arterials as the city grew. A single city-wide fare was introduced with shorter trips subsidizing longer ones, and raised plexiglass tube stations were introduced for easy loading. Today, Curitiba has the third highest rate of car ownership in Brazil, but 70% to 75% of commuters ride the city's 1,902 buses; 89% of the riders are satisfied, and the system is completely unsubsidized.

When the Mayor decided to turn the downtown shopping zone into a pedestrian area, he met resistance from shopkeepers, so he asked for a 30-day trial. When motorists planned a protest, he invited the city's children to come and paint flowers on the street. Today, the scheme is so popular that 26 downtown blocks are car-free; on the Rua das Flores street children tend the gardens.

Resources

African Sustainable Cities Network: www.iclei.org/la21/ascn

Afribike: www.afribike.org

Busways in Curitiba: www.busways.org

Cities for Climate Protection: www.iclei.org/co2/co2.htm

City of Curitiba: www.curitiba.pr.gov.br

Curitiba Transport Planning: www.solstice.crest.org/sustainable/curitiba

Institute for Transportation and Development Policy: www.itdp.org

International Bicycle Fund: www.ibike.org

Midrand Ecological City (South Africa): www.midrand-ecocity.co.za

Model Law for Eco-Municipalities: www.planetdrum.org/modellaw.htm

Mexico Climate Action (5 cities): www.iclei.org/co2/mexicoCAP.html

Philippine Climate Action (5 cities): www.iclei.org/co2/philippinesCAP.html

Sustainable Santiago: www.iclei.org/iclei/sustsant.htm

Curitiba's lawns

Suzanne Hodges / Asterisk Productions

In spite of Curitiba's many initiatives, car-use is increasing as the low-density areas between the radial routes expand faster than the planning for transit and pedestrian neighborhoods. The city has attracted four new automobile plants, and as people become more prosperous they buy cars, clogging the roads. The new mayor is actively pursuing more innovative policies, including local Citizenship Streets where government services, shopping, health services, and daycare are all provided. There is no simple way to build an ecological city; they need to be invented and re-invented month by month.

Create Sustainable Citizenship

The philosophy that Curitiba encourages is one of active citizenship in the Athenian sense, where citizens bear responsibility for the soul of their city. The purpose of the city is to give individuals a better life and create a space where they can flourish. For individuals to function well, the community must also function well.

Curitiba is not the only city in the developing world that is pursuing ecological policies. Cali, Columbia has built a light train system; Lima, Peru is building a bicycle network. In South Africa, the African National Congress is planning to give away 1 million bicycles to rural residents by 2010, inspired by the success of the non-profit agency 'Afribike'. Cajamara, in northern Peru, has decentralized its governance to 12 neighborhood councils and 64 rural councils, created a planning framework for ecological management, and developed an energy efficiency strategy to reduce CO_2 emissions, declaring itself Peru's first ecological municipality. Rabat, Morocco has developed an environmental action plan and is promoting clean technologies for business, transportation, and waste management. Tripoli, Lebanon has done an environmental audit, produced a green strategy for the city region, and is using a strategy of decentralized cooperation to implement a large number of projects. Jinja-Bugembe on Lake Victoria, Uganda's second largest city, has adopted a participatory multi-stakeholder group, including many women, to manage an environmental management program. The International Council for Local Environmental Initiatives (ICLEI) works with cities such as these to help them develop environmental strategies and climate action plans.

If there is a key, it is building a participatory framework that will empower people to take responsibility for their communities and make things happen. Without local organizations, people feel passive and disempowered. When they are stressed, they become angry, making them vulnerable to political forces that foment violence against other religious or ethnic groups. Community organizations enable people to express themselves, and join the shared journey to sustainability.

Urban Creativity in Curitiba

- 90 miles of bicycle paths criss-cross the city.
- 13% of the city's waste is recycled; slum-dwellers exchange collected trash for bus tickets, local farm-grown food, toys, and teaching materials.
- Curitiba has 23 parks full of wildlife, after the city solved its flood problems by creating the parks and using them to divert floodwaters into lakes.
- Orphaned street children are "adopted" by businesses that provide them with a daily meal in exchange for gardening or simple chores.
- Street vendors are organized into a mobile trade fair that circulates through the city's neighborhoods.
- 1.5 million tree seedlings were given to neighborhoods to plant and care for; now there are trees everywhere.
- Builders get a tax break if their projects include green space.

84
Learn from Gaviotas

During 1999, the world's population passed 6 billion. Everywhere, global ecosystems are in retreat under the impact of our activities, and yet our numbers keep growing. With each year that passes, another 75 million people join our population. By the best average estimates, we will reach 10 billion by 2080 before the numbers start to decline.

This means that we'll need to find room for another 4 billion people somewhere on the Earth in a way that is socially and ecologically sustainable. Our children and grandchildren will live in this world, and have every right to enjoy its beauty. So where is everyone going to live? And *how* are they going to live?

In the late 1960s, Columbian activist Paolo Lugari found himself asking these questions. Columbia is a huge country, but if you drive east from Bogota out of the Andes, you come to the rain-soaked savanna of the country's eastern los Llanos region just north of the Equator, where almost no-one lives because the soil is so poor. Go further east and you're in the Amazon.

Flying over the area in 1965, Paolo realized that if people could live here, they could live anywhere. He staked a claim to 25,000 acres at a place called Gaviotas (named after a local bird) and asked his friends at Bogota's universities to help.

Working together, they invented a way to mix soil with cement to make buildings, dams, and drainage pipes. They created a light-weight pump that children can work by riding a see-saw or a swing. After 58 attempts, they designed a windmill that would catch the slightest breeze and run for years without the need for repairs. They designed a solar water heater that works in the rain and is so cheap and effective that they set up a factory in Bogota staffed by street kids, installing heaters all over Bogota. All this without registering any patents; they wanted the world to enjoy their technologies for free.

They manufactured their windmills at Gaviotas, and installed thousands all over the country. They made biogas generators, designed a solar pressure-cooker, and invented a solar kettle that provides safe, clean drinking water. It took six years to perfect, but in a world that is running out of fresh water, the implications are enormous.

They needed food. The soil on the riverbanks was too poor for vegetables, so they grew lettuce, tomatoes, cucumbers, and eggplants in containers made of nutritionless rice hulls, washed by manure tea. By the late 1970s, they had created a third of a square kilometer of hydroponic greenhouses. They built a solar hospital, cooled by the wind and heated by the sun, named by a Japanese architecture journal as one of the 40 most important buildings in the world.

Resources

Gaviotas: A Village to Reinvent the World, by Alan Weisman (Chelsea Green, 1998)

Gaviotas book's home page: www.chelseagreen.com/Gaviotas

Colombia's Model City: www.context.org/ICLIB/IC42/Colombia.htm

Gaviotas slideshow: www.zeri.org/systems.htm

Global EcoVillage Network: www.gaia.org

Columbia Support Network: www.colombiasupport.net

At the end of the 1980s, Gaviotas began to run into trouble. Columbia's embrace of free trade was flooding the market with mass-produced food, undercutting local farmers and driving them to the cultivation of coca. The oil industry was booming and the market for windmills and solar collectors was declining. The Gaviotas pioneers were not the kind to give up, however.

Searching for a plant that could survive the harsh llanos soil, they found a Caribbean pine from Venezuela that would grow if the roots of its seedlings were dipped in a fungus that was missing from the local soil. They planted 20,000 acres.

"Just imagine: There are 250 million hectares of savannas like these in South America alone. There's Africa. The tropical Orient. Places where there's space and sun and water. If we show the world how to plant them in sustainable forests, we can give people productive lives and maybe absorb enough carbon dioxide to stabilize global warming. This is a gift just as important as our pumps and solar water purifiers. Everywhere else they're tearing down rainforests. We're showing how to put them back." — Paolo Lugari

As the pine-forest grew, it provided shade for a host of other seeds, dropped by birds or blown in on the wind, and in front of their eyes, the rainforest started to return, with plants, shrubs, jacarandas, saplings, deer, ant-eaters, armadillos, capybaras, and eagles.

The Gaviotans discovered that they could tap the pines for resin and process it into turpentine, replacing imported petroleum products used in paints, glues, cosmetics, and medicines. They designed a pollution-free factory to distill it (earning a 1997 United Nations World Zero Emissions award), and found a new source of income. They realized that by planting the pines in ever-increasing circles and harvesting the resin, they could restore los llanos to a fertile rainforest with rich, productive soil. Now they are working with the Guahibo Indians to research 250 new species of native plants that have appeared, seeking their ethnobotanical properties.

Today, Gaviotas is a self-sufficient village of 150+ people. Its residents live by their creative endeavors, powered by the energy of the sun and wind. They manage their growing forest and ship clean drinking water to Colombia's many villages. In one of the world's most tormented, violent countries, they have created an oasis of joy, health, and fulfillment filled with music and birdsong. They have shown that it is possible.

What can we learn from Gaviotas? That it is important to let people experiment and work together to discover what is possible; that universities have a lot to contribute if their students and professors are allowed to think outside the normal constraints; that empty land offers an incredible opportunity if people are able to live there and experiment. As Gaviotas founder Paolo Lugari puts it: *"If we can do this in Colombia, there's hope that people can do it anywhere."*

Gaviotas' Gifts to the World

1. A means to restore the rainforest, absorbing carbon.
2. New habitat for wildlife.
3. New habitat for humans.
4. A way to enrich the soil, absorbing more carbon.
5. The technology to create safe, clean drinking water.
6. A whole village powered by biogas, the sun, and the wind.
7. Windmills that produce energy in the lightest of breezes.
8. Solar water heaters that work in cloudy conditions.
9. A zero-emissions industry that makes natural turpentines to replace petrochemicals.
10. Something for the third world, by the third world.

85

Maximize Your Energy Efficiency

As economies mature, they use less energy for each gram of output; their production becomes less energy-intensive. At least, that's what's supposed to happen.

With the exception of China, most countries in the developing world have become stuck in their progress down the energy-intensity curve. The wastage means that they are burning more coal, importing more oil and gas, and producing more greenhouse gas emissions than they need to. Some might say, "What's the rush? The rats have been eating the rice for centuries." But every time an electron from fossil fuel is wasted, a molecule of CO_2 joins its friends in the atmosphere. An inefficient factory today means a village destroyed by a mudslide tomorrow, while income is drained from national economies to pay for fossil fuel imports.

Photovoltaic system at the village school in Ceara, Brazil

Roger Taylor

The road to energy efficiency has so many obstacles that it hardly exists. From Punta Arenas in Chile to Mwaniwowo in the Solomon Islands, people waste energy in homes, factories, utilities, offices, irrigation systems, and vehicles. What would it take to make the process of becoming energy-efficient as easy as buying rice or mangos? In countries where some families spend their entire lives scavenging mountains of garbage, there must be a way to scavenge the mountains of wasted electrons.

The Global Environment Facility (GEF) was set up in 1991, just before the Rio Earth Summit. It brings together governments, development institutions, scientists, the private sector, and NGOs to address biodiversity loss, ocean degradation, climate change, and ozone depletion in the developing world. Thirty-six nations pledged $2.75 billion. From 1991 to 1999, the GEF spent $706 million on projects to promote energy efficiency and renewable energy.

One of these projects, in Thailand, was to find new ways to promote energy efficiency in the residential sector. The project team persuaded Thai lighting manufacturers and importers to improve the efficiency of their products; bought 1.5 million compact fluorescent light bulbs and sold them through a chain of convenience stores; promoted new commercial building codes; carried out demonstration building retrofits; developed labeling standards for household appliances and industrial motors; carried out a consumer

Resources

- American Council for Energy-Efficient Economy (China, Brazil): http://aceee.org
- ASE's International Energy Efficiency: www.ase.org/programs/international
- ENEWS - Energy Efficiency in Developing Countries: www.unicamp.br/nipe/enews
- Ghana Energy Foundation: www.ase.org/ghanaef
- Global Environment Facility: www.gefweb.org
- International Institute for Energy Conservation: www.cerf.org/iiec
- World Energy Efficiency Association: www.weea.org

"Given the right information at the right time, the Ghanaian consumer is capable of reducing electricity consumption by between 25% and 75%. One of the causes of utility-consumer misunderstanding -- inaccurate billing - is also drastically reduced if consumers monitor their consumption regularly." — Ghana Energy Foundation

awareness campaign; and provided incentives for manufacturers and consumers to buy more efficient equipment. Over a short period of time, the entire Thai market of 45 million fluorescent lamps per year was converted to the production and sale of more efficient lamps.[1]

Turning Wasted Energy Into Cash
In China, the GEF trained workers in the existing network of provincial energy efficiency centers how to do energy audits. The centers are now considering turning themselves into energy service companies, earning an income by doing efficiency retrofits, financed by the savings, and establishing a free market system for capturing wasted energy that will attract others into the field.

This is definitely the way to go, accompanied by political champions and by policies and programs that remove the blockages to greater efficiency. What is needed is a spark that will turn energy efficiency into a fashion, causing everyone to rush out and harvest their wasted electrons.

One incentive might be the Clean Development Mechanism (see #90), through which western countries buy carbon credits to make up for their own lack of carbon reduction. We need a system in which a community works to aggregate its efficiency savings. The local utility knows what carbon offsets are worth on the international market, so whenever a business or a household produces a documented energy savings, it pays them in cash and sells their saved carbon to buyers in the north.

The system could work for a business or a middle-class family that keeps records and can validate its savings. But what if you live in a slum in Mumbai, or a squatter settlement in Jakarta? The key probably lies in local community or religious organizations, which could train and organize local businesses to do the retrofits, creating both jobs and income.

The more effective the incentives, the more people will want to use them, giving politicians a reason to get moving. The whole program needs packaging so that a local or regional jurisdiction can pick up the phone and say, "we'll have one." It needs selling in the way that good birth control programs and inoculation programs are sold, creating the excitement and the expectation that everyone will take part. The first developing world country to set up such a program will be inundated with visitors from the north, asking "How did you do it?"

A Blockage-Free Road to Energy Efficiency

(see also #24, #32, #45, #46, #65 & #74)

- The maximum support (e.g. tax-free status) for Energy Service Companies (ESCOs)
- A Public Benefit Charge on all utility bills, creating income for efficiency programs
- Clear efficiency labeling on all appliances and equipment, using local symbols
- Higher building codes and vehicle fuel efficiency standards
- Accurate billing, giving clear consumer information
- Colorful consumer awareness campaigns
- Easy financing for the cost of consumer and industry efficiency upgrades
- The bulk purchase of efficient light bulbs, appliances, and equipment
- Technical training for engineers, manufacturers, and dealers
- Achievement incentives for energy-efficiency managers, officials, and staff
- Shared incentives for efficiency upgrades on rental and lease property
- Public endorsements by film stars, sports heroes, poets and singers
- The use of tax credits, subsidies, and incentives
- Political champions for energy efficiency, locally and nationally
- Local and national NGOs to champion the cause of energy efficiency

86
Embrace Renewable Energy

The step beyond fossil fuels is renewable energy, combined with greater efficiency and sustainable methods of transportation. This will soon become a political as well as a scientific and ecological necessity, as rising oil and gas prices cause political tempers to flare and as the feverish atmosphere inflicts more typhoons, floods, mudslides, and droughts on ill-prepared people. Some individual countries show what is possible.

In Egypt, 68 million people produce 0.5% of the world's carbon emissions (0.5 tonnes per person), 72% from oil and 25% from natural gas. Most of Egypt's electricity comes from gas, the rest from hydroelectricity; 40% of its export earnings come from the sale of oil, which will cease as internal demand grows. If new gas finds create a surplus, there are plans to build a pipeline to Israel, Gaza, Lebanon, Syria, and Turkey.

A solar power station is being built in the south at Toshka, and the Noor Al Salam solar-gas plant is being planned near the Red Sea, to operate on solar during the day and gas at night, providing electricity to Egypt, Israel, and neighboring countries. Egypt is 386,000 square miles in size, most of which is sun-baked desert. If one-ninth of its desert area was covered with solar PV or solar thermal plants, it could provide enough electricity for the entire world. If it built partnerships with Libya, Algeria, Jordan, and Syria, it could use its solar capacity to manufacture enough hydrogen for all of Europe's future cars. So here's a huge potential. What would it take to make it happen?

First, it needs a political atmosphere where non-profit societies can challenge fossil fuel companies and promote sustainable energy paths without fear of retribution. The role of independent NGOs to develop and champion the new energy thinking is critical. Next, it needs policy, taxation, and regulatory changes to remove the barriers and create incentives for renewable energy, similar to those in Denmark and Germany. Finally, it needs the price of solar to fall to $1 per watt, which is coming soon (see p. 30, and #92). It's all within reach; we just have to reach out and take it.

> **"We do indeed need to co-operate more effectively. Too many international and national governmental institutions are failing us. Too many public servants in international and national institutions are so governed by obstructive procedures, that smaller-scale and newer developments are particularly difficult to get off the ground. There are too few incentives and too many disincentives so that renewables usually play on an uneven playing field."**
> **— Michael Jefferson, World Renewable Energy Congress, July 2000**

How to Bring down the Price of Renewable Energy

- Remove direct subsidies to fossil fuel energy (see #72).
- Add environmental taxes to fossil fuel energy (see #72).
- Use the income to provide subsidies and tax credits for renewable energy (see #73).
- Use the Clean Development Mechanism to earn carbon offset credits (see #90).
- Set up revolving loan programs to finance solar villages (see #82).
- Use global buying partnerships to reduce the technology costs (see #92).
- Construct local solar and wind factories to save import costs.
- Twin debt write-off programs with solar gain and carbon offset programs (see #90).
- Tie the whole thing together with political champions and active NGOs.

Roger Taylor

Homes in this Brazilian village each have 50-watt PV systems to provide energy for two fluorescent lights.

In Chile, 15 million people produce 0.2% of the world's carbon emissions (0.96 tonnes per person); 64% from oil. Emissions have doubled since 1982, and continue to grow. Chile imports 93% of its oil, but 80% of its electricity comes from hydroelectricity. Their 1999 drought was the worst in 50 years and there are fears that future droughts caused by El Nino and global-warming will threaten the hydro supply, so new electricity is planned to come from gas, imported by pipeline from Argentina.

The Chilean government has done little to encourage energy efficiency or generate renewable energy. With such a long westerly ocean coastline, however, it should have superb prospects for wind and tidal energy. The extreme south around Punta Arenas could become a world center for the manufacture of hydrogen from wind and tidal energy, for use in South America's vehicles. What's needed? A national commitment to switch from fossil fuels and to remove the barriers to renewables; a full package of taxation and regulatory incentives; and active NGOs to promote renewable energy, help the utilities sell it, and counteract the fossil fuel interests.

In Brazil, 192 million people produce 1.4% of the world's carbon emissions (0.46 tonnes per person); 82% from oil. Brazil produces 95% of its electricity from hydro, and several more plants are under construction. There's a World Bank/GEF program to encourage greater energy efficiency, helped by a 1% Public Benefit Charge on utility bills.

Following the 1973 oil crisis, Brazil turned to ethanol from biomass crops as a fuel for its vehicles; today, ethanol provides 41% of Brazil's transportation fuel, releasing almost no net CO_2 emissions, since the CO_2 released in burning is recaptured by the crops grown to produce more ethanol. Brazils's fuel is a 22% ethanol mix, and 1% of Brazil's new vehicles are designed run on pure ethanol. Biomass is also being used as a fuel for pulp and paper mills, sugar mills, and petrochemical and steel plants. Brazil's Biomass Power Program is using gasified wood in high-efficiency combined-cycle gas turbines, with help from the Global Environment Facility.

For the 20 million Brazilians who have no electricity, the government is investing $25 billion to provide 20,000 MW of renewable energy to schools and health centers; the sale of solar water heaters is also growing. Brazil has nation-wide biomass and solar potential, and high wind-power potential in the north and east. It just needs political champions and the right policies. Brazil also has the Amazon rainforest, 30% of the world's remaining tropical forest, which shelters 10% of the world's plant and animal species and plays such important roles in moderating the climate and in absorbing CO_2 from of the atmosphere (see #97). Policies, programs, and laws are needed to protect the forest, too.

Resources

Clean Energy: www.cleanenergy.de

Country Analysis Briefs for Energy: www.eia.doe.gov/cabs

International Network for Sustainable Energy: www.inforse.dk

International Project for Sustainable Energy Paths: www.ipsep.org

International Solar Energy Society: www.ises.org

World Energy Council: www.worldenergy.org

World Renewable Energy Network: www.wrenuk.co.uk

87 Solutions for India

"While a ten-fold increase in energy consumption to, say, about 4,000 kWh per capita, may appear to be a distant dream (from the present 385), and while we would like to avoid the profligate use of energy as in the industrialized countries, we would like to provide the same levels of comfort to our increasing population by using less energy."
— C.R.Kamalanathan

Picture a billion people chasing fast economic growth in a land bathed by the sun for 300 days a year. A land filled with noise and color, pollution and poverty, mysticism and adventure. A land of ancient traditions and dynamic visions. That's India.

In 2000, India's electrical capacity was around 108,000 MW, with demand growing by 12,000 MW a year; 69% of India's power comes from coal-fired and other fossil fuel plants, encouraged and financed by the World Bank (see #96). Hydroelectricity provides 25% of India's power. 3% comes from nuclear, and 1.6% (1,700 MW) from renewables. 86% of India's villages have electricity, but only 31% of rural households can afford to use it; 76 million rural homes do not use electricity; 80,000 villages are yet to be electrified.[1]

Can India become a climate-neutral nation powered by renewable energy, protecting her forests and managing her farms sustainably, while putting an end to the suffering and ecological devastation caused by large-scale coal projects? The climate crisis threatens India's agriculture (rising heat, falling wheat yields[2]), coastal regions (typhoons, floods), and the entire Ganges river delta (melting Himalayan glaciers). Within 17 to 30 years, all of India's oil will be gone. Her uranium will be gone in 40 years, her gas in 60 years.

India has one of the largest renewable energy programs in the world, championed by the Ministry of Non-Conventional Energy Sources. From 1992 to 1997, India set a target of 600 MW of renewable energy and achieved 1,050 MW. They are planning an additional 3,000 MW by 2002, with the goal of achieving 24,000 MW by 2012 (10% of the assumed total of 240,000 MW).[3] As policy measures, the government uses integrated resource planning, soft loans, concessional customs duties, exemption from excise duty, accelerated depreciation, subsidies, educational programs, and reduced subsidies to fossil fuels. While 10% from renewables is a good start, the goal assumes an additional 108,000 MW from large hydro, nuclear, and fossil fuels, meaning that things will get worse much faster than they get better. New studies are showing that rotting vegetation in large dam reservoirs is a significant source of greenhouse emissions, and the problem appears to be worse in hotter climates. (See p. 35). With these plans, India's greenhouse gas emissions will be three times higher in 2010 than they were in 1990.

In 1990, the south Indian state of Karnataka commissioned a study that showed that the state could eliminate 95% of its CO_2 emissions from fossil fuels, use 40% less energy, and save 66% of its energy costs if it focused on energy efficiency and the generation of electricity from small hydroelectric plants, sugarcane wastes, solar hot water, and a small amount of natural gas. If every state followed a similar path, India would need 40% less energy overall.[4]

If India increased its use of power ten-fold, it would need 1 million MW; using the best efficiency policies and practices, this could be reduced to 500,000 MW. The Ministry of Non-Conventional Energy Sources estimates India's potential for renewable energy capacity at 100,000 MW, but when solar becomes economically competitive, this could increase to 600,000 MW.

Resources

Climate Change in Asia: www.ccasia.teri.res.in

Council of Energy Efficiency Companies of India: www.ase.org/ceeci

Department of Energy Conservation: www.conservinfo.org

India's Ministry of Non-Conventional Energy Sources: http://mnes.nic.in

Karnataka Renewable Energy Development Ltd: www.kredl.com

Indian Renewable Energy Develpoment Agency: www.ireda.nic.in

Renewing India: www.renewingindia.org

Jim Hamm

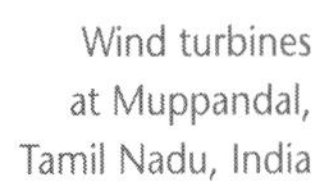

Wind turbines at Muppandal, Tamil Nadu, India

Wind

India's wind power potential has been estimated at 45,000 MW; 1,175 MW has been installed, mostly in the far south, and the potential for a further 5,500 MW at 177 sites has been identified.[5] Between 1993 and 1995, wind power grew by 25 MW a month thanks to a 100% depreciation allowance during the first year. When the policy ended, the investments slowed down rapidly; the government is now considering renewing it.

Microhydro

India has major plans for big hydroelectric dams, some of which are very unpopular, such as the massive Sarovar dam on the Narmada River. The country also has 10,000 MW of potential capacity for micro-hydro, of which 217 MW has been installed and 133 MW is under construction.

Biomass

India is the world's largest producer of sugarcane, the waste of which (known as bagasse) makes a great biomass fuel for use in cogeneration plants within the sugar mills. India could generate a potential 16,000 MW from biomass; they have built 222 MW, including a 6 MW rice husk biomass plant in Madya Pradesh, a 12 MW biomass plant in Tamil Nadu, and a further 332 MW is under construction. Agricultural wastes could also become a source of ethanol for vehicle fuel. India also has 2.75 million biogas plants that generate energy from methane from farm manure, providing 250,000 permanent jobs.

Solar

India has huge solar potential. A 140 MW solar combined cycle plant is being built at Mahania, Rajasthan, with help from the World Bank's Global Environment Facility, which will generate 35 MW from a solar parabolic trough and 105 MW from fossil fuels, confirming (alas) the World Bank's preference for fossil fuels (see #96) . The country's entire electricity needs could be met with PV cells over 4,500 sq km,[6] or solar thermal plants over 2,600 sq km of the western state of Rajasthan (342,000 sq km), 50% of which is desert. India is the third largest producer of solar cells after the US and Japan, and has installed 500,000 small PV systems (58 MW), the equivalent of 750,000 households installing a 75-watt system (0.2% of India's 250 million households). If every household had a 2 kW system, that would yield 500,000 MW;[7] and they're working on it. Calcutta is considering legislation that every multi-storey building must have a solar system. A network of 300 stores has been set up to sell PV devices; 500,000 sq metres of hot water solar collectors have been installed, shaving 500 MW off India's peak load demand.[8] A 200 MW solar chimney is being planned for Rajasthan, similar to a 50-kW system that operated at Manzanares, Spain. And 1,165 MW of grid-connected PV have been installed, with a further 500 MW in the pipeline.
A solar-powered crematorium is being built in Gujarat that will use a 540 sq ft solar dish to cremate a body instead of 600 lbs of firewood.
India has the vision and the capacity to become one of the world's first sustainable energy nations.

What is Needed for India to Achieve 100% Sustainable Energy?

- Cities designed for bicycles, buses, trams, and electric neighborhood cars.
- A transportation system based on electric or hydrogenc buses and railways.
- Global purchasing policies to bring down the price of solar energy (*see #82*).
- The removal of fossil fuel subsidies, and the addition of environmental taxes.
- Legislation requiring the use of solar energy on all multi-storey buildings.
- The use of the Clean Development Mechanism to support renewable energy.
- A complete program of energy efficiency policies (*see #65, #74 and #85*).
- A complete program of renewable energy policies (*see #64 and #75*).

88 Solutions for China

"**The challenge is either to build an economy that is sustainable or to stay with our unsustainable economy until it declines. It is not a goal that can be compromised. One way or another, the choice will be made by our generation, but it will affect life on earth for all generations to come."**
— Lester Brown, Worldwatch Institute[3]

Now we come to China. With 1.3 billion people, it is the world's most populated country. China has just come through an atrocious century of poverty, war, revolution, and upheaval. Its people are finally enjoying a period of relative stability and the economy is growing by about 8% per year.

In 1996, China burned 1.4 billion tonnes of coal, a third of the world's total, and produced 794 million tonnes of carbon, 0.6 tonnes per person. (The US produced 1,469 million tonnes, 5.4 tonnes per person). To reduce its emissions and improve the terrible air quality of its cities, the Chinese government is accelerating a move away from coal to gas and renewables. They've reduced the subsidies to coal, closed 38,000 of the least efficient coal mines, switched 28,000 boilers from coal to natural gas, and are looking into clean coal technologies. They've brought in an Energy Efficiency Act and other initiatives that are capable of delivering a 13% reduction in demand.

In 1998, the Chinese economy grew by 7.8%; at the same time, her carbon emissions fell by 3.7% to 740 million tonnes as a result of increased efficiency and the reduced use of coal, while US emissions rose by 1.7% to 1,494 tonnes. China produces 12.1% of the world's carbon emissions, but her future energy needs are enormous. In 1993, China had 180 GW of electrical generating capacity (180,000 MW). By 2000, it had 310 GW. By 2020 it could need 725 GW, which would require a new 800 MW power plant to be opened every two weeks for 20 years. In 2000, they used 9.6% of the world's energy; by 2020, it could be 16%.

The Chinese are also falling in love with the automobile. In 1986, China had 4.2 million vehicles; by 1996, it had 11 million — one for every 115 people, compared to one for every 1.3 people in the USA. If the Chinese were to drive as many vehicles per person as Americans and Canadians do today, they would need 47% more vehicles than today's entire global fleet and use 18% more oil than today's entire global oil consumption.[1] They would also have to pave the equivalent of 40% of the nation's cropland.[2] It's clearly not possible, which is why China is looking into alternative fuels, electric bicycles and cars, and hydrogen fuel cell vehicles.

Could China's electricity come from renewable energy? In 2000, they obtained 310 GW from coal (68%), hydro (24%), renewables (7%), and nuclear (1%). By their reckoning, China will need 725 GW by 2020. With the best efficiency measures, this could be cut by 50%.

Hydro: 378 GW

China has 378 GW in large-scale hydro resources, of which 41 GW (11%) has been tapped. The goals are 70 GW by 2000 and 125 GW by 2010, but this includes hugely controversial projects such as the Three Gorges Dam (18 GW), which will displace two million people.

Resources

Asia Alternative Energy Program: www.worldbank.org/astae
Beijing Energy Efficiency Center: www.gcinfo.com/becon
China Energy Group: http://eande.lbl.gov/EAP/China
China E-News: www.pnl.gov/china
China Environment: www.chinaenvironment.com
China New Energy: www.newenergy.org.cn
China Renewable Energy Development Project: http://setc-gef.newenergy.org.cn
China Sustainable Energy Program: www.efchina.org
Energy Policy and Structure: www.iceu.net/china
Global Village of Beijing: www.gvbchina.org
Pew Center China Report: www.pewclimate.org
Professional Association for China's Environment:www.chinaenvironment.net

Micro-hydro: 75 GW
China has 75 GW of micro-hydro potential, of which 25 GW has been developed in 4,300 small hydro stations. The goal is to develop 28 GW by 2010.

Tidal and Geothermal: 30 GW
Tidal energy has a 20 GW potential; a dozen stations provide 5 MW today. It also has 3.5 GW potential for hot-water geothermal energy and 6 GW for ground-source heating.

Wind: 253 GW
China has major wind energy potential (253 GW), mainly in Inner Mongolia, 500 km north of Beijing, and along the coast. Some 200,000 small wind turbines operate in rural areas, and 20 large wind farms have been established. The goal for 2000 was 1000 MW, but progress is slow because of a variety of institutional barriers. A massive expansion could provide power to the Beijing area, using compressed air storage to balance windy and calm days.[4] The Chinese used the wind for sailing 3,000 years ago, and for pumping water 2,000 years ago.

Biomass & Biogas: 1-2 GW
There is 700 to 900 MW potential from burning sugarcane wastes, and another 300 million tons of agricultural feedstock available each year as fuel. Five million rural families use methane biogas from agricultural manures, saving two million trees a year.

Solar PV: 800 GW
Two-thirds of China receives at least 2,000 hours of sunshine a year. Assuming the best policies, improved technology, and a competitive price, if China's 400 million households each installed a 2 kW system, this would provide 800 GW. The large western areas of the country could support major solar thermal or solar PV plants for the grid.

Solar Hot Water: 100 GW
China has 5 million sq. metres of solar water heaters (50% of the world market) that are used by over 1,000 factories and 30 million consumers, and the volume is growing rapidly. In India, 1,000 household solar water heaters shave 1 MW off the peak demand. In China that might be 0.5 MW because of cloudier skies. If 200 million households installed systems, this would save 100 GW of peak energy.

Electricity needs in 2020 (with 50% efficiency)**: 362 GW**

Possible renewable energy: 1328 GW

> **"Hold fast to dreams, for without them we are like birds with broken wings."**
> **— Chinese proverb**

The Barriers to Expansion of Wind Energy in China

- High interest charges and short terms on bank loans for wind energy.
- High taxes for imported equipment: sales tax 17%, import tax 12%.
- No depreciation incentives available, as they are for coal.
- No taxes to pay for coal's heavy social and environmental costs.
- Companies reluctant to get involved because of lack of incentives.
- No adequate policy framework.

Source: CICERO, Norway[5]

89

Build Strong Citizens' Organizations

All around the world, from Borneo to Bolivia, developing countries are bursting with the potential to use solar energy, biogas, wind power, biomass, micro-hydro, and other renewables as a means of development. Rural families can earn new income, villages can be solar-electrified, agricultural wastes can be put to good use, and solar appliances can be manufactured. Forests can be protected, trees planted, and farming practices improved. Each of these activities will reduce carbon emissions while enriching people's lives.

One of the keys to realizing this potential, as well as legislation, policies, and a good investment climate, is strengthening local citizens' organizations, giving them the training, encouragement, and resources they need. India has some 25,000 citizens' organizations. In the developing world as a whole, we may be looking at as many as 100,000 organizations, each of which has the potential to embrace the opportunities of renewable energy, to mobilize local people, and to accelerate the sustainable energy revolution.

In the southern Indian state of Kerala, the Kerala Sastra Sahitya Parishad (KSSP) has embraced the mass dissemination of science as its goal, inspiring the foundation of People's Science Movements all over India. As well as their educational and health work, the KSSP's 60,000 members promote local sustainable development and encourage village-level planning. As a result of their activities, half a million homes in Kerala have installed high-efficiency smokeless 'chulhas' (wood-burning ovens), saving an estimated 700,000 tonnes of firewood every year. They also have programs to replace a substantial proportion of the state's 20 million 60-watt light bulbs with compact fluorescent bulbs, to help local governments install micro-hydro stations, and to encourage farmers to adopt agro-forestry, conserving soil and trees. They have led a successful campaign to protect the Silent Valley, a protected evergreen forest that was threatened with flooding by a major hydroelectric project, and campaigned for a rational energy plan for Kerala.

In Kenya, the Green Belt Movement has inspired the creation of 3,000 tree nurseries and the planting of 20 million trees by women and schoolchildren on farms and school and church compounds, protecting the soil against erosion, stopping deforestation and desertification, and creating self-sufficiency in firewood for the women involved, reducing the hardship of their lives. The movement was founded in 1977 by Wangari Maathai and the National Council of Women of Kenya, and has grown to embrace wider community environmental action in East Africa, leading to the establishment of a Pan African Greenbelt Network. When people organize to take practical action in this manner, they cease being the victims of circumstance and become its creators.

In northern India, the Chipko movement emerged spontaneously in the Himalayan region of Uttar Pradesh, as villagers started hugging their trees to save them from being logged and to preserve the soil, water, fuel, and fodder that the trees ensured, following the Gandhian philosophy of satyagraha (the power

Resources

Centre for Science and Environment, India: www.oneworld.org/cse

Chipko Movement: http://iisd1.iisd.ca/50comm/commdb/list/c07.htm

Environmental Development Action in the Third World: www.enda.sn

Global Greengrants Fund: www.greengrants.org

Greenbelt Movement, PO Box 67545, Nairobi, Kenya

Kerala Sastra Sahitya Parishad: www.south-asian-initiative.org/kssp

People-Centered Development Forum: http://iisd.ca/pcdf

Right Livelihood Award: www.rightlivelihood.se

Thai-Danish Cooperation on Sustainable Energy: www.ata.or.th

We the People--50 Communities Awards: http://iisd1.iisd.ca/50comm

Groundwork Trust

The Groundwork Trust's Earthday Celebrations, 2000, Pietermaritzburg, South Africa.

of truth) and non-violence. The movement grew through hundreds of locally autonomous initiatives, mostly led by women, and led to a 15-year ban on cutting in the Himalayan forests. One of its leaders, Sunderlal Bahaguna, undertook a 5,000 km trans-Himalayan march to spread the message, and later fasted for 45 days to prevent the Tehri Dam project, which would have dammed the Ganges River. As well as fighting to save their forest, the Chipko villagers are alarmed that the Himalayan glaciers are receding because of global warming. If nothing is done to stop the warming, they fear the glacier that feeds the Ganges will disappear within 100 years.

In southern Thailand, villagers are organizing to oppose two large coal-fired power plants that are to be built near Prachuap Khiri Khan on the beautiful Gulf of Thailand. The proposed 2-mile-long pier to receive coal from Indonesia will release waste water, harm local fishing grounds, destroy the coral, pollute the air, drive away tourists, and damage the local economy. With help from Thai-Danish Cooperation on Sustainable Energy, the villagers have shown that the coal-fired power could be replaced without extra cost by industrial cogeneration, efficiency savings, and hydroelectric power. Greenpeace Nordic persuaded the Nordic Investment Bank and a Finnish company to withdraw from the project, but it still has Indonesian investors and has yet to be cancelled.

A few years ago, the corporations would have used strong-arm tactics and bribed local officials to get the plants approved, but in 1997, political reform gave Thailand a new constitution that requires public hearings for major projects and allows local people to have their say. The villagers have received assassination threats and their homes have been fired upon, but the campaign continues.

Citizens' action groups like these represent people-centered development. They follow a path to sustainability that is very different to the top-down, fossil-fuel heavy developments that the World Bank and the global corporations try to impose on developing countries. People-centered development reflects the real development needs of local communities rather than the imposed will of distant bankers or corporations.

> **"The solution of present-day problems lies in the re-establishment of a harmonious relationship between man and nature. To keep this relationship permanent we will have to digest the definition of real development: development is synonymous with culture. When we sublimate nature in a way that we achieve peace, happiness, prosperity and, ultimately, fulfillment along with satisfying our basic needs, we march towards culture."**
> **— Sunderlal Bahuguna, Chipko Movement**

International Network for Sustainable Energy

In 1992, citizens organizations from around the world met at the Rio Earth Summit in Brazil and decided to form a network to obtain equitable, sustainable energy consumption from renewable energy and local participation, phasing out nuclear power and fossil fuels.

INFORSE links 200 organizations in 60 countries that are working to promote sustainable energy and social development. It publishes *Sustainable Energy News*, which includes worldwide listings of sustainable energy NGOs. www.inforse.dk

90 Demand Sustainability Adjustment Programs

Beneath the developing world's poverty and debt, beneath the rich world's consumption and greenhouse gas emissions, there is a layer of insight that may offer a solution to these difficult issues.

In the 1960s, the US government printed extra dollars to deal with an unexpected deficit brought on by the Vietnam War. This caused the dollar to fall, reducing the income of the oil-producing nations, who increased their prices and deposited the surplus in the West's banks. This caused interest rates to plummet, so the banks sent youthful employees off to Third World countries offering "can't lose" loans below the rate of inflation.

Many of the countries had corrupt leaders. Some of the money went into private bank accounts, often with the same banks that loaned the money; some was spent on armaments, policing, and megaprojects. The developing nations were advised to grow more food and sell more raw materials, but increased production caused prices to fall. Then interest rates and oil prices rose. Like consumers caught in a credit card trap, they had to borrow to pay the interest. In 1988, they owed $1.2 trillion. By 1993, they had repaid $1.4 trillion, but still owed $1.76 trillion.

When consumers get caught in a credit card trap, they can declare bankruptcy and start again. Nations cannot do this, so with every year, the trap gets worse. Since the 1980s, many nations, especially in sub-Saharan Africa, have seen falling incomes and declining education and health care. The World Bank and the International Monetary Fund have imposed "Structural Adjustment Programs," demanding that the debtor nations reduce their government budgets, cut health care spending, stop subsidizing education, grow more cash crops, exploit their natural resources, and allow western corporations to take over essential functions such as water supply.

Desperate for cash, developing countries have been producing more crops on small areas of land (releasing CO_2 through soil erosion and nitrous oxides from nitrogen fertilizers), selling logging rights to their forests (producing more CO_2 emissions) and chopping down forests (direct loss of CO_2 absorption) to make room for beef cattle (methane-producing) and crop farming. Brazil, one of the world's largest debtors ($112 billion), is also the world's fastest deforester.

Meanwhile, in the wealthy North, the rich are building larger houses, buying larger cars, and releasing more greenhouse gases into the planet's atmosphere, for which they pay no fee. The atmosphere is a global commons, but it has a limited ability to absorb these emissions, so Earth's temperature is rising and causing increasing climate-induced disasters around the world. These strike developing nations with disproportionate severity — by 2025, over half of all people living in developing countries will be highly vulnerable to floods and storms.[1] When Hurricane Mitch hit Central America, the Honduran president commented, "We lost in 72 hours what we have taken more than 50 years to build."[2]

Debt Realities[3]

- Money raised in 1997 by the UK charity Comic Relief for famine in Africa: **$38 million**
- Days of debt-service payments that this represented for Africa: **1.3**
- Money raised by Bob Geldof's Live Aid in 1985: **$200 million**
- Days of debt-service payments that this represents for Africa: **7**
- Money per person per year that Uganda spends on debt-repayment: **$17**
- Money that Uganda spends per person on health-care: **$3**
- Money loaned to the notoriously corrupt Zaire president Mobutu before he was deposed: **$8.5 billion**
- Face-value debt owed by the world's 52 poorest nations: **$376 billion**
- Actual cost to cancel this debt: **$71 billion**
- Cost to citizens of OECD nations to cancel it, over 20 years: **$4 per person, per year**
- Number of people who signed the Jubilee 2000 "Cancel the Debt" petition: **24 million**
- Cost of world military spending: **$800 billion per year**
- Global spending on narcotics: **$400 billion per year.**

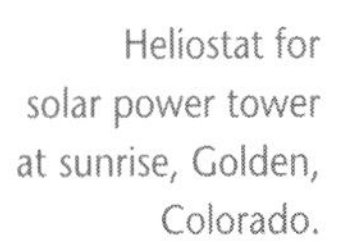

Heliostat for solar power tower at sunrise, Golden, Colorado.

Warren Gretz

Balance Fiscal and Environmental Debt

We are used to financial debt, but the concept of "environmental debt" is new. By taking more than their share of the atmosphere to dispose of their emissions, the wealthy nations are borrowing from those who are not using their share. If there was a World Environment Bank, or an International Emissions Fund charged with ensuring that individual countries remained solvent in their global ecological transactions, they would demand ecological debt repayment from the developed nations, and impose "Sustainability Adjustment Programs" as a precondition of continued use of the atmosphere, requiring a commitment to reduce greenhouse gas emissions and payment for continued use of the atmosphere above the agreed limit. In total, the world's richest nations have accumulated an ecological debt of $13 trillion for their use of the global atmosphere.[4]

So here is the proposition: Can the two debts be balanced? The developed nations could account for their use of the atmosphere at a price between $10 to $50 per tonne of CO_2 ($200-$1,000 per year for each US and Canadian citizen, levied via a carbon tax; see #73). The income could be used to write off developing nation debt on the condition that the retained income was spent on measures to reduce poverty and greenhouse gas emissions, and to fund global greenhouse gas reduction projects anywhere on the planet.

Adopt "Contraction and Convergence" as a Framework

The Global Commons Institute has developed a model called "Contraction and Convergence" that would suit this purpose. It proposes that all nations should contract their emissions towards an agreed-upon global goal, with the per capita emissions of individual nations converging towards an equitable sharing of the atmosphere's ability to absorb pollution. The model allows for an increase of CO_2 in the atmosphere that would peak at 450 ppm in 2025 (compared to today's 370 ppm), and stabilize at an ecologically safe level by 2100. As an associated measure, Third World debt could be traded for commitments to preserve forests (see #97), develop solar energy (see #92), introduce zero-emission vehicles, and reduce methane emissions.

A New Worldwide Campaign

The worldwide Jubilee 2000 Campaign to forgive the debts of the most heavily indebted poor countries by 2000 was pursued by campaigning groups in 60 countries, supported by Desmond Tutu, the Dalai Lama, Bono, Muhammad Ali, Yousou N'Dour, Robbie Williams, George Michael, the Prodigy, David Bowie, and the Pope. It was possibly the largest citizen campaign the world has ever seen and, as a result, most of the G7 nations finally agreed to write off a portion of the debt on the condition that the retained income was used for poverty relief. The campaign ended on December 31, 2000, but now it needs to re-appear in a new form, joining the world's climate change NGOs with the poverty and religious NGOs that gave it such strength. We need to think as one world, and cooperate to solve our problems as one world.

Resources

Drop the Debt: www.dropthedebt.org

Global Commons Institute: www.gci.org.uk

Jubilee 2000 USA: www.j2000usa.org

Jubilee 2000 UK: www.jubilee2000uk.org

Red Cross, Red Crescent: www.ifrc.org

World Disasters Report 2000: www.ifrc.org/publicat/wdr2000

91

Refresh and Renew the Kyoto Protocol

We come now to the global challenge, and the actions that will be needed on the world stage to reduce greenhouse gas emissions by 80% by 2025. This is a big challenge, but the world has met big challenges before when nations agreed to work together.

As authors, we are writing this just after the collapse of the global climate talks that took place in The Hague, Holland, in November 2000, with 182 of the world's governments (also known as COP 6[1]). The talks were to agree on the operating details of the climate change negotiations that started in 1992 and led to the signing of the Kyoto Protocol in 1997, when the developed nations agreed to reduce a basket of the six main gases by an average of 5.2% below the 1990 level by 2008-2012.

By the time you read this book, the world's nations may have reached agreement over the sticking points that caused the talks at The Hague to break down, which were chiefly to do with the ability of carbon sinks to absorb emissions. In terms of world diplomacy, this would be a very significant success. In terms of reducing greenhouse gas emissions, it would be just the first step. Over the years 2001-2005 it will become increasingly clear that we need to embrace a much greater reduction in emissions, and faster than planned. It will also become clear that the carbon-free technologies we need to do it are ready and waiting, and that the treaty should be seen as an exciting invitation to join a transition into the solar-hydrogen age, not a painful exercise in hanging onto the past. The world's scientists say we need a 60%-70% reduction in greenhouse gas emissions *now* if we want to stave off climate disaster. The Dutch government is considering an 80% reduction by 2040. Our goal is an 80% reduction by 2025, which will require a frequent upgrading of the Kyoto Protocol. How can this be achieved?

1. Break Open the Basket

The Kyoto Protocol calls for the aggregated reduction of CO_2, methane, nitrous oxides, HFCs, PFCs, and SF_6, with each gas being assigned a CO_2 equivalence based on its global warming potential over a 100-year period. In reality, most of the discussion has focussed on CO_2, which represents only 53% of the problem. The challenge of reaching the target has been made harder by sidelining the other gases. Our hunch is that the task would be easier if the basket of gases was broken open, and separate goals assigned to each gas. Methane, for instance, has a natural life of 12 years, after which it breaks down into CO_2 and hydrogen. Over 100 years, it has a global warming potential that is 23 times more powerful than CO_2. Over 20 years, it is 62 times more powerful, so a more rapid methane reduction will have a faster and more beneficial effect.

2. Include Black Carbon

Recent studies have shown that black carbon (BC), an aerosol that results from the burning of fossil fuels and biomass, may be responsible for as much as 15%-30% of direct radiative forcing, making it the second most important greenhouse gas after CO_2 (see p. 10). The elimination of BC should be a major feature of a renewed Kyoto treaty.

3. Include Aviation Fuels

Aircraft account for 6% of the total greenhouse effect, yet their emissions have not been included in the Kyoto Protocol. Deal them in! (See #94.)

4. Emphasize HFC Reduction

In the Kyoto Protocol, developed nations are supposed to phase out HFC gases, while in the Montreal Protocol, developing nations are being paid to phase in the same gases

Resources

COP 6, The Hague: http://cop6.unfccc.int

Global Commons Institute: www.gci.org.uk

Global Legislators Organization for a Balanced Environment: www.globeint.org/int

Linkages - COP 6: www.iisd.ca/climate/cop6

UN Framework Convention on Climate Change: www.unfccc.int

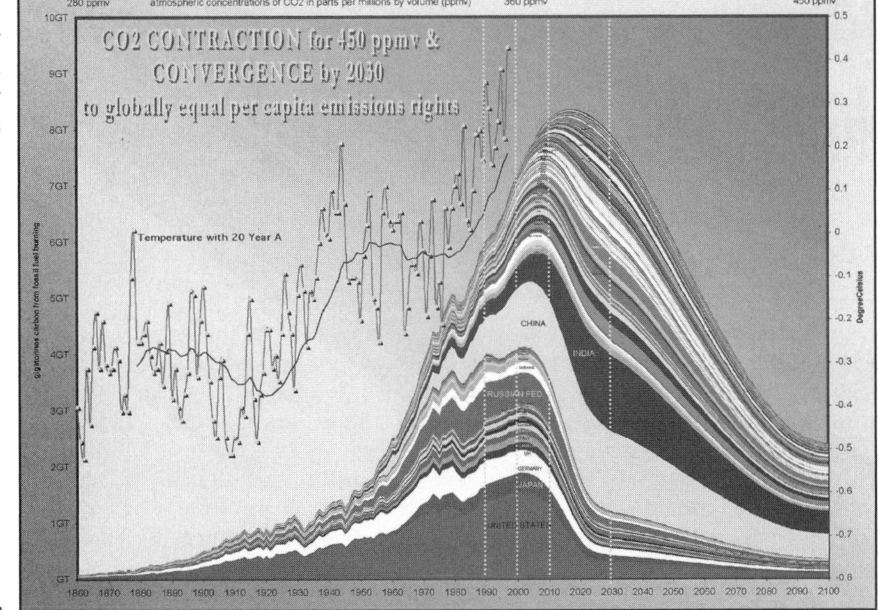

CO_2 CONTRACTION for 450 ppmv & CONVERGENCE by 2030 to globally equal per capita emissions rights

as a substitute for CFCs. This makes no sense — and there are perfectly adequate alternatives to HFCs. Phase them out! (See #95.)

5. Allow Conditional Use of Forest Sinks

The scientific understanding of carbon sequestration by forests, vegetation, and soil is still very immature. Bad management of forests and soils releases CO_2; good management (eco-certified forestry and organic agriculture) absorbs it. We should allow forest and soil sinks to be counted only on the condition that there is reasonable expectation that they will remain stable and hold carbon over a long term. They should be assessed and monitored over a 150-year period, and risk factors including fire, pest damage, and rising heat should be taken into account. The rules for sequestration should not provide an incentive to replace old forests with fast-growing tree plantations. Because of the uncertainties involved, the industrial nations should not be allowed to meet more than 10%-20% of their treaty obligations through sequestration.

6. Use the Clean Development Mechanism Carefully

The Clean Development Mechanism was created to allow the developed nations to finance carbon reduction projects in the developing world, if this proved cheaper than doing the same at home. This could become a huge scam if it allows the developed world to benefit from selling nuclear energy or clean coal technology to India or China. Used carefully, however, by limiting its use to renewable energy and forest conservation projects, it could be a very valuable way to reduce emissions.

7. Include the Developing World

By 2020, under a "business as usual" scenario, the developing world may be producing more emissions than the developed world. It makes more sense for them to leapfrog over the dirty age of fossil fuels than to play catch-up with old technologies that have no place in the 21st century. The best thought-out way to bring them into the treaty process is through the "Contraction and Convergence" model, developed by the Global Commons Institute (see #90), which allows for a fair allocation of the use of the global atmosphere on a per capita basis. Financing will be needed to help the developing world make the transition, creating the need for a Global Climate Fund (See #93).

8. Include a Solar Compact

We will know when we have achieved a successful treaty by the global excitement it generates. The treaty needs to switch from negative to positive by including global compacts to purchase solar and other renewable energy (see #92), and to invest in renewable energy through the World Bank and other multilateral financial institutions. Make it positive. Make it win.

> **"A much tighter and sharper framework of commitments is clearly now needed, one that is focussed on longer term emissions targets that are substantially lower."**
>
> **— Michael Meacher, UK Minister for Environment[2]**

92

Launch a Global Green Deal

A renewed treaty will be a good beginning, but the world needs something far bigger. We need to turn the corner in our imagination, away from the belief that reducing carbon emissions will be enormously painful and difficult towards the realization that the transition to a solar-hydrogen economy will be a thrilling adventure with all sorts of unexpected benefits.

We need a global metamorphosis, casting off our fossil-fueled caterpillar legs for a pair of solar-hydrogen wings. We need what author Mark Hertsgaard calls a Global Green Deal. It begins with an understanding that the two most pressing problems facing humanity — the growing gap between rich and poor and the breakdown of the global environment -- are two sides of the same coin. Solutions can be crafted that address both at the same time and produce benefits for everyone.

The North needs a transition to the new energy technologies we describe in this book. At the same time, the North can help the South to leapfrog over the fossil fuel age and move directly to the new technologies. This idea is at the heart of a Global Green Deal. It will provide an insurance policy against climate destabilization, and by building the global market for new technologies such as photovoltaics, it will reduce the price for everyone.

The technology is ready; the people are ready -- you can tell by the smiles that greet a wind turbine or a solar roof. We are stuck inside a planetary chrysalis, awaiting the day when we will burst out of our cocoon and fly. The coal and oil companies, and the governments and banks that support them, are doing all that they can to prevent the cocoon from opening, but many executives privately admit that these days are numbered. We can either wait until climate chaos forces us to change, or we can choose to breathe together, give an enormous shove, and launch the world into the solar-hydrogen age.

What might a Global Green Deal entail? It would include a series of global compacts to purchase renewable energy wherein everyone agreed to push together — governments, corporations, banks, municipalities, NGOs, and ordinary citizens. It would include agreements by companies, corporations, and banks to embrace codes of global ethics, and agreements to phase out fossil fuel subsidies and apply ecological taxes, using part of the income to retrain fossil fuel workers. It would involve the creation of a World Environment Organization to offset the powers of the World Trade Organization. It is like the Renaissance: the only limits are in our imagination.

> **"The idea is to renovate human civilization from top to bottom in environmentally sustainable ways. Humans would redesign and renovate everything from our farms to our factories, our garages to our garbage dumps, our schools, shops, houses, offices, and everything inside them, and we would do so in the wealthy North and the impoverished South. The economic activity such renovating would generate would be enormous."**
> **— Mark Hertsgaard[1]**

Companies for Global Responsibility

- Business for Social Responsibility (USA): www.bsr.org
- Canadian Business for Social Responsibility: www.cbsr.bc.ca
- Coalition for Environmentally Responsible Economies: www.ceres.org
- Copenhagen Centre: www.copenhagencentre.org
- Global Compact: www.unglobalcompact.org
- Global Reporting Initiative (CERES): www.globalreporting.org
- Global Sullivan Principles: www.globalsullivanprinciples.org

Wind turbines in Holland

Jim Hamm

Build a Global Corporate Responsibility Compact
Most corporations treat the Earth as a scavenging yard for materials and a dumping yard for wastes. Some of the more progressive ones are signing social responsibility compacts, such as the CERES Global Reporting Initiative, which commits them to move towards social responsibility. To accelerate this process, a Global Green Deal would encourage *all* major corporations to commit themselves to this type of compact.

Build a Global Solar Compact
Solar energy is ready, but the price is still too high for lift-off. BP Solar has been told that if it were to build a manufacturing plant producing 500 MW per year, the price of PV modules would fall fourfold to $1.25 per watt. The world production of PV in 1999 was 201 MW. As part of the Global Green Deal, the world's governments, solar corporations, banks, and NGOs should sign a compact to introduce policies, programs, and legislation that would guarantee a global solar lift-off:

(1) Lower the price of PV modules to $1/watt through mass production.
(2) Phase out the direct and indirect subsidies that fossil fuels receive.
(3) Introduce net-metering, allowing people to sell their surplus energy to the grid.
(4) Provide 25-year solar mortgages.
(5) Use NGO community marketing to persuade people to install solar PV (see #47).
(6) Upgrade building codes, requiring the use of PV and solar thermal on all new buildings.
(7) Governments, banks, and corporations agree to install solar PV on all their buildings.
(8) G-8 governments and banks agree to write off developing nation debts in exchange for the creation of solar development funds and programs.

The same approach should be applied to energy efficiency, solar-thermal, micro-hydro, wind turbines, and ground-source (geothermal) heat. These things are perfectly possible. In December 2000, the 15 members of the European Union signed a compact to increase the market share of renewable electricity from 14% to 22% by 2010. Germany, Denmark, and the EU Parliament want the compact to be legally binding; Austria and Luxembourg want it to be non-binding.[2]

Build a Global Zero-Emission Vehicles Compact
A similar compact could build the market for zero-emission vehicles (ZEVs), following California's model. An agreement to introduce legislation around the world requiring that 5% of all new vehicles be ZEVs by 2010 would give a massive boost to the production of fuel-cell, clean electric, and compressed air vehicles. This could grow to 25% by 2015, 50% by 2020, and 100% by 2025.

Build a Global EcoForestry Compact
Create a phased-in agreement that governments would not allow the import of timber products from forests that have been not certified by the Forest Stewardship Council. It doesn't matter if not all governments sign on; the important thing is to get the ball rolling.

Establish a World Environment Organization
Under current rules, the World Trade Organization would probably oppose these compacts on the grounds that they restrict free trade. We need a World Environment Organization to balance the needs of trade with the needs of the planet. A Global Green Deal will require a huge grassroots movement to lobby for it, to counteract the power of the fossil fuel corporations and their lobbyists. This is a central component of Earth's "enormous shove" (see #98 and #99).

93

Establish a Global Climate Fund

For countries in the developing world, and many in the former Soviet Union, the call to reduce greenhouse gas emissions sits at the bottom of a huge shopping list. Spending on education, health care, stopping the spread of AIDS, stemming corruption, repaying foreign debts, fighting crime, controlling population growth, dealing with floods and famine, investing in sewage treatment, preventing ethnic conflicts, coping with water shortages, and protecting endangered species all clamor for funds, as well as tackling poverty and hunger.

Without support from a Global Climate Fund, the solutions that can reduce greenhouse gas emissions are not going to receive the attention they need. The developing nations are going to suffer disproportionately from floods, hurricanes, droughts, and rising sea levels. The developing countries need a Climate Adaptation Fund to help them cope with climate-related disasters, and take preventative measures.

A global endeavor of this size needs a huge commitment of funds, similar to the funds that financed the Marshall Plan to rebuild Europe after the devastation caused by World War II. We need to think globally, and we need to tax ourselves globally. The two funds will probably need to provide $100 billion a year. By contrast, between 1991 and 1999, the World Bank's Global Environment Facility provided around $100 million a year for renewable energy and energy efficiency initiatives.[1]

Various methods are being proposed to build such a fund. The simplest involves direct contributions from the developed nations, but this is vulnerable to changes of government, the disappearance of budget surpluses, and public burn-out. The Climate Fund will need to last at least 25 years - longer if an 80% phase-out of fossil fuels by 2025 is unsuccessful. The Adaptation Fund will need to last for 100 years or longer.

A Global Oil Tax

The world uses 76 million barrels of oil a day, 28 billion barrels a year. With oil prices fluctuating between $25 and $35 a barrel, $1 per barrel could be taken as tax without much political fall-out. A barrel is 42 US gallons (35 Canadian/Imperial gallons), so the cost to oil-users would be 2.3 cents a gallon (0.06 cents per liter for Canadians). This would generate $28 billion a year if the oil producing nations agreed to participate, which would appear unlikely. At the 2000 Hague Climate Conference, Saudi Arabia (which sits on a huge solar potential) even demanded compensation for the loss of its oil revenues if the world agreed to reduce greenhouse gas emissions.

> **"We are running an uncontrolled experiment on the Earth, and it continues as we go about our daily lives."**
> **— Dan Becker, Sierra Club**

Resources

Association for the Taxation of Financial Transactions for the Aid of Citizens (France): www.attac.org

Center for Economic & Policy Research: Global Taxes: www.cepr.net/globalization/speculation

Global Taxes - Bibliography of Resources: www.ceedweb.org/iirp

Global Taxes: Policy Instruments & Revenue Sources: www.igc.org/globalpolicy/socecon/glotax

Halifax Initiative (Canada): www.web.net/halifax

Network Institute for Global Democratization (Finland): www.nigd.u-net.com/campaign.html

Tobin Tax Campaign (UK): www.tobintax.org.uk

Tobin Tax Q & A: www.waronwant.org/tobin/faqf.htm

War on Want Tobin Tax Campaign: www.waronwant.org/tobin/faqf.htm

World Economists Call for a Tobin Tax: www.attac.org/fra/asso/doc/doc18en.htm

World Parliamentarians Call for Tobin Tax: http://tobintaxcall.free.fr

"If it were only a few degrees, that would be serious, but we could adapt to it. But the danger is the warming process might be unstable and run away. We could end up like Venus, covered in clouds and with the surface temperature of 400 degrees. It could be too late if we wait until the bad effects of warming become obvious. We need action now to reduce emission of carbon dioxide."
— Stephen Hawking, Physicist, on *Larry King Live*, Dec 25, 1999

A Global Aviation Tax

There are no taxes on aircraft fuel; aviation was left out of the Kyoto Protocol, which is both an oversight and an opportunity. There is a clear logic in using an aviation tax for climate change purposes. The atmosphere is a global commons that belongs to everyone; without an atmosphere, aircraft would not be able to fly. The developed world is mainly responsible: 63% of all flights are within or between Europe and North America.[2] If aircraft paid for the cost of eliminating 80% of the CO_2 and other greenhouse gases that they produce through a fuel tax, this would generate $79 billion a year. (See #94.)

A Global Tax on Currency Dealing

In 1972, the Nobel prize-winning economist James Tobin proposed levying a 1% tax on foreign exchange transactions "to throw some sand in the well-greased wheels" of global currency trading and mitigate the predominance of speculation over enterprise.[3] The proposal has become known as the Tobin Tax and is being promoted as a way to provide stable funding for the United Nations,[4] tackle world poverty, and finance the transfer of clean energy technologies to developing world.[5] We tax tobacco and alcohol, so why not tax financial speculation, that has so few redeeming benefits? In 2000, international currency trading totaled over $2 trillion per day, of which only 5% was needed for financing trade in goods and services. All the rest was speculative activity. Back in 1975, 80% of these transactions were trade-related,[6] but not any more. In 1998, Britain's National Westminster Bank made $625 million profit from currency trading; in 1997, the Hong Kong and Shanghai Bank boasted that it made $3.3 million a day from currency speculation.[7]

Assuming 240 trading days per year, that 20% of the transactions were tax-exempt, that 20% evaded the tax, and that the volume of currency trading fell by 50% because of the tax, a 0.25% tax would generate $360 billion a year — enough to pay for the UN; the climate change funds; the worldwide elimination of AIDS, leprosy and diphtheria; the elimination of Third World debt[8]; the elimination of the worst forms of poverty;[9] and the dismantling of all nuclear warheads. A widespread movement is growing in support of such a tax (see box). Four hundred world parliamentarians and 300 world economists have signed petitions in support of the tax. In March 1999, after substantial campaigning, Canada's Parliament voted to enact a tax on financial transactions in concert with the international community. This proposal is not going away!

Solar Electric Light Fund

A Brazilian family gathers to celebrate their new solar home system.

How Can You Promote the Tobin Tax?

- Spend an hour learning about the tax (see Resources, especially the War on Want site).
- Write to your senator, representative, or MP about it. Ask to meet them to talk about it when they are next in town.
- Create a petition, and invite your friends and neighbors to sign it.
- Write an Op-Ed piece for your local paper.
- Persuade a local environmental group, church, school, or student union to take it up as a campaign.
- Write a short information sheet and send a copy to every politician in Washington, Ottawa, or your capital city.

94 Include Aviation

NASA's Centurion Solar Plane in test flight with internal wing structure visible

"Calling all passengers for American Airlines Flight 196 from San Francisco to Boston. Please have your carbon tax receipts ready with your boarding passes."

It's not a reality yet — but it could be soon. In 1999, 1.5 billion passengers took scheduled services on 16,000 commercial jet aircraft, and the volume is increasing by 5%-6% a year as more and more people reach for their vacation catalogs.[1] It is estimated there are 10,000 aircraft in the air at any given moment.[2] We also shipped 27 million tonnes of freight to keep the world's shopping malls full, a tonnage that's growing by 6%-7% a year.[3]

Every flight is fueled by kerosene, which is a fossil fuel. Globally, commercial jet aircraft generate 600 million tonnes of CO_2 per year, 3% of the world's CO_2 emissions. Aircraft also release nitrogen oxides (NOx) directly into the troposphere, 5 to 8 miles high, where they oxidize into troposhere ozone. Higher up, the ozone would be helpful, but at this level it acts as a greenhouse gas that represents 12% of the overall problem. When you combine the CO_2 and the ozone, the combined impact comes to 6% of the greenhouse effect,[4] or a possible 10%.[5] Aircraft contrails (carrying water vapor) can also lead to the formation of high-level cirrus clouds, which add further to global warming by trapping heat within the atmosphere.

Aircraft are the world's fastest growing source of greenhouse gas emissions, but they were left out of the Kyoto Protocol (along with fuels for ocean shipping) because the nations could not agree on how to allocate the release of CO_2, since aircraft fly between different nations. No reduction targets have been set for the aviation or ocean shipping sectors (known collectively as bunker fuels).

The biggest shocker is that aircraft fuel is not taxed, as gasoline is. The absence of a fuel tax gives aircraft an unfair advantage over cars, buses, and trains, making tickets cheaper and encouraging people to fly more often. The aviation industry also receives subsidies from duty-free sales, government expenditures on surface access roads, and freedom from sales taxes. In Europe, the aviation industry receives an estimated $42 billion each year in direct and indirect subsidies.

To those who love it, flying is an adventure of the very soul — but it should at least pay its way. And let us pray for wings that will celebrate our planet, not pollute it.

The High Cost of Flying[6]

	Miles	CO_2 eq per person	Carbon Tax $50/ton CO_2[7]
Los Angeles - London	5380	6.4 tons	$320
San Francisco - Beijing	5903	7.0 tons	$350
Winnipeg - Miami	1850	2.5 tons	$125
Seattle - New York	2375	3.0 tons	$150
Vancouver - Toronto	2099	2.7 tons	$135

Source: Ben Matthews

"Air travel has an unfair advantage over other transport modes such as the car, bus, and train because airlines don't pay tax on aviation fuel. The absence of a fuel tax or an emissions-based levy allows airlines to charge artificially low ticket prices as the cost of pollution is passed on to society, and not the passenger."
— Friends of the Earth, London, May 2000

Establish A Global Aviation Tax
In 2000, the world's airlines flew 3,000 billion passenger kilometres and carried 27 million tonnes of freight. Altogether, human activities produced 33 billion tonnes of CO_2 equivalents. Reducing this by 80% at $50 per tonne/$CO_2$ would cost $1.65 trillion per year. Aviation is responsible for 6% of this, i.e. $79 billion. If the burden was split 50:50 between passengers and freight, the cost for passengers would be $39 billion, so tickets would need to rise by 1.3 cents per kilometre. A 8,663 km trip from Los Angeles to London would cost an additional $112.[8] The cost for freight would need to rise by $1,444 per tonne, or $1.50 per kilogram. The yearly income for the Global Climate Fund and the Climate Adaptation Fund would be $79 billion per year.

Impose Fuel Efficiency Charges
The call for an aviation tax is a very hot topic in the European Union (EU), where finance ministers would willingly tax jet fuel if there was the international will to do so. If Spain and Ireland had not objected, they would have already imposed a kerosene tax within Europe's borders. Norway recently abandoned an internal aviation tax because it put its airlines at a disadvantage. Lacking agreement on a fuel tax, the EU is proposing to encourage greater fuel efficiency by means of a ticket surcharge, a charge based on distance and emissions per mile, and a take-off and landing charge. The OECD has shown that a cumulative fuel levy that increased the price of aviation fuel by 5% a year would cause a 30%+ reduction in energy use by 2020. If today's passenger-miles growth-rate continues, aircraft will consume 250% more fuel by 2020,[9] resulting in a net 175% increase in fuel-use.

Start Aviation Emissions Trading
Give airlines a fixed and declining CO_2 emissions allowance under the Kyoto Protocol, and allow them to take part in global carbon trading to help them achieve the reductions, passing on the cost to their passengers as a fuel tax. Some passengers already pay for their flights to be carbon neutral (see #35). This way, all passengers would be required to do so over time.

The Future of Flying
We must end the era of cheap flying. The cost to the atmosphere is far too heavy. NASA has flown automatic solar aircraft and, in 1981, a solar plane with a pilot flew from Paris to London, but solar aircraft will never carry 1.5 billion passengers a year. Teleconferencing can be substituted for many business trips; for distances up to 400 miles, trains are quicker. Aviation's future may lie with low-temperature liquid hydrogen. If the hydrogen is obtained from algae or renewable energy, there will be no emissions, but the impact of water vapor from hydrogen exhaust on contrail formation will need to be considered. In 2000, DaimlerChrysler Aerospace Airbus announced plans for a liquid hydrogen commuter jet, the "cryoplane," with 33 partners in 11 countries. It hopes to start series production by 2010.

Resources

Aviation and the Global Atmosphere (IPCC Report): www.grida.no/climate/ipcc/aviation

Aviation Environment Federation (UK): www.aef.org.uk

Ben Matthews - CO_2 emissions from any flight: www.chooseclimate.org/flying

Cryoplane (DaimlerChrysler): www.aero-net.org/about/relproj/21-cryoplane.htm

Friends of the Earth, UK (Aviation): www.foe.co.uk

Hydrogen and Aviation (Cryoplane): http://dom.decollage.org

International Air Transport Association: www.iata.org

International Civil Aviation Organization: www.icao.org

Right Price for Air Travel Campaign: www.milieudefensie.nl/airtravel

95 Accelerate the Phase-Out of CFCs, HCFCs, and HFCs

Before there was an ozone layer, there was no life anywhere on the Earth outside the oceans. As life in the oceans gradually released oxygen (O_2), the molecules rose up into the stratosphere, where light rays broke their bonds and produced ozone (O_3), creating the ozone layer. Only three molecules in every ten million are ozone, yet this is enough to protect all life on Earth from UV-B radiation.

And so life continued for many millions of years.

Then, in 1928, Thomas Midgley, a chemist with General Motors Research Division, was asked to develop a refrigerant to replace sulfur dioxide, methyl chloride, and ammonia. Midgley's solution was the chlorofluorocarbons (CFCs), a family of chemicals that are neither flammable nor poisonous. DuPont started to manufacture them and soon they were being used everywhere, as air-conditioning coolants, spray-can propellants, foam-blowing agents, and circuit-board cleaners.

In 1974, the chemists Sherwood Rowland and Mario Molina theorized that the chlorine molecules in the CFCs were probably destroying the ozone layer, but companies like DuPont ridiculed the idea and continued to produce them. In 1985, a gaping hole in the ozone layer was discovered over Antarctica, where the CFCs gather before they attack the ozone molecules. The consequences were potentially horrendous. As the ozone hole spread, harmful UV-B rays would penetrate deeply into water, leaves, and skin. There would be increased skin cancer and eye cataracts among humans and animals, a weakening of immune systems in all creatures, reduced crop yields, damage to forests, reduced fish yields, and decreased phytoplankton populations in the oceans, affecting both the marine food chain and the ability of the oceans to store carbon.

The nations of the world met in Montreal in 1987 and agreed to phase out CFCs and some other ozone-depleting substances. Industrial countries have already phased out some, but developing countries have been given longer. Later treaties have speeded up the process, but of the two main substitutes, HCFCs (hydrochlorofluorocarbons) are still ozone destroyers (though to a lesser degree), while HFCs (hydrofluorocarbons) are powerful greenhouse gases, so we're "out of the frying pan and into the fire," as the Antarctic scientist Joe Farman put it.

Halons (halogen family chemicals used in fire-fighting) and methyl bromide (used as a soil fumigant on non-organic strawberries, tomatoes, and peppers) are two more ozone destroyers that are proving difficult to eliminate because of institutional and corporate resistance. A significant number of countries, including Armenia, Bulgaria, China, India, Nigeria, the Philippines, Romania, and Russia have not ratified the most recent treaties. There's also a significant amount of CFC smuggling going on. Meanwhile, the ozone hole over Antarctica that appears every September when the sun re-appears over the horizon is still growing; it is also appearing over the Arctic each spring. All in all, it's not a pretty picture.

Resources

- EPA Methyl Bromide Phase-Out: www.epa.gov/ozone/mbr
- EPA Ozone Depletion: www.epa.gov/ozone
- Environment Canada Stratospheric Ozone: www.ec.gc.ca/ozone
- Friends of the Earth - Healing the Atmosphere: www.foe.org/ptp/atmosphere
- Greenpeace Ozone Campaign: www.greenpeace.org/~ozone
- NASA's Ozone Information: http://jwocky.gsfc.nasa.gov
- Ozone Action: www.ozone.org/page19.html
- Ozone Depleting Substances: www.epa.gov/ozone/ods.html
- United Nations Ozone Secretariat: www.unep.org/ozone
- University of Cambridge Ozone Hole Tour: www.atm.ch.cam.ac.uk/tour

Accelerate the CFC and HCFC Phase-out

The reasons given for such slow phase-out periods (e.g. 2040 for HCFCs in developing countries) are not technical — they're financial. The countries say, "we can't afford it." Since 1991, the richer countries have contributed just over $100 million a

Gases that Mess Up the Atmosphere

	Life span in years	GWP* in years	Developed world phase-out	Developing world phase-out	Share of warming	Global atmospheric status
CFCs	45 - 1700	4,600-10,600	1996	2010		Falling
Halon 1301	65	4,900	1994	2010		Rising
Methyl Bromide	1.3	5	2005	2015		Rising
HCFCs	1.4 - 18.5	120-2,300	2020	2040	12%	Rising
HFCs	1 - 264	1456	None	None	altogether	Rising
PFCs	3,000 - 50,000	8000	None	None		Rising
SF_6	3200	23,900	None	None		Rising

*GWP: Global Warming Potential

year into a Multilateral Fund administered by the World Bank to help developing countries change to HCFCs and HFCs. In the period 1985-1995, however, after the ozone hole was discovered, companies such as DuPont, Elf-Atochem, ICI, Great Lakes Chemicals, Dead Sea Bromide, Solvay, and Allied Signal produced and sold over 17 million tonnes (worth nearly $30 billion) of ozone-depleting substances. So far, they have not contributed a cent towards the cost of a global phase-out or towards reparations for the worldwide damage their chemicals have caused.

Phase Out HFCs

HFCs are potent greenhouse gases that should be phased out rapidly, separately from their inclusion in the bundle of six gases covered by the Kyoto Protocol. If we continue to use them as CFC and HCFC substitutes, global production (currently 101,000 tonnes a year) could reach 1.35 million tonnes by 2040, equivalent to 15% of current fossil fuel emissions.[4] Over a 20-year period, HFC-134a has a huge global warming potential of 3,200. Safe natural refrigerants are already in use, including hydrocarbons, CO_2, water, air, and ammonia[5] (see Greenfreeze fridge, #5), but chemical companies are marketing HFCs and HCFCs to the developing world in spite of the fact that HFCs cost more, leak more, produce toxic waste, and are unstable in very humid climates. The companies also use their influence within the World Bank (see #96) to ensure that the Multilateral Fund finances HCFC and HFC projects which lock developing nations into a toxic, antiquated technology, when it would be more rational to embrace natural refrigerant technologies with a promising future.

End CFC Technology Exports

The developing nations complain that chemical companies produce CFC-based fridges for sale overseas and dump ozone-depleting products in the developing world. We can't afford to play with the ozone layer and the Earth's climate in such an irresponsible manner. If there was a World Environmental Court (see #92), these practices would immediately be challenged.

Eliminate PFCs and SF_6

PFCs (perfluorocarbons) are industrial chemical greenhouse gases with high global warming potentials and very long atmospheric lifetimes; 59% comes from the production of aluminum, 15% from plasma etching in semiconductor manufacturing, and 26% from solvents and unknown sources. New smelting technology avoids PFCs altogether, allowing for total phase-out.[6] Sulfur hexafluoride (SF_6) is the most potent greenhouse gas the IPCC has ever evaluated (GWP 23,900). It is used as a dielectric in electrical transmission and distribution systems. Emissions come from leaks and from servicing substations and circuit breakers, especially from older equipment. It is also used as a protective covergas in the casting of molten magnesium.

> **"It does not seem very logical to try to replace them (CFCs) with another family of related halocarbons, the HFCs, which are equally foreign to nature. It must obviously be preferable to use natural compounds that are already circulating in quantity in the biosphere and are known to be harmless."**
>
> **— Professor Gustav Lorentzen, Trondheim, Norway[2]**

96

Reform the Global Financial Institutions

All around the world, oil pipelines, coal-fired power plants, open-cast coal mines, and other fossil fuel megaprojects are still being built, even as floodwaters and tropical storms sweep away people's homes. The world's scientists,[1] business leaders,[2] and national governments[3] have agreed that climate change is perhaps the biggest crisis facing the planet — and yet the fossil fuel construction boom continues. What is going on? The captains on the bridge have issued orders to switch fuels and change course, but down in the boiler-room, nothing is happening. Is there a mutiny? Are the captains not in control?

Correct, and correct. The captains could be in control if they wanted to be, but as soon as the media's backs are turned, the leaders of the G-7 nations (USA, Japan, Germany, France, Britain, Italy, and Canada) use their majority votes in the World Bank, the International Finance Corporation, the Inter-American Development Bank, the European Investment Bank, the European Bank for Reconstruction and Development, and other financial institutions to support fossil fuel projects that benefit their corporations and allow their export credit agencies[4] to do the same. It's business-as-usual out there, and the world's leaders are not stopping it.

Since the 1992 Earth Summit, the World Bank has spent 25 times more money on fossil fuel projects in countries such as India, China, Morocco, Chad, Bolivia, and Thailand than it has on renewable energy projects. Coincidentally, 90% of these projects gave contracts to corporations from the G-7 nations.[5] In 1995, Larry Summers, undersecretary in the US Treasury, told Congress that for every dollar the US government contributes to the World Bank's coffers, it gets $1.30 in procurement contracts for US corporations.[6]

How can this be, when the World Bank is supposedly part of the democratically governed United Nations? Back in 1946, when the World Bank and the International Monetary Fund were created as part of the Bretton Woods agreement, the lions seized control of the cage and pushed the gate-keepers out. Instead of democratic voting, they set up a system whereby each member gets 250 votes, plus one vote for each $100,000 they contribute. In a throwback to the pre-democratic era when only land-owners were allowed to vote, the wealthy nations contribute more money, so they control how the money is used. The developing and transitional countries make up 83% of world's nations but control only 39% of the votes. The G-7 nations also make sure that the top jobs all go to Europeans and Americans.

In other words, they've got it all sewn up in the name of the big fossil fuel corporations, which happen to be among the biggest US political campaign donors, and happen to have been members of the Global Climate Coalition, which opposes the Kyoto Protocol (see p. 50). Call it a conspiracy or whatever you want, but what it means for the overheated planet is that the world's financial institutions are piling on the coal.

Destroying Orissa[8]

The Indian state of Orissa is being devastated by its World Bank financed coal mining projects. Rivers carry toxic effluent through villages where people rely on the blackened waters for bathing, drinking, and washing their clothes; the black waters of the Nandira, once a life-sustaining river, are slowly poisoning and killing people, animals, fish, and plants as far away as 50 miles downstream. Agricultural productivity has dropped for farmers who depend on its water; fishing communities have been wiped out; and the incidence of cancer, bronchitis, and skin diseases is soaring. All for the sake of coal, financed by the World Bank. If Orissa were to invest the money in solar and biomass energy, it could trigger a statewide shift towards sustainability and the elimination of poverty.

Source: Institute for Policy Studies, 1997. www.seen.org

"If the people we entrust to manage the global economy — in the IMF and the US Treasury Department — don't begin a dialogue and take their criticisms to heart, things will continue to go very, very wrong."
— Joe Stiglitz, Former World Bank chief economist and vice-president, April 2000[7]

What's to be done?
What might be possible if the World Bank, the Export Credit Agencies, and the other public financial institutions used their resources to expand solar, wind, energy efficiency, and other sustainable energy technologies, instead of fossil fuels? Instead of the $540 million that the World Bank is paying to subsidize Exxon's oilfields in Chad, where the pipeline will destroy tropical rainforest and risk creating an all-out civil war, the World Bank could launch a micro-loan banking program to establish village-based solar energy systems throughout the country. In China, instead of investing $1.3 billion to subsidize four massive coal-fired power plants, it could invest in wind turbines on the Mongolian plateau or the coastal region. The Bank and its colleagues could be major players in a Global Green Deal (see #92), instead of a business-as-usual Brown Deal.

It's Democracy, Stupid!
There is one change that would make a big difference: democracy — fair voting, open debate, and transparent decision-making within the World Bank and the other international financial institutions. But how to achieve it? History shows that undemocratic rulers rarely give up their power without a struggle. The worldwide campaign for reform is building momentum; sooner or later, change must come. The Bank's current position — that the fight against poverty must take precedence over climate change — is transparently thin. If we do not stop climate change, it will undo everything, in the Third World and everywhere.

It's Sustainability, Dumbo!
The World Bank, the Export Credit Agencies, and the other international financial institutions should create a sustainability screen for all their investments and loan guarantees. They should pledge not to use public resources to finance projects that increase greenhouse gas emissions, destroy rainforests, pollute local rivers, destroy people's livelihoods, or undermine human rights. The world's NGOs and leading corporations could write a Sustainability Pledge, produce a score-card, and build a worldwide campaign to persuade every financial institution to sign on.

It's the New Economics, Rambo!
Forget the Rambo-style free market policies that economists in the International Monetary Fund and elsewhere like to impose on the countries they deal with. Their top-down policies and structural adjustments have proved disastrous in Asia, Russia, and other debt-ridden nations. A more sane, sustainable approach to economics is emerging from groups such as the New Economics Foundation (London, UK), which advocates restoring social and environmental values and using community-based approaches for local decision-making.

Resources

Bank Information Center: www.bicusa.org

Bankwatch: www.bankwatch.org

Berne Declaration: www.evb.ch/bd/bdfin.htm

Bretton Woods Project: www.brettonwoodsproject.org

Business Council for Sustainable Energy: www.bcse.org/itfinance.html

Carbon Watch Bulletin: www.seen.org/bultin.asp

Debt and International Institutions: www.foei.org/LINK/indexlink94.html

Export Credit Agencies Watch: www.eca-watch.org

Fifty Years is Enough: www.50years.org

Friends of the Earth campaign: www.foe.org/international/finance

Jakarta Declaration: www.environmentaldefense.org/pubs/Filings/jakarta_dec.html

New Economics Foundation: www.neweconomics.org

Real World Bank: www.realworldbank.org

Sustainable Energy and Economy Network: www.seen.org

World Bank: www.worldbank.org

World Bank-Climate Change: www-esd.worldbank.org/cc

97
Stop Global Deforestation

As planetary citizens, we are making a terrible mess of our forests. We have cut, burned, or cleared four-fifths of the Earth's original forests, and we're still at it. The developed world has a voracious appetite for timber and wood products (including paper). Developing countries such as Brazil, Guyana, Indonesia, Gabon, Cameroon, Malaysia, Borneo, and Papua New Guinea have an equal appetite for the dollars that can be earned by cutting down the forests. In poorer countries, settlers burn forests to clear land for agriculture; in Canada, forest companies clearcut the ancient rainforests of British Columbia and the boreal forests of Alberta, Saskatchewan, Ontario, and Quebec.

8,000 years ago, the world had around 43 billion acres of forests.[1] By 2000, a UN forest assessment showed that we were down to 9.5 billion acres. The World Resources Institute estimates that we are losing natural forest in the tropics at the rate of 40 million acres a year. That's 110,000 acres per day, or 76 acres a minute. If an acre has 200 trees, the loss is an astonishing 2 million trees a day, every day of the year. Europe, India, Bangladesh, and Sri Lanka have already lost their rainforests; the Ivory Coast's forests are nearly gone; the Philippines, Thailand, and Indonesia have lost almost half of their forests in the last 25 years.

Most of this destruction is taking place in the Earth's rainforests, home to two-thirds of the world's plant and animal species and thousands of indigenous tribes who have lived among them for tens of thousands of years. Forests are also a store for billions of tonnes of carbon.

As a result of our activities, we humans are releasing over ten billion tonnes of carbon into the atmosphere a year. 7% of this (700 million tonnes) comes from deforestation, so if we can stop this forest loss, we will remove an important part of the problem.

Scientists estimate that the world's forests and vegetation currently absorb 25%-33% of the excess carbon we are releasing, mostly in the northern mid-latitudes, not the tropics. Planting new forests in northern latitudes will absorb more carbon, but as the world's temperature rises, the forests will become stressed by the heat and begin to release their carbon faster than they absorb it, leading to further warming.[2] By 2050, if temperatures continue to rise, the whole Amazon forest will begin to die as soils dry out and rainfall patterns change, bringing increased fire and insect attack and the release of the stored carbon.

Global Carbon Emissions	Biomass burned (million tonnes dry matter/year)	Carbon released (million tonnes carbon/year)	% of global carbon
Savannahs	3,690	1,660	16%
Agricultural waste	2,020	910	9%
Tropical forests	1,260	570	5%
Firewood	1,430	640	6%
Temperate & boreal forests	280	130	1%
Charcoal	20	30	0.3%
Global biomass sub-total	8,700	3,940	38%
Fossil fuels, cement manufacture, gas flaring	-	6,518	62%
Global total	-	**10,458**	**100%**

Source: GEO-2000 Global Environment Outlook

"There is no one-size-fits-all solution to protecting the world's forests. The world's forests and their circumstances are too diverse. If we search for a single program, a single framework, or a single institution to comprehensively address this problem, we'll be disappointed."
— David Sandalow, U.S. Assistant Secretary of State for Oceans, Environment and Science[5]

Resources *(see also #37, #69, #79)*

Amazon Watch: www.amazonwatch.org
Forest Conservation Archives: www.forests.org
Forest Stewardship Council: www.fscoax.org
The Forest Industry in the 21st Century: www.panda.org/forestandtrade
GEO-2000 - Forest fires and biomass burning: www.grida.no/geo2000/english/0040.htm
Global Forest Watch: www.globalforestwatch.org
Greenpeace International Forests Campaign: www.greenpeace.org/~forests
Pachamama Alliance: www.pachamama.org
Rainforest Action Network: www.ran.org
World Rainforest Movement: www.wrm.org.uy
World Resources Institute—Forests: www.wri.org/forests
Forests for Life: www.panda.org/forests4life

Canada is looking at a 50% increase in the extent of forest lost to fire over the coming 50 years;[3] the boreal forest is already losing carbon, not from its trees but from its soil.[4] The global ecological unravelling has begun.

So what is to be done? Many environmentalists think that trying to negotiate a global forest treaty would take forever, and that the final agreement would probably be very weak. There are other avenues, however, that show promise.

Monitor what's Happening
The World Resources Institute has established Global Forest Watch as a way to put up-to-date digital information on the state of the world's forests into the hands of the public. By 2005, it will span 21 countries and cover 80% of the world's remaining frontier forests, helping NGOs to shine a spotlight on abuses, develop campaigns, and negotiate forest-saving initiatives. This data is critical for any future success.

Swap Debt for Forests
The 1998 U.S. Tropical Forest Conservation Act allows developing countries to reduce their debts to the U.S. government in exchange for setting up a trust fund to pay for the protection of tropical rain forests. In 2000, the U.S. government allocated a very modest $13 million to the fund, and Bangladesh became the first beneficiary with a write-off that will free up $8.5 million for forest protection over 18 years. All northern countries should pass similar legislation to speed the process of debt forgiveness and forest conservation.

Require Eco-Certification for All Timber Products
The best long-term solution for forests that are not being set aside for permanent protection is eco-certification by the Forest Stewardship Council (see #37). In Malaysia, whose forest companies have developed a terrible reputation for clearcutting in the tropics, the World Wide Fund for Nature (WWF) is negotiating to set up a national working group to start the process of certifying selected Malaysian forests. The WWF is hoping to establish similar groups in Indonesia, Papua New Guinea, Cambodia, and Vietnam. In Canada, federal or provincial legislation requiring forest eco-certification would end the clearcutting of forests in British Columbia, Alberta, and Ontario. (The Ontario government justifies 10,000-ha clearcuts on the spurious basis that forest fires sometimes reach that size.) Governments could announce that all imported timber and wood products will require eco-certification by a certain date.

Adopt Environmental Investment Standards
The process of deforestation is often underwritten by government-run export credit agencies and investment guarantee programs. In the U.S., the Overseas Private Investment Corporation and the Export-Import Bank have adopted policies that prohibit logging, mining, and infrastructure projects in primary tropical forests, but few governments have followed suit. There is an urgent need for multilateral environmental standards, especially among the European Union, the G-8 countries, the World Bank, and the IMF, which have been pressuring countries to sell their forests to repay their debts.[6]

Limit the Consumption of Timber-Products
Global deforestation is being fueled by our enormous appetite for wood and paper products (see #10, #26, #37, #69, #79). One way to reduce the consumption would be to eliminate the subsidies that timber companies receive, and apply ecological taxes. The income could be used to retrain loggers, build value-added industries, and support forest conservation. It's just good natural capitalism.

98 Build a Global Movement

Joachim Sefzick
www.save-our-world.org

It all sounds so easy when it's put together in a book like this. There are so many good initiatives underway that it's possible to believe there's a huge global movement happening to replace fossil fuels with clean, renewable energy. There *is* a movement happening - but it's not huge. Not yet.

The technologies exist to make the transition into the solar-hydrogen age. The policies that can accelerate them have been pioneered in places as far afield as Denmark, India, Sacramento, and Freiburg. The piece that is missing is a worldwide movement in every village and every city, from London to Los Angeles and from China to Chile, that will create the political will to make the transition happen.

There is strenuous opposition to these changes by many coal, oil, gas, and auto corporations, by electricity utilities, and by ordinary people who feel they are being targeted by environmental groups and resent having to change their lifestyles. One way to counter their fears is to pool the resources of NGOs, schools, universities, churches, and scientific establishments to build a huge global movement, so that people gain a better understanding of the dangers and become more comfortable with the solutions.

Global climate change is such a complex issue that many people become befuddled and discouraged. Many ask: What could be the significance of my small actions, in the face of something so massive? This perception needs to be turned on its head. To paraphrase Margaret Mead, it is only by small, everyday actions that we can change the world and turn things around. The world is in our backyard, and every solution we implement has significance. A rising tide of individuals and institutions making commitments to reduce their emissions can turn into an unstoppable force, leading to national and global commitments.

In preparation for the 2000 Climate Summit in The Hague, the world's environmental organizations joined forces as "Climate Voice." As a result of their efforts, 11 million people wrote, faxed, or emailed their leaders demanding action to reduce greenhouse gas emissions and embrace the solutions. Four million messages were sent from Holland; one million from the South Pacific.

It was clearly not enough. The American, Canadian, Japanese, Australian, and New Zealand governments listened to the voices of industry, not to the voices of concerned citizens. By insisting on the use of loopholes such as forest sinks, instead of making a commitment to reductions, they prevented the rest of the world from reaching an agreement. Eleven million voices were not enough to counteract the influence of the fossil fuel producers.

Global climate change is going to affect all of us. As well as NGOs, we need the world's schools, colleges, churches, scientists, economists, writers, musicians, astronauts, bird-watchers, Nobel Prize winners, women's organizations, health care organizations, doctors, nurses, aid agencies, parents, grandparents, farmers, philosophers, labor unions, progressive business organizations, lawyers, social investors, parliamentary groups, and senior citizens organizations to join the NGOs in stressing how urgent this is, and call for solutions. (See also #90).

Climate Voice

Climate Voice established a new level of co-operation between organizations around the world that wanted to see action taken against global warming at the Climate Summit in The Hague in November 2000. Using the Internet, 11 million people sent messages to their leaders, sent electronic postcards to their friends, and became Climate Voice Ambassadors, spreading the word through their schools and workplaces. The initiative was coordinated by the World Wide Fund for Nature, Greenpeace, and Friends of the Earth. Even with all this effort, however, it still wasn't enough. www.climatevoice.org

"Never doubt that a small group of thoughtful, concerned citizens can change the world. Indeed, it is the only thing that ever has."
— Margaret Mead

Climate Voice Members

Alliance of Religions and Conservation: Religious and environmental groups working together to care for the environment. www.religionsandconservation.org

Center for International Environmental Law: Public interest environmental law firm works to strengthen international and comparative environmental law. www.ciel.org

Clean Air Network (USA): Local, state, and national environmental, health, and religious organizations working for clean air. www.cleanair.net

Climate Action Network: Global network of 289 NGOs that share information and coordinate strategies. www.climatenetwork.org

Climate Action Network (Australia): groups working together on climate change. www.climateaustralia.org

Climate Alliance of European Cities: 900 cities work with indigenous peoples in the Amazon to reduce greenhouse gas emissions and preserve rainforests. www.klimabuendnis.org

Climate Network (Europe): 76 environmental groups in 19 countries working to limit climate change to ecologically sustainable levels. www.climnet.org

Climate Solutions (USA): Works to make the Pacific Northwest a world leader in practical and profitable solutions. www.climatesolutions.org

David Suzuki Foundation (Canada): Science-based and solutions-oriented, protecting the health of people, other species, and the atmosphere. www.davidsuzuki.org

Earth Day Network: 5,000 organizations, 184 countries. www.earthday.net

Earthsystems.org: Environmental education and information. www.earthsystems.org

Ecologistas en Acción: 300 Spanish environmental groups. www.ecologistasenaccion.org

Euronatura: Non-profit research on environmental law and policy, based in Lisbon. www.euronatura.pt

France Nature Environnement: A network of 3000 French environmental NGOs. fne.energie@magic.fr

Friends of the Earth International: Active local groups in 61 countries. www.foei.org/campaigns/ClimateChange/indexcc.html

Forum Umwelt & Entwicklung: German umbrella organization that lobbies for the implementation of decisions passed at Rio in 1992. www.gmh.uni-mannheim.de/forum

Germanwatch: Directs German policy towards sustainability for the South as well as the North. www.germanwatch.org

Greenpeace: Campaigns globally for governments to address the climate change crisis. www.greenpeace.org/~climate

Grupo de Acção e Intervenção Ambiental (Portugal): "We bet that we can save 8% CO_2 in 8 months instead of 8 years." http://students.fct.unl.pt/gaia

HELIO International: A network of global energy researchers promoting a better contribution by the energy sector. www.helio-international.org

National Environmental Trust (USA): Public education, scientific studies, and public opinion research. www.environet.org

Natural Resources Defense Council (USA): 400,000 members. Protects the planet's wildlife and wild places. www.nrdc.org

Ozone Action (USA): Grassroots, legal, direct action, and corporate campaigns on global warming to weaken corporate opposition to common-sense solutions. www.ozone.org

Physicians for Social Responsibility (USA): 18,000+ physicians and health care providers works nationally and overseas to promote global environmental health. www.psr.org

QUERCUS (Portugal): Portugal's largest environmental citizens' organization. http://www.quercus.pt

Save our World (UK): Protects the natural world through awareness, inspiring changes of attitude, habit, and lifestyle. www.save-our-world.org

WISE (Holland): Grassroots organization works on nuclear energy issues to counter the nuclear industry promoting itself as the solution to climate change. www.antenna.nl/wise

Worldwide Fund for Nature: 4.7 million supporters in 96 countries. Works to ensure that the industrialized nations reduce their CO_2 emissions. www.panda.org/climate

99

Form a Global Ecological Alliance

When nations faced crises of this magnitude in the past, they formed alliances to work together against the common threat. In the past, the threat was usually a military one, requiring a shared commitment to stop the invader. Today, it is a threat to the ecological balance of the entire planet. The threat has a name: unsustainable development. Overpopulation, pollution, water shortages, habitat loss these are all aspects of unsustainable development. Global climate change is its biggest manifestation, bringing the potential for catastrophic disaster to cities, farmlands, forests, fisheries, glaciers, rivers, coral reefs, and species around the world.

In a sane world, where the wisdom of political and business leaders was not dulled by self-importance and too narrow an understanding of self-interest, the world's nations would quickly agree on a common approach to the threat and put measures in place to remove it. Some political commentators say that the age of national governments is past and the future belongs to global corporations that transcend national boundaries. This is part of the rhetoric, however, designed to weaken the will of governments to control the corporations.

An additional problem is that a few of the big boys are refusing to join the world team. The USA, and a few other countries are unwilling to join the global treaty-making, and seem determined to support the needs of the fossil fuel corporations over the needs of the planet. What can the other nations do when these nations refuse to cooperate? They can form their own alliance and press ahead without them. In 1991, Homero Aridjis, the founder of Mexico's leading environmental movement, and Gabriel Garcia Marquez, a Nobel Prize laureate in literature, proposed the creation of a Latin American Ecological Alliance at the first Ibero-American Summit, but the proposal fell on deaf ears.[1]

The glue that binds such an alliance together would be a shared commitment to reduce greenhouse gas emissions, both generically and through a series of compacts to introduce solar energy and zero-emissions vehicles (see #92), and to encourage the spread of organic farms and ecologically certified forests (see box). The members of such a Global Ecological Alliance (GEA) might agree to work together to:

- conduct shared climate and ocean sciences research;
- promote a global tax on currency speculation;
- promote a global standard for energy efficiency;
- require all corporations to return annual social and environmental impact reports;
- promote a World Environment Organization, with teeth;
- work together on environmental issues at the United Nations;
- share information on the best policies for renewables and energy efficiency;
- introduce airport landing taxes and promote a global aviation tax;
- end fossil fuel subsidies and introduce ecological taxes;
- foster a huge increase in environmental youth exchanges between members; and
- build a sense of hope, as citizens, businesses and governments work together for a peaceful, sustainable world.

Arctic Resources

Arctic Circle: http://arcticcircle.uconn.edu

Arctic Climate Impact Assessment: www.acia.uaf.edu

Arctic Council: www.arctic-council.org

Arctic Environmental Atlas: www.maps.grida.no/arctic

Inuit Circumpolar Conference: www.inuitcircumpolar.com

Inuit Observations on Climate Change: www.iisd.org/casl/projects/inuitobs.htm

Northern Climate ExChange: www.taiga.net/nce

"What we human beings are all living now, whether we are volunteers or not, is an extraordinary but exceptionally dangerous adventure. We have a very small number of years left to fail or to succeed in providing a sustainable future to our species."
— Jacques Cousteau

Possible GEA Policy Commitments	2010	2020	2030
• Greenhouse gas reduction (from 1990 level) (developed nations)	10%	40%	85%
• (developing nations)	1%	15%	30%
• Timber to be FSC eco-certified	10%	50%	100%
• Trees planted per citizen per year	2	4	5
• Solar PV energy installed per citizen	0.1kW	0.5kW	1kW
• Solar thermal energy on new buildings	25%	100%	100%
• Solar thermal energy on existing buildings	2%	30%	60%
• New electricity to come from renewables	100%	100%	100%
• Fossil-fuel electricity shifted to renewables	10%	40%	100%
• Farmland managed organically	10%	40%	100%
• Landfills capturing methane	80%	100%	100%
• New buildings that are super-efficient	100%	100%	100%
• Existing buildings upgraded to super-efficient	10%	40%	80%
• Zero-emissions new vehicles	10%	65%	100%
• Minimum fuel efficiency for new cars (mpg)	40	75	120
• Applications not using CFCs, HCFCs, HFCs	30%	100%	100%

Who Might Join Such an Alliance?

Europe can be counted on to provide a strong contingent, with membership from most countries. African members might include Morocco, Ethiopia, Kenya, Uganda, Tanzania, Mozambique, Madagascar, Zambia, Botswana, South Africa, Namibia, Senegal, and Gambia. Jordan might join, along with India, Nepal, Bhutan, and Bangladesh. So might Thailand, East Timor, the Philippines, the Alliance of Small Island States, and even China (Tibet would love to join, too). From Latin and Central America, we might see Cuba, Brazil, Honduras, Nicaragua, Costa Rica, El Salvador and Guatemala. If states and provinces could have affiliate status, we might see California, Oregon, Washington, Minnesota, Wisconsin, Vermont, Rhode Island, Massachusetts, New Jersey, New York, Florida, Hawaii, the Yukon, Nunavut, the Northwest Territories, and Quebec.

Form an Alliance of Arctic Jurisdictions

An alliance of countries and jurisdictions is urgently needed in the Arctic, where rising temperatures are causing the summer sea-ice and the tundra to melt, threatening wildlife and a whole way of life. The Arctic is the canary in our global coalmine, and the Arctic jurisdictions need to ring the alarm bells all over the world. Members would include Greenland, Iceland, Norway, Sweden, Finland, northern Russia, Alaska, the Yukon, the Northwest Territories, and Nunavut.

Support the Alliance of Small Island States

The Maldives, Philippines, Tuvalu, Tonga, Marshall Islands, and 35 other small island nations have formed the Alliance of Small Island States (AOSIS) to help them to prepare for the impact of rising sea levels and make their voices heard on the world stage. The members do valuable work, sending a clear alarm signal to the rest of the world. If the small island states and the Arctic jurisdictions (the tropical and the polar) could join hands and work together, maybe the rest of the world would begin get the message.
AOSIS: www.sidsnet.org/aosis

100

Declare a Century of Ecological Restoration

In 1950, the historian AJP Taylor spoke for many when he wrote: *In the first half of the century, Western man has achieved every ambition that he set before himself since the Renaissance. He has conquered space, disease, poverty. The scientific method he has perfected guarantees that he can do anything that he wishes.*[1]

Fifty years later, the United Nations Global Environment Outlook 2000 contained a stark warning about looming water shortages, global warming, and worldwide nitrogen pollution. Klaus Topfer, executive director of the UN environment rogram said, *A series of looming crises and ultimate catastrophe can only be averted by a massive increase in political will. We have the technology, but we are not applying it.* Rudd Lubbers, former Dutch Prime Minister and President of the World Wide Fund for Nature added, *You just can't simply go on with the technologies and the way of life we have today. Nature and the Earth are going to retaliate against humankind.*[2]

The promise of 50 years ago has vanished. *In the past half century, the world has lost a fourth of its topsoil and a quarter of its forest cover. At present rates of destruction, we will lose 70% of the world's coral reefs in our lifetime. In the past three decades, one third of the planet's resources, its "natural wealth," has been consumed. We are losing freshwater ecosystems at the rate of 6% a year, marine ecosystems by 4% a year...The decline in every living system in the world is reaching such evels that an increasing number of them are starting to lose their assured ability to sustain the continuity of the life process. We have reached an extraordinary threshold.*[3]

Jane Lubchenco has captured the essence of this unique moment in human evolutionary history: *During the last few decades, humans have emerged as a new force of nature. We are modifying physical, chemical, and biological systems in new ways, at faster rates, and over larger spatial scales than ever recorded on Earth. Humans have unwittingly embarked upon a grand experiment with our planet. The outcome of this experiment is unknown, but has profound implications for all of life on Earth.*[4]

We have applied the scientific method and created an economic miracle — but only for a minority of the world's population, and at a huge cost to our planet's ecosystems, with much of the cost still to be revealed. For the first time in centuries, perhaps since the Renaissance, today's younger generation believes that the future will be worse than the present.

In March 2000, after input from 200 NGOs and thousands of people all over the world, The Earth Charter was released. It begins with the words: *We stand at a critical moment in Earth's history, a time when humanity must choose its future...The protection of Earth's vitality, diversity, and beauty is a sacred trust.*

The vision that drove the past 500 years is over. If we do not want to surrender our spirits to pessimism, we must create a new vision.

> **"The future belongs to those who believe in the beauty of their dreams."**
> **— Eleanor Roosevelt**

Resources

Restore the Earth: www.restore-earth.org

The Earth Charter: www.earthcharter.org

The Earth Charter, USA Campaign: www.earthcharterusa.org

United Nations Millennium Assembly: www.un.org/millennium

World Scientists' Warning to Humanity: www.ucsusa.org/about/warning.html

"There can be no purpose more enspiriting than to begin the age of restoration, reweaving the wondrous diversity of life that still surrounds us."
— Edward O. Wilson, author of *The Diversity of Life*

What are our dreams? In September 2000, the leaders of 189 nations met in New York for the United Nations Millennium Summit, where they agreed "to free all of humanity, especially children and grandchildren, from the threat of living on a degraded planet whose resources would no longer be sufficient to meet their needs."

Restore the Earth: A Vision of Hope for the 21st Century

From Scotland, Alan Watson Featherstone is spreading the idea that the United Nations should declare the 21st century a "Century of Ecological Restoration." He is proposing that an Earth Restoration Service be created, enrolling volunteers from all over the world to restore the mangroves to Vietnam's Mekong Delta, the cedar forests to Lebanon, the Caledonian pine forest to Scotland; to clean polluted beaches; to remove unnecessary roads from wild places; to take down the dams that block wild rivers; to clean toxic waste sites; to restore the forest to the African Sahel; to heal the holes in the ozone layer; to restore the atmosphere to its previous balance; to fill the world's cities with trees and gardens; to restore the Black Sea and the Aral Sea to their former health; to return the tiger to India and the wild buffalo to the American prairies; to return the beaver to Scotland, the wolf to Japan, the Arabian oryx to the wild Oman.

Establish a new Global Village Parliament

Our world is one village. We breathe the same air, we drink from the same well. We have to cope with each other's wastes. How can we create an effective Global Village Council where we can address the issues that matter to us, and act on them? The United Nations should be our village council, but it has been pushed to the margins by the larger powers and excluded from the important decisions on trading and finance that have been taken over by the global corporations, their bankers, and the World Trade Organization. When confronted with the evidence of climate change, they offer a few platitudes, then continue to buy and sell fossil fuels. Meanwhile, the carbon level keeps rising.

There is nothing wrong with global trade - but it must be fair trade, not free trade untrammeled by social responsibility. There is nothing wrong with science - but its miracles must be used with caution. The lands of which we are ignorant are vastly larger than the lands of which we have knowledge. We need a new system of governance for the global village, a new definition of "progress", a new set of rules, and a new vision to fill our hearts with purpose.

The Earth Charter

We stand at a critical moment in Earth's history, a time when humanity must choose its future. As the world becomes increasingly interdependent and fragile, the future at once holds great peril and great promise. To move forward we must recognize that in the midst of a magnificent diversity of cultures and life forms we are one human family and one Earth community with a common destiny. We must join together to bring forth a sustainable global society founded on respect for nature, universal human rights, economic justice, and a culture of peace. Towards this end, it is imperative that we, the peoples of Earth, declare our responsibility to one another, to the greater community of life, and to future generations...Let ours be a time remembered for the awakening of a new reverence for life, the firm resolve to achieve sustainability, the quickening of the struggle for justice and peace, and the joyful celebration of life. (Excerpt)

101
Now, Think!

By whatever standard you choose, we are living through a most remarkable period in humanity's evolution on this planet. Within this last tiny handful of years, we have waged the most destructive and painful war the world has ever seen; we have rid the world of most dictatorships and tyrannies; we have removed many of the barriers that prevent women from participating in the affairs of the world; we have launched ourselves into space and looked back on our planet for the first time; we have doubled our population; we have increased our use of the Earth's resources tenfold; we have poured a vast burden of pollution into Earth's oceans, soils, and atmosphere; we have eliminated more species than at any other period of evolution; we have moved beyond nationalism and tribalism to create a truly global society; we have created a global communications system and a global brain. We take these things for granted — but all this has happened since 1940, after 100,000 years of modern human existence. Something very remarkable is happening.

Some argue this has no meaning, since evolution is just about genes replicating themselves.[1] Some argue that technology will solve our outstanding problems. Others argue that human evolution has taken a terrible turn, and we should never have taken up the tools of agriculture and imposed our will on the rest of creation.[2] Pessimism has become very fashionable of late.

Our perspective is different. We see humans as having great capacity for creativity, and for compassion, imagination, and action, as well as for selfishness and ignorance. Once we were tribes; then we were nations. Now we are learning to love the Earth as a whole. Who knows what surprises evolution may yet unfold? Can we stop burning fossil fuels and cease destroying our soils and forests? Of course we can.

We have stumbled into a crisis that invites us to act. Global climate change comes as the greatest of teachers. As perhaps nothing else, it underscores the four laws of ecology elaborated by Barry Commoner:

1. Everything is connected to everything else.
2. Everything must go somewhere.
3. Nature knows best.
4. There is no such thing as a free lunch.

If we can learn the lessons that climate change is placing before us, then for all the suffering we will experience as a result of undertaking this uncontrolled experiment with our planet's atmosphere, we will have gained greatly. A profound leap in human understanding, insight, and compassion awaits us if we open ourselves to the fullness of our situation and commence the transformational response for which it calls.

Within every human, a hero waits to appear. If you are a dancer, dance your concern. If you are a teacher, teach it. If you are a householder, demonstrate it. If you are an employee, share it. We have done our best to lay the solutions before you. At this point, we stand aside: the stage is yours.

Resources

Center for a New American Dream: www.newdream.org

Institute for Deep Ecology: www.deep-ecology.org

New Dimensions Broadcasting Network: www.newdimensions.org

Turning Point Project: www.turnpoint.org

YES! Magazine, A Journal of Positive Futures: www.yesmagazine.org

Carolyn Herriot

Glance at the sun. See the moon and the stars. Gaze at the beauty of the Earth's greenings. Now, think.

— Hildegard von Bingen (1098-1179)

Driven by the forces of love, the fragments of the world seek each other, so that the world may come into being. Love alone is capable of uniting living beings in such a way as to complete and fulfil them, for it alone takes them and joins them by what is deepest in themselves.

— P. Teilhard de Chardin (1881-1955)

Our task must be to widen our circle of compassion to embrace all living creatures, and the whole of nature in its beauty.

— Albert Einstein (1879-1955)

If the world is to be healed through human efforts, I am convinced it will be by ordinary people, people whose love for this life is even greater than their fear. People who can open to the web of life that called us into being.

— Joanna Macy

We are at this moment participating in one of the very greatest leaps of the human spirit.

— Joseph Campbell (1904-1987)

How I lead my life speaks a prayer for the world I want to create.

— Shannon Service

Listen carefully this, you can hear me. I'm telling you because earth just like mother and father or brother of you. That tree same thing...Tree working when you sleeping and dream.

— Bill Neidjie, Kakadu Australian Aboriginal

Wei Chi: Before Completion[3]

The conditions are difficult. The task is great and full of responsibility. It is nothing less than that of leading the world out of confusion back to order. But it is a task that promises success because there is a goal that can unite the forces now tending in different directions. At first, however, one must move warily, like an old fox walking over ice.

— I Ching, *Hexagram 64*

Notes

Introduction

Time to Get Started

1 Thinning of the Arctic Sea-Ice Cover, by D.A. Rothrock, Y. Yu, and G.A. Maykut, University of Washington, Seattle, Washington, Geophysical Research Letters, Vol. 26, No. 23, Pages 1-5, December 1, 1999.

Our Story

1 Based on *The Cosmic Walk*, developed by Sister Miriam Therese McGillis, based on the works of Thomas Berry and Brian Swimme, and adapted by Ruth Rosenhek.

2 There is no consensus on a date for the first deliberate use of fire. See www.beyondveg.com/ nicholson-w/hb/hb-interview2c.shtml

3 The Eemian period may have begun anywhere between 132,000 and 140,000 years ago. See *Sudden climate transitions during the Quaternary*, by Jonathan Adams et al, www.esd.ornl.gov/projects/qen /transit.html

4 The Lunar Orbiter 1 carried an Eastman Kodak camera which took the first ever photo of the Earth from a spacecraft in the vicinity of the Moon on August 23rd 1966, showing a cloud-covered, crescent Earth above a lunar landscape. For the actual photo, see http://cass.jsc.nasa.gov /expmoon/orbiter/orbiter-images.html

Planetary Weaning

1 *The Medieval Machine: The Industrial Revolution of the Middle Ages*, by Jean Gimpel. (Pimlico, 1992)

The Greenhouse Effect

1 'Atmospheric carbon dioxide concentrations over the past 60 million years' by Paul N. Pearson & Martin R. Palmer, *Nature*, Aug 17th 2000.

2 'Oceanography: Stirring times in the Southern Ocean' by Sallie Chisholm (MIT), *Nature*, Oct 12th 2000

3 *Effects on carbon storage of conversion of old-growth forests to younger forests*, by M.E. Harmon, W.K. Ferrell and J.F. Franklin, 1990. *Science* 247: 699-702.

4 *A Note on Summer CO_2 Flux, Soil Organic Matter, and Microbial Biomass from Different High Arctic Ecosystem Types in Northwestern Greenland*, by M. H. Jones, J. T. Fahnestock, P. D. Stahl, and J. M. Welker. *Arctic, Antarctic, and Alpine Research*, Feb 2000. www.colorado.edu/INSTAAR/arcticalpine

5 *Acceleration of global warming due to carbon-cycle feedbacks in a coupled climate model*, by Peter M. Cox et al. *Nature*, Nov 9, 2000.

The Greenhouse Gases

1 In 2000, it was listed on Roschem Pacific Group's chemical inventory (www.roschem.com/pravail.html) in Australia, but it has since been removed.

2 *Global Warming in the 21st Century: an Alternative Scenario*, by James Hansen et al (NASA Goddard Institute for Space Studies), 2000. (www.giss.nasa.gov/gpol /abstracts/2000.HansenSatoR.html)

Greenhouse Gases Chart

1 The data on concentrations and annual growth rates for the long-lived industrial gases comes from James Elkins, at the Halocarbons and other Atmospheric Trace Species Group, Oceanic and Atmospheric Research Climate Monitoring and Diagnostics Laboratory, NOAA, Boulder CO.

2 The different gases do not suddenly disappear at the end of their lifetime, so these figures are generally approximate.

3 GWP is an index defined as the cumulative radiative forcing between the present and some chosen time horizon (typically 100 years) caused by a unit mass of gas emitted now, expressed relative to a reference gas such as CO_2, including any indirect effects of the emitted gases.

4 The data for radiative forcing comes from the IPCC's Third Assessment Report of Working Group 1 (January 2001). The figures represent increased forcing since 1750.

5 The percentages are based on the radiative forcings of the different gases, excluding the aerosols.

6 It is likely that total atmospheric water vapour has increased by several percent per decade over many regions of the Northern Hemisphere (IPCC).

7 The increased water vapor is caused by increased ocean precipitation, which is caused indirectly by human activities. There is uncertainty about the extent to which it causes increased radiative forcing due to an increase in high level cirrus cloud cover (which traps heat), or decreased forcing due to increased albedo on lower stratus clouds (which reflect heat back into space).

8 Water vapour feedback approximately doubles the warming for fixed water vapour (IPCC).

9 Overall global methane sources and methane sinks are not well understood. Methane emissions are continuing, but the accumulation of methane in the atmosphere has slowed since 1990 for reasons that are not clear. Methane's radiative forcing has been falling since 1987.

10 Identified in the January 2001 IPCC Report as an additional cause of methane increase.

11 James Hansen et al (see above) suggests that the radiative forcing of the CFC replacements (HCFCs and HFCs) will be similar to that of the CFCs they replace. If HFC-134a were phased out, and the stockpiles of CFC-21 were destroyed, the overall forcing might fall to 0.25W/m2. Range of uncertainty for radiative forcing: +/-0.5W/m2

12 Yoshio Makide and a team of researchers at Tokyo University reported in 1998 that HFCs are accumulating rapidly as they are used to replace CFCs, though they are still at a very low level. In 1995 they were 2 ppt (parts per trillion); by 1997 they had reached 6 ppt. The figure of 12 for the year 2000 is based on an increase of 2ppt per year.

13 We were not able to locate data for the growth rate of tropospheric ozone.

14 The principal emissions that cause the accumulation of tropospheric ozone (known as the 'precursors') are volatile organic compounds and nitrogen oxides (NOx), which result from transportation, power plants and industrial processes. It is also caused by the break-up of methane in the atmosphere.

15 Research on the different aerosols is difficult, partly because they are so short-lived.

16 The direct radiative forcing of the different aerosols is estimated by the IPCC to be -0.4 Wm2 for sulfate, -0.2 Wm2 for biomass burning aerosols, -0.1 Wm2 for fossil fuel organic carbon and +0.2 Wm2 for fossil fuel black carbon. The radiative forcing of indirect aerosol effects (eg on clouds) is very uncertain, and ranges from 0 to - 2.0.

The Enhanced Greenhouse Effect

1 "Climate and atmospheric history of the past 420,000 years from the Vostok ice

core, Antarctica." by JR Petit et all. *Nature*, June 3, 1999.

2 "Atmospheric carbon dioxide concentrations over the past 60 million years," by Paul Pearson and Martin Palmer. *Nature*, August 17, 2000.

3 "How Warm Was the Medieval Warm Period?" by Thomas Magnuson et al, *Ambio*, Feb 1, 2000; also "Historical Trends in Lake and River Ice Cover in the Northern Hemisphere," by John J. Magnuson et al, *Science*, Sept 8, 2000.

4 "A doubling of the Sun's coronal magnetic field during the past 100 years," by M. Lockwood et al. *Nature*, Vol 399, p 437, 1999

5 "The record breaking global temperatures of 1997 and 1998: Evidence for an increase in the rate of global warming?" by Thomas Karl et al, *Geophysical Research Letters*, March 1, 2000

6 See "Forests and soils may speed up global warming," Hadley Centre for Climate Research, UK, Nov 8, 1998.

The Climate Impacts

1 "Warming of the World Ocean," by Sydney Levitus et al, *Science*, March 24, 2000.

2 Alan Strong, National Oceanic and Atmospheric Administration, July 2000.

3 "Growing season extended in Europe," by Annette Menzel and Peter Fabian. *Nature* Vol 397 No 6721 (1999).

4 Barnaby Briggs, Birdlife International.

5 World Wide Fund for Nature (Climate): "An Early Spring."

6 For references, see www.climatehotmap.org

7 National Oceanic and Atmospheric Administration news release, Washington Post, April 19, 2000

8 Story by David Roberts, Globe and Mail, April 3, 2000.

9 David Welch, Fisheries Canada, paper in Journal of Fisheries and Aquatic Science, 1999.

10 *Global warming and terrestrial biodiversity decline,* WWF 2000.

11 *Destructive storms drive insurance losses up: will taxpayers have to bail out insurance industry?* by Seth Dunn & Christopher Flavin, Worldwatch Institute, 1999.

12 Munich Re, Dec 29th 2000.

The Runaway Greenhouse Effect

1 *Sudden climate transitions during the Quaternary,* by Jonathan Adams et al. www.esd.ornl.gov/projects/qen/transit.html

2 "Biotic feedbacks: Will Global Warming Feed Upon Itself?" by Bruce Johansen. In *Global Warming Desk Reference,* Greenwood Press, 2000.

3 "Arctic effect could multiply global warming," Reuters/Planet Ark 2 March, 1999. Ohio State University research team.

4 The arctic tundra has been a carbon sink for the last 9000 years, but since 1982 its role has reversed and it has become a source. Walter C. Oechel, director of the Global Change Research Group, San Diego State University. Rachel's Environment & Health Weekly #664, August 19, 1999.

5 Shawn Marshall, University of Calgary, personal correspondence.

6 "The Threat of Rising Seas," by Grover Foley, *The Ecologist,* Vol. 29, No 2, March/April 1999.

7 "Gas Blasts: Methane once frozen under the seafloor may help heat up the climate,"' by Prof. Erwin Suess, Research Center for Marine Geosciences, Kiel, Germany. *Scientific American*, Dec 20, 1999.

8 James Kennett, oceanographer, quoted in *The Guardian,* Nov 11-17, 1999.

9 Lisa Sloan and Gerald Dickens, report to American Geophysical Union, 1999. See Note 1.

10 *Forests and soils may speed up global warming,* Hadley Centre for Climate Research, UK, Nov 8th 1998.

11 Quote by Wallace Broecker, whose work has awoken the world to this scenario. See 'Thermohaline circulation, the Achilles heel of our climate system: will man-made CO_2 upset the current balance?' by Wallace Broecker, *Science* 278, 1582-1588, 1997.

12 'Risk of sea-change in the Atlantic' by Stephan Rahmstorf, Potsdam Institute for Climate Impacts Research. *Nature*, 388: August 28th, 1997.

13 Bill Turrell, Scottish Executive's Marine Laboratory, Aberdeen. *New Scientist*, Nov 27th, 1999.

14 Svein Sterhus, University of Bergen. *New Scientist,* Nov 27th, 1999.

15 For a good general description, see *Sudden climate transitions during the Quaternary*, Note 1.

The Counter-Arguments

1 Research by Knud Lassen and Eigil Friis-Christensen, Danish Meteorological Institute. See "Don't blame the Sun," *New Scientist,* May 9, 2000.

2 IPCC Third Assessment Report of Working Group One, January 2001.

3 "Sun studies may shed light on global warming," *Washington Post,* Oct 9, 2000.

4 "Satellite climate record in error," *Nature Science Update,* August 13, 1998.

5 *Reconciling Observations of Global Temperature Change,* National Academy of Sciences, January 2000.

An 80% Reduction by 2025

1 Quoted by David Malakoff in "Thirty Kyotos Needed to Control Warming," *Science* 278, Dec 19, 1997.

2 For a blow-by-blow description of the lobbying, see *The Carbon War: Global Warming and the End of the Oil Era*, by Jeremy Leggett (Routledge, 2001).

3 57% of the total radiative forcing from fossil fuels comes from 75% of the CO_2 (1.095 Wm2), 21% of the methane (0.096 Wm2), 70% of the N_2O (0.0315 Wm2), and 100% of the ozone (0.35 Wm2), for a total of 56.56% of the total 2.78 Wm2; 8% for farming comes from 25% of the methane (0.12 Wm2) and 70% of the nitrous oxide (0.015 Wm2), for a total of 8.09%.

4 *Global Warming in the 21st Century: an Alternative Scenario,* by James Hansen et al (NASA Goddard Institute for Space Studies), 2000. (www.giss.nasa.gov/gpol/abstracts/2000.HansenSatoR.html)

Sustainable Energy for a Sustainable Planet

1 World Energy Perspectives: www.iiasa.ac.at/cgi-bin/ecs/book_dyn/bookcnt.py

2 Assuming 7.5 billion people, and 3 - 4 people per household. This is a purely hypothetical calculation.

3 The World Energy Assessment's Chapter 7: Renewable Energy Technologies (p237), says that the area needed to supply the world's present electricity consumption (12,000 TWh) is 217,000 square miles.

4 World Energy Assessment, Chapter 5, p 165.

5 World Energy Assessment, Chapter 5, pp. 165-6.

6 Times Atlas of the World, 1983. The UN's FAO has estimated the world's total land availability at 50.5 million square miles.

Building a More Efficient World

1 IPCC Scenario Database compiled by Morita and Lee, 1998: www-cger.nies.go.jp/cger-e/db/ipcc.html

2 *Global Energy Perspectives,* International Institute for Applied Systems Analysis & World Energy Council, 1998. www.iiasa.ac.at/cgi-bin/ecs/book_dyn/bookcnt.py
3 *Factor Four: Doubling Wealth, Halving Resource-Use.* 1997. Ernst Von Weisacker, Amory Lovins, and L. Hunter Lovins, Earthscan, London.
4 *Natural Capitalism.* 1999. Paul Hawken, Amory Lovins, and L. Hunter Lovins. Little, Brown.
5 The Carnoules Declaration was signed by a group of 16 scientists, economists, government officials, and businesspeople who met at the Wuppertal Institute in Germany in 1994, and subsequently formed themselves into the Factor Ten Club, calling for a ten-fold leap in efficiency and resource-use. See *Natural Capitalism,* p. 11.
6 Association to Save Energy, 2000.
7 *State of the World 1998,* Worldwatch, p.121.

Solar Energy

1 In 1990, the Worldwatch Institute said 10% of Arizona. To check this, we did our own calculation using solar radiation data from the Union of Concerned Scientists (www.ucsusa.org/energy/brief.solar.html):
At noon on a cloudless day, sunlight yields 1,000 watts (1 kW) per square meter.
Arizona gets 6 kWh a day; Boston 3.6 kWh; Seattle in December 0.7 kWh. (For a solar radiation map of the US, see www.homepower.com/solmap.htm)
At 15% solar efficiency, 1 square meter in Arizona yields 1 kWh per sunny day.
A square meter in Boston yields 0.6 kWh per sunny day.
Average Boston + Arizona = 0.8 kWh/day.
50% sunny = 0.4kWh/day = 146 kWh/year.
1 square solar meter = 146 kWh per year
1 million square meters = 1 square kilometer.
1 square solar kilometer =146 million kWh per year.
US electricity demand in 1997: 3,570 terawatt hours.
Land needed for 3,570 terawatt hours = 24,452 sq. km (9440 square miles).
USA is 9,363,132 sq. km. Arizona is 293,985 sq. km (113,508 square miles).
US electricity demand could be met 382 times from total US;12 times from Arizona.
1 terawatt hour needs 2.64 square miles. With 50% efficiency, divide the land needed by 2.
2 The USA has 103 million dwellings. If each dwelling uses 50 sq metres for PV coverage, this produces 5150 million square meters = 5150 square kilometers of PV. In Holland, non-residential roof area is estimated at 30% of residential roof area. If the same proportion applies in the USA, this increases the area to 6,695 square kilometers, which is 27% of 24,452 square kilometers. Result: 27% of US electricity needs could come from solar roofs. If efficiency reduced demand by 50%, this would increase to 54%.
3 In Sacramento, SMUD has a long-term purchase agreement that reduces the price to $2 per watt.
4 The Siemens Solar Energy Payback Study is at www.siemenssolar.com. From PV Power Resource www.PVpower.com: "According to the article by J. Nijs (et al) in their paper *Energy Payback Time of Crystalline Silicon Solar Modules (Advances in Solar Energy),* ed. K. Boer, ASES, Boulder, CO USA; Vol. 11, 1997. pp. 291-328), conservative calculations for the payback time of crystalline silicon PV modules varies from 2.58 years for multicrystalline silicon and 2.66 years for single-crystal silicon in the sunbelt, to 4.92 years and 5.07 years, respectively, for these same materials in less sunny areas. Projections for additional manufacturing improvements indicate improvements to 1.4 years (sunbelt) and 2.67 years (less sunny areas) in the mid-term future, and 1.22 years and 2.33 years, respectively, for longer-term improvements. Estimates for amorphous silicon are just more than a year for making up their energy cost."
5 1 quad = 293.1 terawatt hours. The conversion of capacity to energy differs for each energy source by the number of hours per day electricity can be produced. Some energy sources (e.g., biomass) also produce heat or liquid fuel, which cannot be measured in TW. The data here assumes that all energy is converted to electricity.
6 "Solar Energy: From Perennial Promise to Competitive Alternative." (KPMG, 1998). See www.greenpeace.org/~climate/renewables/reports/kpmg8.pdf
7 Strategies Unlimited, quoted in *How the Northwest Can Lead a Clean Energy Revolution,* p.5, by Patrick Mazza. Climate Solutions, WA, 1998.
8 Avi Brenmiller, CEO of Solel Solar Systems, Israel, has suggested that a 5-6 cents/kWh tax may be needed.
9 *Gaviotas — A Village to Reinvent the World,* by Alan Weisman, p.129 (Chelsea Green).

Wind Energy

1 With thanks to Ramakrishna, who wrote: *The winds of grace blow all the time. All we need do is set our sails.*
2 *Global Wind Energy Market Report, 2000* www.awea.org/faq/global2000.html
3 In the US classification system, Class 3 = an average wind power density exceeding 250-300 watts per sq meter 50 meters above the ground.
4 World Energy Assessment 2000, *Energy and the Challenge of Sustainability,* Chapter 5: Energy Resources, by Hans-Holger Rogner, et al.
5 "Wind Energy: Resources, Systems, and Regional Strategies" in *Renewable Energy: Sources for Fuels and Electricity.* Island Press, 1993.
6 *Wind Energy Potential in the United States,* by D.L. Elliott and M.N. Schwartz, Sept 1993. Pacific Northwest Laboratory. www.nrel.gov/wind/potential.html
7 US DOE study, quoted in *There's Something in the Wind,* by Lester Brown, Worldwatch Institute, September 2000.
8 See "Harvesting Clean Energy for Rural Development. Part One: Wind" by Patrick Mazza. Climate Solutions, 2001, and American Wind Energy Association.
9 Source: *"Wind Energy Potential in the United States"* by D.L. Elliott and M.N. Schwartz, Sept 1993, updated with 1998 electricity data from www.eia.doe.gov/emeu/iea/table63.html

Geothermal, and Hydroelectric Energy

1 *World Energy Assessment 2000,* Ch 5, "Energy and the Challenge of Sustainability," p. 165.
2 *Global warming impacts of ground-source heat pumps compared to other heating and cooling systems.* Caneta Research, NRCan Renewable and Electrical Energy Division, April 1999.
3 *China, Land of Discovery and Invention,* by Robert Temple, p.54 (Patrick Stevens, 1986).
4 *World Energy Assessment 2000,* Chapter 5, p 154.

Biomass Energy

1 Union of Concerned Scientists estimate.

2 US Department of Energy estimate, quoted by the Union of Concerned Scientists, *How Biomass Energy Works*

3 *Why We Should Support Ethanol*, by David Morris, Institute for Local Self-Reliance, May 1997.

4 "Ottawa may pave the way for ethanol," by Cameron Smith," *Toronto Star*, March 4, 2000.

5 Biomass is hard to measure because it includes firewood consumption all over the world. According to the World Energy Assessment, the world produced 41 quads of biomass energy, or 12,000 TWh, in 1999.

Hydrogen

1 Speech by Fritz Vahrenholt, board member of Deutsche Shell, May 2000. For the scenario study (in German), see www.deutsche-shell.de and click on "Publikationen."

2 This is Plug Power's estimate of the price.

3 *Fuel Cells — a Green Solution*. David Suzuki Foundation and Pembina Institute, 2000.

4 These three final options were not included in the Suzuki Foundation study.

5 *Natural Capitalism*, Paul Hawken, Amory Lovins, L. Hunter Lovins (Little, Brown, 1999), p. 36.

6 BBC News Online, Nov 8, 1999, and *Nanotubes for Electronics*, Scientific American, December 2000.

7 For a full description of the hypercar, see www.hypercar.com

8 Option E gives a 72% reduction. A 3-fold increase in fuel efficiency would increase this to 91%.

9 Natural Capitalism (above), pp 32-39. The US fleet is 200 million vehicles. The US electrical grid's capacity in 1998 was 779 GW. Generating capacity per vehicle: 20-40kW.

Natural Gas and Nuclear Energy

1 The global warming potential of natural gas was updated by the IPCC in January 2001 from 21 to 23 over 100 years, and from 50 to 62 over 20 years.

2 The data used in this chart comes from the US Energy Information Administration (www.eia.doe.gov).
Col. 1: Energy Consumption by Source (www.eia.doe.gov/pub/energy.overview/aer1999/txt/aer0103.txt)
Col. 2: Carbon emissions by sector by source. (www.eia.doe.gov/pub/energy.overview/aer1203.txt)
Col. 3: Col. 2 divided by Col. 1.
Col. 4: Methane emissions over 100 years in million metric tonnes of carbon equivalent. (www.eia.doe.gov/pub/energy.overview/aer1205.txt). Updated using the IPCC's new methane GWP of 23 over 100 years, instead of 21. To convert methane to MMTCE, multiply by 23, then divide by 3.667.
Col. 5: Add the carbon and methane, over 100 years. Col. 2 + Col. 4.
Col. 6: Divide the total (Col. 5) by the energy consumed (Col. 1)
Col. 7: Methane emissions over 20 years, using the IPCC's new methane GWP of 62, instead of 56.
Col. 8: Add the carbon and methane, over 20 years. Col. 2 + Col. 7.
Col. 9: Divide the total (Col. 8) by the energy consumed (Col. 1).

3 US Department of Energy forecast

4 Estimate by Ray Niles, natural gas analyst with Salomon Smith Barney, quoted in *Soaring natgas prices revive interest in coal.* Reuters 8 Dec 2000.

Carbon Sequestration

1 "Effects on carbon storage of conversion of old-growth forests to younger forests," by M.E. Harmon, W.K. Ferrell and J.F.Franklin. *Science* 247:699-702, 1990.

2 "Climate Change: Managing Forests After Kyoto," by Ernst-Detlef Schulze, Christian Wirth, Martin Heimann. *Science*, 22 September 2000.

3 *Forests and soils may speed up global warming*, Hadley Centre for Climate Research, UK, 8 Nov 1998.

4 Here's how the calculation goes:
(1) US carbon emissions from fossil fuels, 1998: 1495 million metric tonnes = 1,647 million tons.
(2) American Forests data: 1 acre of forest will absorb 1.25 tons of carbon per year.
(3) 1,647 million tons divided by 1.25 tons per acre = 1,318 million acres = 2.059 million square miles.
(4) US land area: 3.53 million sq. miles.
(5) Carbon divided by land = 58% of the land to be planted with forest every year. Canada produces 136 million tonnes of carbon from fossil fuels (150 million tons). This would need 120 million acres of trees to be planted every year, covering 187,000 square miles, which is 5% of Canada's total land area (3.56 million square miles).

5 "Soil Ecology, Agriculture and the Greenhouse Effect," by Ranil Senanayake. *Australian Journal of Soil and Water Conservation* Vol. 6 No 1, 1993.

6 "Legume-based cropping systems have reduced carbon and nitrogen losses," by Laurie Drinkwater, Peggy Wagoner and Marianne Sarrantonio, *Nature*, November 18, 1998. pp. 262-265.

7 Based on 1994 emissions data.

Wild Cards

1 "On the Fringe," by Erik Baard and Frank Byrt, *Wall Street Journal*, 13 Sept 1999.

2 *Fossil Fuel Revolution Begins*, BBC News, 23 November 1999.

3 "Little gem - it will cut your car's emissions and make diamonds as you go," *New Scientist*, 7 October 2000.

4 *Policy Implications of Greenhouse Warming: Mitigation, Adaptation, and the Science Base* by the Panel on Policy Implications of Greenhouse Warming, National Academy of Sciences, National Academy of Engineering, Institute of Medicine. www.nap.edu/catalog/1605.html

What Will It Cost?

1 *Global Warming: The High Cost of the Kyoto Protocol* by Wharton Econometric Forecasting Associates, 1998; www.wefa.com/download/NATL1998A.pdf. Another study, *Economic Implications of the Adoption of Limits on Carbon Emissions from Industrialized Countries*, by Charles River Associates, 1997 (www.crai.com), predicts that implementing Kyoto will cost 10 times more than President Clinton's White House estimated. The author, W. David Montgomery, previously received funding from the American Petroleum Institute to develop a model that would assess the economic costs of climate change.

2 "Impacts on Canadian Competitiveness of International Climate Change Mitigation," by Christopher Holling and Robin Somerville. *Conservation Ecology* [online] 2(1): 3, 1998; www.consecol.org/vol2/iss1/art3. Thanks to *The Gallon Environment Letter* (cibe@web.net)

3 *The Benefits of Reduced Air Pollutants in the US from Greenhouse Gas Mitigation Policies* by Dallas Burtraw and Michael Toman. Resources for the Future, 1998. www.rff.org/issue_briefs/summaries/ccbrf7.htm

4 *Scenarios of US Carbon Reductions: Potential Impacts of Energy Efficiency and Low Carbon Technologies by 2010 and Beyond.*

This study was completed by five US Department of Energy laboratories and peer-reviewed by industry and academic experts, but it appears to have been removed from the relevant websites. (www.ornl.gov and www.eren.doe.gov).

5 *Solving the Kyoto Quandary: Flexibility with No Regrets* by Dr. Florentin Krause. www.ipsep.org
6 *America's Energy Choices*, Association to Save Energy, 1991.
7 *A Small Price to Pay: US Action to Curb Global Warming is Feasible and Affordable*, Union of Concerned Scientists and Tellus Institute, July 1998.
8 *Fortune Magazine*, 24 May 1999. www.economics.harvard.edu/faculty/mankiw/fortune/may99.html
9 Paul Kovacs, analyst with the Insurance Board of Canada and head of the Institute for Catastrophic Loss Reduction, quoted in *Wild Weather: Our Changing Climate*, Vancouver Sun, 30 December 1998.
10 "Destructive storms drive insurance losses up: will taxpayers have to bail out the insurance industry?" *Worldwatch News* brief, 26 March 1999.
11 *Global warming to cost $300 bn a year — UN Report*, Reuters/Planet Ark, 4 Feb 2001.
12 "*America's Global Warming Solutions*", Tellus Institute and World Wide Fund for Nature, 1999.
13 "Creating Jobs, Preserving the Environment" by Michael Renner, in "State of the World, 2000", p. 173. *Worldwatch Institute.*
14 "Energy Efficiency and Job Creation" by Howard Geller, John DeCicco and Skip Laitner. American Council for an Energy-Efficient Economy, 1992.

The Carbon Barons

1 For details, see *The Heat is On-Line* by Ross Gelbspan (www.heatisonline.org)
2 *Media Moguls on Board* by Norman Solomon. Extra (FAIR), Jan/Feb 1998. www.fair.org/extra/9801/cato-media-moguls.html
3 Quoted on *What's Up with the Weather?* (PBS) www.pbs.org/wgbh/warming/debate/palmer.html
4 *Carbon War: Global Warming and the End of the Oil Era* by Jeremy Leggett (Routledge) 2001. Don Pearlman works as an attorney for the law firm Patton Boggs LLP, which received $180,000 from the Global Climate Council in 1998. (www.opensecrets.org/lobbyists/98profiles/5077.htm) He previously worked for the US National Coal Association, and was Chief of Staff to the US Secretary of Energy.
5 Personal communication from Andrew Weaver, IPCC lead author from the University of Victoria's Climate Modelling Group.
6 For details of all political campaign donations, see www.opensecrets.org and www.commoncause.org

A Global Effort

1 US GHGs: www.eia.doe.gov/oiaf/1605/ggrpt
2 Canadian GHGs: www2.ec.gc.ca/pdb/ghg/English/docs/eGHGTable_trendsByGasSector.htm
3 The US and Canadian data has been updated to reflect the IPCC's upgrading of methane's global warming potential (GWP) from 21 to 23 over 100 years, and from 50 to 62 over 20 years; and N_2O's GWP from 310 to 296 over 100 years.

The 101 Solutions

Solution 3

1 The Canadian data show vehicles achieving 20% higher mileage than the US data because the Canadian gallon is 20% (0.76 liters) larger than the US gallon.

Solution 5

1 Throughout this book, we use the US average of 8 cents/kWh for electricity and 1.6 lbs of CO_2 generated by each kWh. The figures will not always be accurate for individual states or provinces.
2 *Natural Capitalism*, p. 167
3 *Natural Capitalism*, p. 96.

Solution 6

1 Solar Energy Industries Association
2 Canadian Earth Energy website, www.earthenergy.com (prices converted to $US)
3 Data from BP Solar's Clean Power Estimator. "The actual price may be substantially more or less...depending on...the complexity of your solar installation, the distance you live from the distribution agent, local permits and connection fees...These calculations are in no way binding on BP Solar and are made in good faith."

Solution 7

1 All calculations assume an average 1.6lbs of CO_2 per kWh.

Solution 8

1 Ed Ayres, 'Will We Still Eat Meat?', *TIME Magazine*, November 8, 1999
2 As above.

Solution 9

1 *Solar Energy: from perennial promise to competitive alternative.* KPMG Bureau voor Economische Argumentatie, Holland, August 1999.

Solution 10

1 Making a car requires one-eighth of the energy that it uses during a 9-year lifetime; 9 years @ 11,500 miles per year @ 25 mpg = 4176 gallons over 9 years; 1/8th @ 20lbs CO_2 per gallon = 5.22 tons of CO_2.
2 At the sawmill, 50% of the tree becomes lumber, 50% becomes chips and sawdust. At the pulpmill, 50% becomes pulp, 50% becomes fuel. Energy needed at mills for the virgin fibers: 0.3 kWh = 109.5 kWh/yr = 175 lbs CO_2. Energy needed for the 50% recycled fibers: 0.15 kWh = 54.75 kWh/yr = 88 lbs CO_2. Energy needed to make the ink, print, deliver, and recycle: unknown. Total CO_2 emissions: 263 lbs per year. From *Stuff: The Secret Lives of Everyday Things.*
3 EPA data.

Solution 13

1 National Renewable Energy Laboratory, Summersp@tcplink.nrel.gov
2 Design for Workplace Productivity, www.cool-companies.org

Solution 17

1 *Car-Sharing and Mobility — An Updated Review*, by Daniel Sperling, Susan Shaheen, and Conrad Wagner. www.calstart.org/resources/papers/car_sharing.html

Solution 25

1 Study by Ryan Jensen, University of Florida, July 2000

Solution 26

1 *Natural Capitalism*, Ch 3. The 2 tons includes 1 ton of water and 370 lbs of rock tailings. The dry weight total is 123 lbs per day, almost the same as the data in Solution #10.

Solution 28

1 *Transportation Cost Analysis Summary*, Victoria Transportation Policy Institute, 2000 (www.vtpi.org), describes 25 different studies on auto subsidies. The figure of $1,700 ($2,500 Canadian) comes from KPMG and Litman, *The Cost of Transporting People in the British Columbia Lower Mainland*, Transport 2021/Greater Vancouver Regional District, March 1993.

2 245 days a year, 10 miles a day = 2,450 miles at 25mpg = 98 gallons at 20 lbs of CO_2 per gallon = 1 ton of CO_2

3 League of American Cyclists

Solution 29

1 Eben Fodor, *Better Not Bigger*, New Society Publishers, 1999.

2 Sierra Club, Report *The Costs of Sprawl*,

3 Peter Newman, *YES! Magazine*, Summer 1999.

Solution 32

1 *Natural Capitalism*, p. 245.

2 *Cool Companies*, p. 5.

3 *Cool Companies*, p. 7.

4 *Cool Companies*, p. 47.

5 *Cool Companies*, p. 63.

6 EPA, Green Lights.

Solution 33

1 Joseph Romm, *Cool Companies*, p. 197.

Solution 34

1 *Cool Companies*, p. 128.

Solution 35

1 Price to plant 1 tree = $1. Plant 3 trees for 1 to survive and absorb 1 ton of CO_2 over 40 years. Cost per ton/CO_2 = $3. Cost per ton of carbon = 3 x 3.666 = $11. Source: American Forests.

2 Price to plant 1 tree = £4. Plant 5 trees for 1 to survive, and absorb 1 tonne of CO_2. Cost per tonne/CO_2 = £20 = $30US. Cost per tonne of carbon = $30 x 3.666 = $110. Convert to tons = $99US. Source: Future Forests.

Solution 36

1 The total radiative forcing from greenhouse gases is 2.78 watts per square meter. Methane contributes 0.48 w/m^2, of which livestock digestion is 18%, rice paddies 17%, and animal manure slurry 7%: total = 0.2016 w/m^2. Nitrous oxide contributes 0.15 w/m^2, of which poor soil management is 70% = 0.105 w/m^2. 0.2016 + 0.105 w/m^2 = 0.3066 w/m^2, which is 11% of 2.78 w/m^2.

2 Traces of dimethyl sulphide (DMS) have been detected in cows' breath in a study by Dr. Phil Hobbs (Institute of Grasslands and Environmental Research, UK) and Dr. Toby Mottram (Silsoe Research Institute, UK). DMS reacts with oxygen to form compounds that could trigger cloud formation. Rough calculations imply that farmed animals could account for about a tenth of the global total of DMS produced by living organisms.

3 *Natural Capitalism*, p. 199.

4 Lester Brown, Worldwatch Institute.

Solution 37

1 British Columbia = 350 cubic meters per hectare (142 cubic meters per acre). 1000 board feet = 4.344 cubic meters. 750 trees per hectare (300 trees per acre) = 2 cubic meters (460 board feet) per tree. 1 acre (200 trees) over 40 years = 50 tons of carbon (American Forests). 3.8 billion board feet = 8.26 million trees at 300 per acre = 27,533 acres @ 50 tons of carbon per acre = 1.37 million tons of carbon = 5 million tons of CO_2. 9.6 billion board feet = 70,000 acres = 12.6 million tons of CO_2.

Solution 38

1 National Renewable Energy Laboratory, Jan 23rd 2000 press release.

2 Ibid.

3 *Natural Capitalism*, p. 85.

4 Association to Save Energy.

5 *Natural Capitalism*, p. 87.

6 "Integrating Ecology and Real Estate," in *Green Developmer*, p. 24.

7 *Natural Capitalism*, p. 92.

Solution 39

1 US Census, 1997

2 Association to Save Energy

3 EPA Greenhouse Gas Emissions and Sinks, 1990-1998

4 Appraisal Journal

5 Pacific Northwest National Laboratory

Solution 42

1 *Making Connections: Case Studies of Barriers to Interconnection of Distributed Power*, www.eren.doe.gov/distributedpower

Solution 44

1 See the Green Power Network, www.eren.doe.gov/greenpower

Solution 45

1 The 1998 total was 1,494 million metric tons, or 1,643 tons.

2 *Profiting from a Nuclear-Free Third Millennium*, by Amory Lovins, Rocky Mountain Institute.

3 With no disrespect to the excellent programs run by many other utilities, such as Pacific Power in Oregon!

Solution 48

1 *Out of Breath: Health Effects from Ozone in the Eastern United States*. Abt Associates for Clear the Air, 1999.

2 *Lethal Legacy: The Dirty Truth about the Nation's Most Polluting Power Plants*. US PIRG, April 2000.

Solution 49

1 *Japan Times*, July 3, 2000.

Solution 50

1 *Cool Companies*, p. 207.

Solution 51

1 Greenpeace report on Shell's 1999 annual figures, May 9, 2000.

Solution 53

1 *Scientific American*, February 2000.

Solution 54

1 *HyWeb Gazette*, 1st quarter, 2000.

2 Tasios Melis, Liping Zhang, Michael Seibert, Maria Ghiardi, Marc Forestier. *Plant Physiology*, Jan 2000.

Solution 55

1 *Kingpins of Carbon: How Fossil Fuel Producers Contribute to Global Warming*, NRDC, 1997.

Solution 56

1 1 gallon = 20 lbs CO_2. 3.667 lbs CO_2 = 1 lb carbon. 74 million barrels per day @ 42 gallons per barrel = 3,108 million gallons = 31 million tons CO_2 = 11,344 million tons CO_2 per year = 3.093 million tons of carbon = 2.812 billion tonnes.

2 US EIA, *Emissions of Greenhouse Gases in the U.S.*, 1997, Appendix B.

Solution 57

1 In European terms, the agreement was to increase efficiency from 7.7 to 5.8 liters per 100 kilometers.

2 In European terms, 80mpg.

3 In European terms, it achieved 97.1 mpg. The European and Canadian gallon is 20% larger than the U.S. gallon.

4 A study done by P. Ahlvik and A. Brandberg, Ecotraffic, Sweden, for the Centre for Science and Environment, New Delhi, showed the cancer potency of

diesel particulates is much higher than the total effect of all carcinogenic compounds from regular gasoline vehicles. A similar German study found the cancer potency of diesel engines to be 10 times greater than gasoline engines (CSE, May 4th 2000). See www.cseindia.org/html/au/au4_20000504.htm.

5 Quoted in the Oregon Climate Change Action Plan.

6 All data quoted in American mpg.

Solution 60

1 *The Carbon War*, by Jeremy Leggett (Allen Unwin).

2 *Common Cause*, "Some Like it Hot," November 30, 1999.

3 *Tomorrow Magazine*, September 1999.

Solution 61

1 *Guardian Weekly,* February 10th, 2000.

Solution 64

1 *The Economic Impacts of Renewable Energy Use in Wisconsin*, Wisconsin Department of Administration Energy Bureau, April 1994.

Solution 65

1 David Morris, Institute for Local Self-Reliance, Minneapolis-St. Paul.

2 Maryland has also waived the sales tax on Energy Star appliances.

Solution 68

1 August 2000 data from Dr Kim Holmen, associate professor of global change studies and greenhouse gases in the Department of Meteorology at Stockholm University, who chairs the greenhouse gas monitoring program of the World Meteorological Organization.

2 Emissions of Greenhouse Gases in the US, 1998, Energy Information Agency.

3 Environment Canada Greenhouse Gas Division, 1997 data.

Solution 69

1 US data from US Greenhouse Gas Emissions and Sinks, 1990 — 1998. In 1998, agriculture produced 144.5 MMTCE out of a total of 1,834 MMTCE = 7.9%.
Canadian data from Environment Canada's Greenhouse Gas Division, and *The Health of our Air: Towards Sustainable Agriculture in Canada* (http://res2.agr.ca/research-recherche/science/Healthy_Air/2d6a.html). This study estimates that Canada agro-ecosystems produce 59 Mt of CO_2 equivalent (18.3 MMTCE), compared to Canada's overall production of 682 Mt (186 MMTCE) = 8.65%.

Solution 70

1 *Living Planet Report 2000*, World Wide Fund for Nature. The Living Planet Index is a measure of the state of natural ecosystems based on the population of animal species in the world's forests, freshwater, and marine ecosystems. Between 1970-1999, these declined, on average, by 12%, 50%, and 35% respectively, with the most severe deterioration in the tropical and southern temperate regions.

2 For the full analysis of this "two more planets" metaphor, look under "Ecological Footprint" on the "Redefining Progress" page at www.rprogress.org

3 British Columbia's Green Economy Initiative was started under the New Democratic Party government in 1998. It is not clear whether the initiative will survive under the Liberal government that replaced it in 2001.

4 "Going to Work for Wind Power," by Michael Renner, *Worldwatch Magazine*, Jan/Feb 2001.

5 *Saving the Environment: A Jobs Engine for the 21st Century,* by Michael Renner, Worldwatch Institute, 2000.

Solution 72

1 *Voltaire's Bastards,* by John Ralston Saul, pp 390-393.

2 *Paying for Pollution*, Friends of the Earth, 2000

3 *Special Report on Global Climate Change: Economic and Market Mechanisms* by Richard L. Ottinger & Mindy Jayne, Pace University School of Law, White Plains, New York. January, 2000. www.solutions-site.org/special_reports/sr_global_climate_change.htm

4 *The Subsidy Scandal: the European clash between environmental rhetoric and public spending*, Greenpeace special report, 1997.

5 For details, see www.EcoAction.ca

6 *Fueling Global Warming*, Greenpeace, 1998, using US government data

7 *Paying for Pollution*, Friends of the Earth, 2000

8 *Gas Guzzler Loophole: SUVs and Other Light Trucks Drive Off with Billions*, Friends of the Earth, August 2000

9 *Public-health impact of outdoor and traffic-related air pollution: a European assessment*, Lancet 2000; 356: 795 - 801. "Air pollution caused 6% of total mortality or more than 40000 attributable cases per year. About half of all mortality caused by air pollution was attributed to motorized traffic, accounting also for: more than 25,000 new cases of chronic bronchitis (adults); more than 290,000 episodes of bronchitis (children); more than 0.5 million asthma attacks; and more than 16 million person-days of restricted activities."

Solution 73

1 "Environmental Tax Shifts Multiplying," by David Roodman *Vital Signs 2000*, Worldwatch Institute.

2 23-30 Euros, rising to 76 Euros.

3 *Taking Stock of Green Tax Reform Initiatives*, by Carola Hanisch, Environmental Science and Technology, December 1998.

4 *The Natural Wealth of Nations*, by David Roodman (Worldwatch Institute, 1998).

5 Poll conducted by International Communications Research, for Friends of the Earth.

6 $50 per ton of CO_2 comes to $13.65 per ton of carbon. EcoSecurities, which invests capital in carbon offsets, estimates that carbon credits will sell for $10 - $15 per tonne. (www.ecosecurities.com)

7 The Sky Trust is being proposed by Peter Barnes.

8 Each gallon releases 20 lbs of CO_2, so 100 gallons = 1 ton.

Solution 74

1 An excellent summary of the market barriers and initiatives to overcome them can be found in *Climate: Making Sense and Making Money*, by Amory and Hunter Lovins, Rocky Mountain Institute, 1997.

2 Lovins, 1997

3 Information from the office of Jeanette Fitzsimons, MP, New Zealand.

Solution 75

1 Sustainable Energy Coalition 5th Annual Survey, conducted by the Republican polling firm of Research/Strategy/Management Inc with 1,003 registered voters, Sept 22-28, 1998. 62% want a high priority for DoE funding of renewable energy; 78% want a 10% Renewable Portfolio Standard; 81% (45% strongly, 36% somewhat) support tax incentives to support renewable energy and energy efficiency.

2 As of May 2001.

Solution 77

1 Solution 28, footnote 1.
2 Canadian Urban Transit Association nation-wide survey, November 1999.
3 US fleet of 137 million vehicles.

Solution 78

1 Welsh Institute of Rural Studies, Aberystwyth.

Solution 79

1 Society of American Foresters "Forest Facts," 2000.
2 Canadian Forest Service, 1995.
3 Sierra Club of Canada "Forest Fires and Climate Change", 1996.
4 See *Carbon budget of the Canadian forest product sector* by M.J. Apps et al, *Elsevier Environmental Science & Policy* 2 (1999) 25-41, and *A 70-year Retrospective Analysis of Carbon Fluxes in the Canadian Forest Sector* by W. Kurz and M.J. Apps, *Ecological applications* 9(20), 1999, pp 526-547.
5 *Perverse Habits: the G8 and Subsidies that Harm the Forests & Economies*, World Resources Institute, 2000.
6 Environmental Systems of America, Inc. Rainforest Factoids.

Solution 81

1 Thanks to Lee Eng Lock, in Singapore, for these insights.
2 Sajeeda Choudhury, Bangladesh environment minister, January 2000
3 *Developing Counties and Global Climate Change: Electric Power Options for Growth.* Pew Center on Climate Change, June 1999
4 These are the ingredients recommended by Jaime Lerner, past Mayor of Curitiba, now Governor of Parana, Brazil
5 Quoted in *Flirting With Disaster: Global Warming and the Rising Costs of Extreme Weather*, US P.I.R.G, October 1999.

Solution 82

1 Solar Electric Light Company data (SELF's affiliate).

Solution 84

1 Many thanks to Alan Weisman, Donella Meadows, and Tom Athaniasou.

Solution 85

1 *Promoting Energy Efficiency and Renewable Energy: GEF Climate Change Projects and Impacts*, by Eric Martinot and Omar McDoom, October 1999.

Solution 87

1 *Renewable Energy for India: A Future Challenge*, by C.R. Kamalanathan, Secretary, Ministry of Non-Conventional Energy Sources, 1998.
2 If CO_2 doubles, India's wheat yields could fall by 28-68%. Rao and Sinha, *Impact of Climate Change on Stimulated Wheat Production in India.* EPA, 1994.
3 As Note 1.
4 *Energy for the Developing World*, by A.K.N. Reddy and J. Goldemberg, *Scientific American*, September 1990.
5 Ministry of Non-Conventional Energy Sources, Annual Report, 1999-2000.
6 Estimate from Herman Scheer, President of Eurosolar, June 1999.
7 This is the authors' estimate, assuming a competitive price.
8 See 5, above.

Solution 88

1 *China at the Crossroads: Energy, Transportation and the 21st Century*, by James Cannon (1998), INFORM Inc. (www.informinc.org).
2 *Seven Wonders*, by John Ryan, p.11.
3 *Daily Yomiuri*, Japan, May 1st 2000
4 *Industrial-Scale Wind Power in China*, by Debra Lew, Robert Williams, Xie Shaoxing and Zhang Shihui. Center for Energy and Environmental Studies, Princeton University, USA, and Chinese Ministry of Electric Power, 1996.
5 Article by Lin Gan, CICERO, *Sustainable Energy News*, Sept 1998.

Solution 90

1 *World Disasters Report*, 2000. International Federation of Red Cross and Red Crescent Societies.
2 *Red Cross warns on climate*, by Alex Kirby, BBC News, June 28, 2000.
3 All data from the UK Jubilee 2000 website.
4 *World Disasters Report*, 2000, Chapter 7.

Solution 91

1 COP stands for "Conference of the Parties," and refers to the parties (nations) that signed the 1992 UN Framework Convention on Climate Change; COP 6 means it was their sixth meeting.
2 Letter to Tom Spencer, congratulating him for being awarded the Best Member of the European Parliament Award in the 1999 Global Green Ribbon Political Awards, and for promoting the "Contract and Converge" approach to treaty-making.

Solution 92

1 "The Global Green Deal," *Earth Island Journal*, Summer 2000.
2 "Europe: Brussels in motion." *New Energy No 3*, 2000: www.wind-energie.de/new_energy/0600/3.html

Solution 93

1 *Promoting Energy Efficiency and Renewable Energy: GEF Climate Change Projects and Impacts*, by Eric Martinot and Omar McDoom, GEF, October 1999.
2 World Air Transport Statistics, 2000. IATA.
3 James Tobin first proposed the levy in his Janeway Lectures at Princeton University in 1972; he first published the proposal in the Eastern *Economic Journal*, 4, 1978.
4 *The United Nations: Policy and Financing Alternatives* (eds. Harland Cleveland, Hazel Henderson, and Inge Kaul), Global Commission to Fund the United Nations, 1995.
5 "The Climate Crisis and Carbon Trading," by Ross Gelbspan, *Foreign Policy In Focus*, July 2000; and "Rx for a Planetary Fever," by Ross Gelbspan, *American Prospect*, May 2000.
6 *All you ever wanted to know about the Tobin Tax*, War on Want (UK), 2000.
7 War on Want, Ibid.
8 Jubilee 2000 has estimated the cost of eliminating the South's unpayable debt at $160 billion.
9 In 1997, the UN and the World Bank estimated that wiping out the worst forms of poverty and providing basic environmental protection would cost around $225 billion a year.

Solution 94

1 In 1992, the breakdown between civil and military aviation fuel was 82% and 18%, respectively. By 2015, this is expected to change to 93% and 7% (IPCC). In 2001, the world's airlines flew 3,000 billion passenger kilometres.
2 *Some Interesting Civil Aviation Statistics*, ICAO, Dec 1994.
3 World Air Transport Statistics 2000, ICAO.
4 *Aviation and the Global Atmosphere*, IPCC, 1999.
5 *Science* (Vol 279, p 49). Using data from NASA, Paul Wennberg of the Californi
a Institute of Technology (an IPCC author) found that the upper troposphere contains much more hydroxyl than expected

— a sign of faster radicalization of the NOx from airplanes into tropospheric ozone, which might raise the figure as high as 10%. The chair of the group that wrote the IPCC report, Dick Derwent, described Wennberg's paper as "absolutely excellent," but added that some researchers have suggested that the reactions involving NOx emissions from aircraft destroy methane, a potent greenhouse gas, which would reduce the overall impact. (*New Scientist,* April 11, 1998).

6 Data from Ben Matthews' site www.chooseclimate.org/flying, with these working assumptions. Fuel per passenger as for a B-747. Fuel = [7840 + 10.1 x (distance-250km)] x 2 for return trip. Take off+climb+descent =7840 kg. Cruising = 10.1 kg/km. Passengers = 370 (x 1.5 for business class). The combined warming effect of CO_2, ozone (made by NOx), water vapor and contrails is about three times greater than CO_2 alone.

7 $50 per tonne of CO_2 is $13.65 per tonne of carbon. EcoSecurities, a company that raises capital to invest in carbon offsets, estimates that carbon credits will sell for $10 - $15 per tonne.

8 The cost differs from the price in the box because this method of calculation is different from the one used by Ben Matthews.

9 OECD study quoted in Environment NGO response to *Aviation and the Global Atmosphere* (IPCC).

Solution 95

1 For a summary of sources for the various environmental effects, see www.ozone.org/page12.html

2 *The Use of Natural Refrigerants, A Complete Solution to the CFC/HCFC Predicament.* 1994 International Institute of Refrigeration, Commission B2, Hanover, Germany.

3 Average GWP of HFC emissions in European Union in 1998. Source, see footnote 5. HFC-134a, the most common, has a GWP of 1300 (EPA).

4 *Fluorocarbons and SF6: Global emissions inventory and options for control* by C. Kroeze. Report No. 773001-007, RIMV Bilthoven, Holland, 1995. See also *Potential Effect of HFC Policy on Global Greenhouse Gas Emissions in 2035*, by C. Kroeze. National Institute of Public Health and Environmental Protection, Holland, 1994. Air Directorate, project #773001. The Alliance for Responsible Atmospheric Policy (representing the chemical industry) claims that "realistic projections show that emissions will be less than 3% in 2050" (*HFCs: an Energy Efficient Solution*, 1998).

5 *How to Limit HFC/PFC/SF6 Emissions? Eliminate Them.* A Greenpeace Position Paper submitted to the Joint IPCC/TEAP Expert Meeting, Holland, 1999, by John Maté, Greenpeace International Greenfreeze Project Coordinator. Updated Feb 2000.

6 *Primary Aluminum Production: Climate Policy, Emissions and Costs,* by Jochen Harnisch et al. MIT Joint Program on the Science and Policy of Change, 1999.

Solution 96

1 Intergovernmental Panel on Climate Change (IPCC), with 2,500 scientists.

2 World Economic Forum, Switzerland, February 2000: www.weforum.org

3 Witness the world's governments' agreements at the 1992 Rio Convention and the 1997 Kyoto Protocol.

4 *The Race to the Bottom: Creating Risk, Generating Debt, and Guaranteeing Environmental Destruction,* ECA NGO Campaign, 1999.

5 *The World Bank and the G-7: Still Changing the Earth's Climate for Business,* 1998, Sustainable Energy and Economy Network and International Trade Information Service.

6 As in footnote 4.

7 *What I Learned at the World Economic Crisis,* by Joseph Stiglitz. New Republic Online, April 2000.

8 *The World Bank's Juggernaut: The Coal-Fired Industrial Colonization of the Indian State of Orissa and the G-7, the World Bank, and Climate Change.* Institute for Policy Studies, 1997.

Solution 97

1 FAO's Global Forest Resources Assessment. Estimates range from 27 to 37 million acres a year. A new report is due by the end of 2000. This is a net loss, which accounts for replanting.

2 *Global Change: That Sinking Feeling,* by Jorge Sarmiento, Nature, 408: 6809, Nov 9, 2000.

3 Mike Flanagan, Canadian Forest Service, Edmonton, *Globe and Mail,* April 3, 2000.

4 Goulden et al., *Sensitivity of boreal forest carbon balance to soil thaw,* Science 279 (5348): 214-217

5 Speech to World Resources Institute, Jan 6, 2000.

6 WWF/WRI Report by Dominiek Plouvier and Nigel Sizer, 1997. See *Report on Forests Suppressed,* Guardian Weekly, June 1, 2000.

Solution 99

1 *Sierra Magazine* interview 'Getting it Right', Jan/Feb 2000.

Solution 100

1 Written in the *Manchester Guardian,* quoted by Stefan Collini in "Is this the end of the world as we know it?" *The Guardian Weekly,* Jan 6th 2000.

2 Interview, Planet Ark, Nov 21, 2000.

3 *Natural Capitalism,* by Paul Hawken, Amory Lovins, and L. Hunter Lovins, p. 4 (Little, Brown, 1999).

4 Jane Lubchenko is the past chair and president of the American Association for the Advancement of Science.

Solution 101

1 With the dominance of biology as the leading science, and the dominance of the "Selfish Gene" as the leading paradigm within evolutionary biology, this has become — for the moment — the dominant belief system for science as a whole. See *The Selfish Gene* and *The Blind Watchmaker* by Richard Dawkins.

2 The leading thinker for this belief is Daniel Quinn, author of *Ishmael* and *Beyond Civilization.* His books have inspired many people to become critically aware of the terrible damage our civilization creates and to seek a way of life that is less harmful.

3 Hexagram 64 in the *I Ching,* Richard Wilhelm translation.

Index

E

F

G

H

I

J

K

About the Authors

Guy Dauncey is an author, speaker, organizer and consultant who specializes in developing a positive vision of a post-industrial, environmentally sustainable future, and translating that vision into action. He has been self-employed for the past 25 years, working in the fields of positive social, economic and environmental change in Britain and Canada.

He has traveled to 25 countries, and co-established a number of initiatives including The British Unemployment Resource Network, the UK Education for Enterprise Network, the UK Social Investment Forum, *EcoNews*, the Victoria *Green Pages*, and The Victoria Car Share Co-operative. His passion lies in creating a positive vision for the future, and showing how it can be implemented in practical ways. He is presently engaged in work to establish a car-free ecovillage on the coast of British Columbia, and in founding and directing a new organization — The Street Volunteers — which will draw neighbors together on the streets and in the apartments where they live. He is also very involved in promoting solutions to global climate change through a Canadian Broadcasting Company (CBC) film, its website, and a future book.

Guy's previous publications include: *Earthfuture: Stories from a Sustainable World* (New Society, 1999); *After the Crash: The Emergence of the Rainbow Economy*, (Greenprint, UK: 1988) — awarded the Green Book of the Year award in Britain, 1988; *The Unemployment Handbook* (NEC, UK: 1981); *Nice Work if You Can Get it — How to be Positive About Unemployment* (NEC, UK: 1983), and several books for schools.

Patrick Mazza has a background of over 20 years in advocacy journalism centered on ecological sustainability issues. His work has woven around the theme of bioregional response to global change, with a focus on the Pacific Northwest. Mazza has written about and participated in efforts to save ancient forests, restore watersheds and build sustainable cities. He has published in mainstream journals and helped to build independent media. In the mid-90s, he mounted an early website devoted to the concept of Cascadia as a global sustainability model, Cascadia Planet (www.tnews.com).

In recent years, he has worked as researcher-writer for Climate Solutions, an Olympia, Washington, group that aims to make the Northwest a leader in global warming solutions. At Climate Solutions he has focused on public understanding of climate science, climate-friendly land use and transportation, and emerging clean energy industries and applications. His most recent Climate Solutions papers include a series on *Harvesting Clean Energy for Rural Development* and *Accelerating the Clean Energy Revolution: How the Northwest Can Lead*, available at www.climatesolutions.org.

New Society Publishers

ENVIRONMENTAL BENEFITS STATEMENT

New Society Publishers has chosen to produce this book on New Leaf EcoBook 100, recycled paper made with 100% post consumer waste, processed chlorine free, and old growth free.

For every 10,000 books printed, New Society saves the following resources:[1]

89	Trees
8,087	Pounds of Solid Waste
8,898	Gallons of Water
11,606	Kilowatt Hours of Electricity
14,701	Pounds of Greenhouse Gases
63	Pounds of HAPs, VOCs, and AOX Combined
22	Cubic Yards of Landfill Space

[1] Environmental benefits are calculated based on research done by the Environmental Defense Fund and other members of the Paper Task Force who study the environmental impacts of the paper industry.

For more information on this environmental benefits statement, or to inquire about environmentally friendly papers, please contact New Leaf Paper – info@newleafpaper.com Tel: 888 • 989 • 5323.